MATERIALS SCIENCE AND TECHNOLOGIES

ORGANIC SEMICONDUCTORS: PROPERTIES, FABRICATION AND APPLICATIONS

MATERIALS SCIENCE AND TECHNOLOGIES

Additional books in this series can be found on Nova's website
under the Series tab.

Additional E-books in this series can be found on Nova's website
under the E-books tab.

ENGINEERING TOOLS, TECHNIQUES AND TABLES

Additional books in this series can be found on Nova's website
under the Series tab.

Additional E-books in this series can be found on Nova's website
under the E-books tab.

MATERIALS SCIENCE AND TECHNOLOGIES

ORGANIC SEMICONDUCTORS: PROPERTIES, FABRICATION AND APPLICATIONS

MARIA A. VELASQUEZ
EDITOR

Nova Science Publishers, Inc.
New York

For permission to use material from this book please contact us:
Telephone 631-231-7269; Fax 631-231-8175
Web Site: http://www.novapublishers.com

Library of Congress Cataloging-in-Publication Data

Organic semiconductors : properties, fabrication, and applications / editor,
Maria A. Velasquez.
p. cm.
Includes bibliographical references and index.
ISBN 978-1-61209-391-8 (hardcover : alk. paper)
1. Organic semiconductors. I. Velasquez, Maria A.
TK7871.99.O74O74 2011
621.3815'2--dc22

2011002582

Published by Nova Science Publishers, Inc. † New York

CONTENTS

PREFACE

This new book presents topical research in the study of organic semiconductors with a focus on the magnetic and electronic properties of organic polymer semiconductors; organic photovoltaic semiconductors and devices; solution processed polymers; organic field-effect transistors and the development of new organic semiconductors for application in organic electronics.

Chapter 1 - Organic photovoltaic semiconductors are presenting great attraction for application in organic photovoltaic cells (OPVs) due to their strong light absorption and simplified fabrication on flexible substrates at low cost. Recently, several innovative researches with exciting improvement at power conversion efficiency approaching business standard 7% of OPVs are reported by modifying functional group of organic materials. Thus, to introduce and conduct the fabrication of high efficiency OPVs, this review would systemically summarize optical and charge transporting properties of organic semiconductors on both small molecules and conjugated polymers. Then, we discuss the physical processes including exciton generation and dissociation, charge transportation and collection that lead to photocurrent generation in OPVs. Moreover, effective OPV architectures as donor-acceptor heterojunction for efficient exciton dissociation, exciton blocking layer, mixed or bulk heterojunction and their photovoltaic process are also presented. Finally, important issues relating to the pivotal improving principles of open-circuit voltage, short-circuit current density and fill factor are also concluded.

Chapter 2 - The prerequisite of the further progress in organic electronics is the understanding of fundamental processes governing the overall performance of organic devices, so to speak, the controlling the properties of both organic films and involved interfaces, the entities which constitute the organic devices.

Among other techniques, photoemission characterization directly assesses the electronic structure of both the molecular films and the associated interfaces, and it thereby belongs to the most popular surface and interface analytical techniques.

This work focuses on certain problems encountered while applying the direct photoemission technique for the basic characterization of both the vacuum-sublimed molecular films and the interfaces formed between the films and the metal, organic, and inorganic substrates. Fundamental electronic parameters, such as the work function, the ionization energy, and the interfacial energy level alignment, and their interplay affected by the film structure and morphology will be addressed in particular, with the reference to the molecular orientation in the films. It will be demonstrated that the work function of molecular films is an essential parameter governing the energy level alignment between the film and the substrate. A simple model for the prediction of the injection interfacial barrier based on the work functions of materials brought into the contact is proposed. Also, it will be demonstrated that neglecting to consider the work function may obstruct an unambiguous interpretation of electronic properties of the involved interface, and it may even implicate a spurious detection of the band bending in the molecular film.

Chapter 3 - Solution processed polymer materials are extensively used at various stages of micro/nano fabrication processes either as structural materials or as sacrificial layers like photoresists, release layers etc. due to their special chemical, electrical, thermal and mechanical properties catering to different applications. With the discovery and development of highly conductive organic polymers ("polyacetylene" class) in 1977 by Alan G. MacDiarmid, Hideki Shirakawa, and Alan J. Heeger a new area of research came into existence - polymer electronics. For this revolutionary discovery they were jointly awarded the 2000 Nobel Prize in Chemistry. The electronic properties of these functional polymers are derived from their chemical structure, which contains the so-called "conjugated polymer chains", consisting of a strictly alternating sequence of single and double bonds resulting in delocalized electrons giving conducting/semiconducting properties to these organic materials [1 - 4]. The conducting/semiconducting properties of these materials can be controlled by adding/removing different functional groups to them. Various researchers all over the world have shown tremendous interest in using these organic semiconducting materials as active semiconductors in different organic electronic components and circuits. Results of all such activities have classified the organic semiconductor based electronic area into four broad categories.

Chapter 4 - Fullerene C 60 and its derivatives are recently expected as very useful organic semi- conductors for organic devices such as the organic field-effect transistor (OFET). In this Chapter, effect of chemical addition on carrier transport properties of C 60 is theoretically estimated and systematically discussed by the density functional theory (DFT) calculation, taking fullerene derivative C 60 Xn as examples. Based on the Mar- cus theory, carrier transport properties are related to the reorganization energy. Hole- transport property of C 60 derivatives is strongly dependent on the position and the number of added groups, but is almost independent of chemical properties of added groups. On the other hand, electron-transport property of C 60 is little influenced by the chemical addition. These results are discussed from viewpoints of geometric and electronic structures. It is found that the values of reorganization energies are almost proportional to the degree of geometrical relaxation upon the carrier injection. Delo- calization of frontier molecular orbitals on C 60 sphere results in small reorganization energy and fast carrier transport. From these analyses, specific guidelines for efficient design of useful organic semiconductors are proposed.

Chapter 5 - The processing characteristics of organic semiconductors make them potentially useful for electronic applications where low-cost, large area coverage, and structural flexibility are required.Contrary to amorphous silicon, which is widely used in solar cells and flat screen displays, organic materials offer the benefits that they can be deposited on plastic substrates at low temperature by employing solution-based printing techniques. These deposition techniques would, therefore, reduce the manufacturing costs dramatically. The challenge now lies on finding organic semiconductors which are processable, stable and, at the same time, exhibit high enough mobilities (μ>0.1 cm2/Vs) and ON/OFF current ratios (>106) to be used for applications in modern microelectronics.

Chapter 6 - Electronic and optoelectronic devices using organic materials as active elements are attractive because they can take advantage of organic materials such as easy of functionality, light weight, low cost, and capability of thin-film, large-area, flexible device fabrication. Among them, organic field-effect transistors (OFETs), which consist of organic semiconductors, dielectric layers, and electrodes, are expected to be a promising technology for application in displays, sensors, and memories. Organic semiconductors play a key role in determining the device characteristics. Recent technological advances in OFETs have triggered intensive research into molecular design, synthesis, device fabrication, thin film morphology and transport of holes and electrons and so on. New organic semiconductors are currently being intensively studied, resulting in the development of a growing number of high-performance semiconducting materials with higher mobility than that of amorphous silicon (0.1 cm2V−1s−1). Here, firstly, an introduction to OFET principles and history, as well as the charge transport mechanisms, film alignment and morphology and crystal growth processes are presented. Then we discussed the structural design/realization of recently developed high performance p- (hole-transporting) and n-channel (electron-transporting) semiconductors for OFETs. A survey of the reported molecules and correlations between their structure and transistor performance are presented. The device structures and dielectric gate insulator materials are also described. Besides, the influence of the device fabrication process, organic semiconductor/dielectric layer interface, and organic layer/electrode contact on the device performance was reviewed.

In: Organic Semiconductors
Editors: Maria A. Velasquez

ISBN: 978-1-61209-391-8

Chapter 1

ORGANIC PHOTOVOLTAIC SEMICONDUCTORS AND DEVICES

Junsheng Yu*, Jiang Huang and Yadong Jiang
State Key Laboratory of Electronic Thin Films and Integrated Devices, School of Optoelectronic Information, University of Electronic Science and Technology of China (UESTC), Chengdu, P. R. China

ABSTRACT

Organic photovoltaic semiconductors are presenting great attraction for application in organic photovoltaic cells (OPVs) due to their strong light absorption and simplified fabrication on flexible substrates at low cost. Recently, several innovative researches with exciting improvement at power conversion efficiency approaching business standard 7% of OPVs are reported by modifying functional group of organic materials. Thus, to introduce and conduct the fabrication of high efficiency OPVs, this review would systemically summarize optical and charge transporting properties of organic semiconductors on both small molecules and conjugated polymers. Then, we discuss the physical processes including exciton generation and dissociation, charge transportation and collection that lead to photocurrent generation in OPVs. Moreover, effective OPV architectures as donor-acceptor heterojunction for efficient exciton dissociation, exciton blocking layer, mixed or bulk heterojunction and their photovoltaic process are also presented. Finally, important issues relating to the pivotal improving principles of open-circuit voltage, short-circuit current density and fill factor are also concluded.

1. INTRODUCTION

In the 21st century, human kind confronts with a great challenge to realize economical and sustainable social development. Currently, some energy problems gradually become critical. First of all, high-speed social development requires massive energy consumption of

* To whom all correspondence should be addressed. E-mail address: jsyu@uestc.edu.cn. Tel: 86-28-83207157.

limited resources, e.g., coal, petroleum and natural gas. Secondly, human beings are suffering from many kinds of diseases caused by environmental pollution as a result of burning fossil fuels. Thirdly, the greenhouse effect causes the global climate extremely changing, and the urgent reduction of carbon dioxide emission is imperative. Therefore, we can only depend upon the advanced technology and renewable energy resources to solve the above energy problems.

Specifically, one of most attractive solutions is to make the most of solar energy due to its natural properties: inexhaustible, locally available, and environmental friendly. Although well developed inorganic solar cells based on silicon materials possess high light conversion efficiency, organic photovoltaic materials and devices have attracted more and more attention and researching forces. It's mainly attributed to their highly potential advantages such as large-area fabrication, light weighted, flexible on plastic substrates, low cost and substantially ecological.[1][2] Many western countries have already established research programs to encourage the adoption of this new technology[3] making them to be a highly energy-efficient, low carbon economy. A broad prospect with the improvement of efficiency and reduction of cost in the near future can be expected.

Since the initial research of small-molecular organic photovoltaic cell (OPV) with an electronic donor (D)/acceptor (A) planar heterojunction was reported by Tang in 1986,[1] further improving researches with various materials and device structures have been reported.[2][4] Most notably, Xue et al. pioneered a hybrid planar-mixed donor copper phthalocyanine (CuPc) : acceptor fullerene C_{60} heterojunction (PM-HJ) structure to achieve high *PCE* =5.0% in 2004. One year later, the highest *PCE*=5.7% of small molecular OPVs with a double stacked PM-HJ texture was achieved.[5] In order to further optimize *PCE*, novel materials with lower highest occupied molecular orbital (HOMO) level than CuPc, e.g., chloroboron subnaphthalocyanine (SubNc) [6][7] and chloroboron subphthalocyanine (SubPc) [7] have adopted to increase open-circuit voltage (V_{OC}) near 0.8 V and 0.92 V, respectively. When SubPc/C_{60} and SubNc/C_{60} planar cells are connected in series, high V_{OC}=1.92 V and *PCE*=5.15% were obtained by Cheyns et al.[8]

The polymeric solar cells based on polyacetylene polymer were firstly reported in the 1980s.[9] In the 1990s, the first bulk heterojunction (BHJ) architecture OPVs showed low *PCE* around 1%.[10][11] During last 10 years, one of the most intensely studied BHJ OPVs is poly (3-hexylthiophene) (P3HT) and fullerene [6,6]-phenyl-C_{61}- butyric acid methyl ester (PCBM) blend system, which have reaching its bottle neck values *PCE*=5.3 % under the standard solar spectrum, AM 1.5G.[12] The highest *PCE* of tandem BHJ OPVs was reported as 6.5% by Kim et al. Then, various novel polymer materials by modifying functional group have emerged to further optimize photovoltaic performance. For instance, BHJ OPVs based on poly[4,8-bis-substituted-benzo [1,2-b:4,5-b0] dithiophene-2,6-diyl-alt-4-substituted-thieno [3,4-b]thiophene-2,6-diyl]- alkyl-fluorine (PBDTTT-CF) : PCBM blend film presented the highest *PCE*=7.73%. [14] Currently, the highest PCE at present has arrived 7.9% by Y. Yang group [15].

Although impressive progress has obtained, many aspects relating absorption range for photon-electron conversion efficiency, and charge carrier mobility of light-harvesting polymers requires further optimization to approach theory limits *PCE*>10% [16] for single cell and *PCE*>15% for tandem cells.[17] Furthermore, fabricating large area devices without significantly losing efficiency while maintaining long device lifetimes remains challenging. The purpose of this chapter is to introduce and conduct the fabrication of high efficiency

OPVs. Thus, this review is organized as follows: Section 2 provides a brief description on the photovoltaic process of OPVs embracing exciton generation and dissociation, charge transportation and collection that lead to photocurrent generation; Section 3 would systemically summarize the optical and charge transporting properties of promising organic semiconductors including small-molecular materials, conjugated polymers and fullerene for the near future high efficiency application; Section 4 presents typical small molecular OPVs with high-efficiency architectures, e.g., PM-HJ, inverted planar HJ, tandem OPVs based on PM-HJ, p-i-n structure and multi-charge separation (MCS) structure; Section 5 firstly elaborates the crucial film morphology controlling treatment for bulky film, e.g., thermal annealing, solvent annealing and solvent mixture, then, optimal strategies to optimize single, inverted and tandem BHJ OPVs will be summarized and concluded.

2. Photovoltaic Theory

In this part, photovoltaic process of OPVs embracing exciton generation and dissociation, charge transportation and collection will be systematically presented. First of all, the analyses would begin from the calculation of major photovoltaic parameters to judge the power conversion qualities of OPVs. As well known, power generation by an OPV cell is a process to convert light energy into electrical energy. The incident light on the cell results in the separation of charge carriers, which ultimately yields to photocurrent, and then provide power supply to an external load. The current density (J) –voltage (V) curve is to show the basic characteristics for an OPV cell. A typical PV cell in the dark and under incident illumination is shown in Fig. 1.

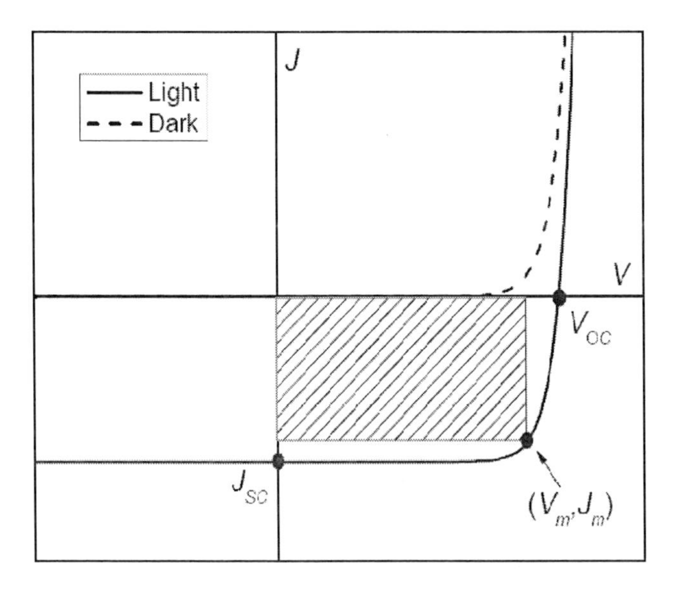

Figure 1. Typical current density (J) versus voltage (V) curves of a solar cell in the dark and under illumination. The various device parameters include open-circuit voltage (V_{OC}), and short-circuit current density (J_{SC}), and the point (V_m, J_m) on solid line corresponds to maximum output power $P_{max}=V_m{\cdot}J_m$, shown as the rectangular region in the forth quadrant.

Major crucial photovoltaic parameters derived from *J-V* curve and their physical denotation are listed as follows:

a) V_{OC}: open-circuit voltage. The free carriers generated from dissociated exciton at Donor/Acceptor heterojunction are all accumulated at two electrodes to compensate the original contact barrier, then, maximum photo-generated voltage can be detected. The operating range of OPVs is within $0 < V < V_{OC}$, where the device generates power to an external load;
b) J_{SC}: short-circuit current density. The free charge carriers generated from dissociated exciton at the heterojunction cannot be accumulated at anode and cathode, and therefore, flow through the external circuit;
c) P_{max}: the maximum output power, defined as $P_{max}=V_m \cdot J_m$, where V_m and J_m corresponds to a maximum power on *J-V* curve under illumination;
d) *FF*: fill factor. It indicates the energy loss due to the device resistances, and practically reflects the quality of a *J-V* curve. *FF* is calculated as:

$$FF = \frac{J_m V_m}{J_{SC} V_{OC}} \tag{1}$$

e) (e) *PCE*: photoelectric (or power) conversion efficiency, defined as below, where P_0 is the incident light intensity:

$$PCE = \frac{J_m V_m}{P_0} = \frac{J_{SC} V_{OC} FF}{P_0} \tag{2}$$

2.1. Exciton Generation

The principle of solar cell is based on the photovoltaic effect resulted from the heterojunction of a semiconductor, or the interface between metal and semiconductor. The first step is the light absorption of various layers, and then excitons are generated with a generation rate of η_A. From the view of the electromagnetic theory of light, the internal optical electric field distribution due to an incident plane wave can be calculated with the assistance of complex indices of refraction and layer thickness of semiconductive materials. [18][19][20] Recently, optical transfer matrix theory have been widely applied to effectively optimize device structure to maximum light distribution in active layers. The total electric field at an arbitrary position inside the organic layer *j* is given in terms of the electric field of incident wave by

$$E_j(x) = E_j^+(x) + E_j^-(x) = (t_j^+ e^{i\xi_j x} + t_j^- e^{-i\xi_j x}) E_0^+ \tag{3}$$

where, $\xi_j=2\pi(n_j+ik_j)/\lambda$, t_j^+, and t_j^- are optical parameters of organic layer j. Most importantly, the effects of optical spacer on the enhancement of light absorption in active layers have been disclosed, as shown in Fig. 2.[21][22] For instance, by inserting a ~10 nm exciton blocking layer, e.g., bathocuproine (BCP) layer [21] or hole blocking layer titanium oxide (TiOx) [22] between acceptor and cathode would obviously redistribute more light density in active films, and higher J_{SC} and PCE were obtained accordingly.

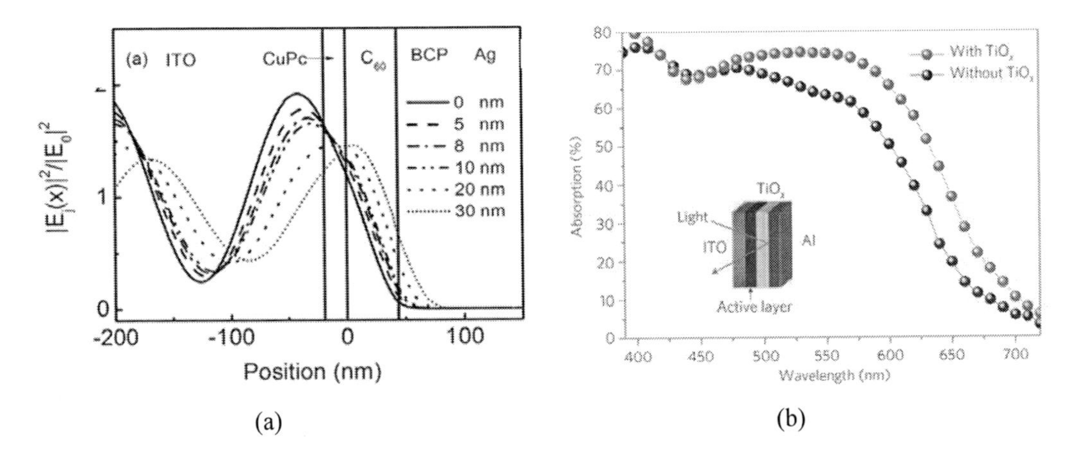

Figure 2. (a) The normalized electric field $|E_j(x)|^2 / |E_0(x)|^2$ distribution of the OPVs with a structure of ITO/CuPc (20 nm)/C_{60} (40 nm)/BCP (0, 5, 8, 10,20, 30 nm)/Ag (100 nm), adopted from [21]; (b) Total absorption in the active layer measured in a reflection geometry with the TiOx layer (red symbols) and without the TiOx layer (black symbols). The inset shows a schematic of the device structure, adopted from [22].

2.2. Exciton Diffusion

The next step is exciton diffusion with a diffuse length of L_D, and η_{ED} gives the percentage of photo-generated excitons that diffuse to a D/A interface. Firstly, continuum steady state exciton diffusion equation should be solved for planar structure with separated donor and layers [19][21]

$$L_D^{j\,2}\frac{d^2e}{dx^2}-e+\tau_j G_j=0 \tag{4}$$

where j corresponds to the sequence of active layer, $e(x)$ is the exciton density in active layers, $L_D=(D\tau)^{1/2}$, D is the exciton diffusion constant, τ is the mean lifetime of the exciton, $G(x)$ is the exciton generation rate. Assuming that donor/acceptor and acceptor/cathode interfaces act as perfect sinks for the excitons, i.e., all excitons can either dissociate into free charges or recombine at the interfaces resulting in boundary conditions at the interface that $e(x)$=0. If the exciton blocking layer (EBL) or hole blocking layer is added, excitons would not quench at acceptor and EBL interface, thus, boundary condition de/dx=0. As a result, the photocurrent current density at the interface $x=x_{DA}$ can be derived as:

$$J_{Exc} = \eta_{CT}\eta_{CC} D \left|\frac{de}{dx}\right|_{x = x_{DA}} \tag{5}$$

where η_{CT} is exciton dissociation efficiency at D/A interface into free charges, η_{CC} is charge collection efficiency at two electrodes.

2.3. Exciton Dissociation

Researchers have pointed out that a thin film organic PV cell composed of a D/A HJ with η_{CT}=0.95% [1] shows much higher efficiency than that of η_{CT}≈0.01 % [23] in single layer due to the efficient dissociation of excitons at the D/A interface. A schematic energy diagram of the D/A HJ is shown in Fig. 3 presenting ionization potential IP (or highest occupied molecular orbital (HOMO)) and electron affinity EA (or lowest unoccupied molecular orbital (LUMO)) of donor and acceptor. The exciton binding energy (E_B) of each material is equal to the difference between transport gap (E_{tran}) and optical gap (E_{opt}). The vacuum energy (E_{vac}) is the point of zero energy. After excitons arrive at D/A interface, the electron transfer rate from E_i in donor to LUMO energy EA_A of acceptor E_f can be explained by nonadiabatic Marcus theory as K_{if} [24][25]

$$K_{if} = \left(\frac{4\pi^3}{h^2\lambda_{if}K_BT}\right)^{1/2} V_{if}^2 \exp\left(-\frac{\left(E_{if}+\lambda_{if}\right)^2}{4\lambda_{if}K_BT}\right) \tag{6}$$

where i and f represent initial and final energy levels, V_{if} is the electronic coupling matrix element, λ_{if} is the molecular reorganization energy, and E_{if} is the free energy difference between frontier energy levels i and f, expressed as $E_{if}=E_i-E_f$. Then, coulombically bound polaron pairs would form, sometimes termed as an “exciplex”.[26] Meanwhile, the polaron pairs can also recombine with an energy transfer rate K_{CVD} from EA_A to IP_D-$E_{B,D}/2$. As deduced in previous research, exciton dissociation efficiency in donor layer can be simplified as $\eta_{CT}=K_{CCD}/(K_{CCD}+K_{CVD})$.

Recently, Kodama et al. investigated how the charge carrier separation proceeds between C_{60} and zinc phthalocyanine (ZnPc) at different intermolecular distances by using Ehrenfest dynamics simulation on the basis of time-dependent density functional theory. [27] The origo (*p*-phenylenevinylene) (OPV1) periphery was attached to the side of ZnPc molecule to utilize its light-harvesting property.

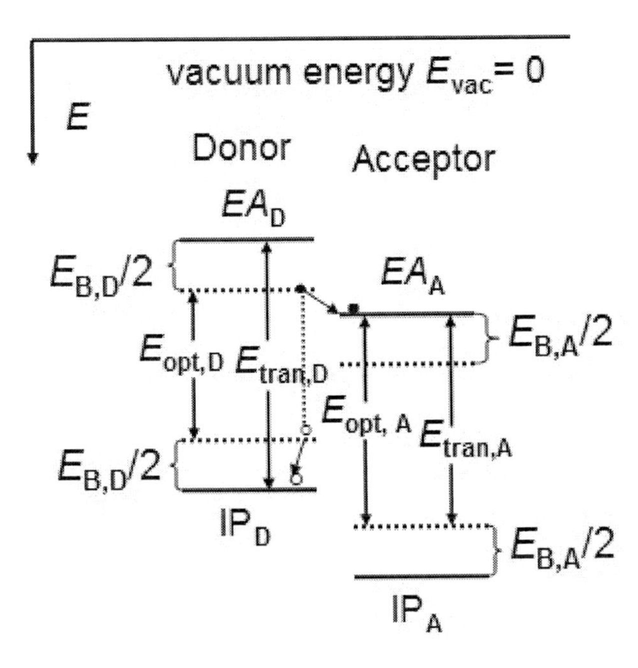

Figure 3. Schematic energy level diagram of an organic heterojunction between a donor (D) and an acceptor (A) layer. The process of charge transfer from D→A is also illustrated. Adopted from [25].

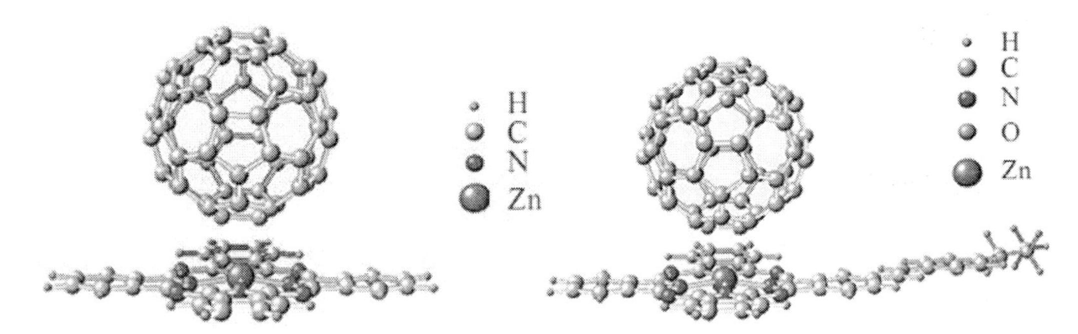

Figure 4. Optimized structure composed of C_{60} and ZnPc or C_{60} and ZnPc-OPV1. Adopted from [27].

Fig. 4 shows the optimized structure of a system composed of C_{60} and ZnPc or C_{60} and ZnPc-OPV1. For both C_{60}-ZnPc and C_{60}-ZnPc-OPV1 systems, one single bond of C_{60} faces the zinc atom, being parallel to one N-Zn-N bond axis (OPV1 is attached in this direction) and perpendicular to the other N-Zn-N bond axis. For C_{60}-ZnPc, the distance between zinc atom and carbon atoms belonging to the C_{60} single bond is 2.42 Å. For C_{60}-ZnPc-OPV1, this carbon-zinc distance is 2.41 Å in the OPV1 side and 2.45 Å in the other side. The dependence of energy level on the intermolecular distance (d) is also shown in Fig. 5.

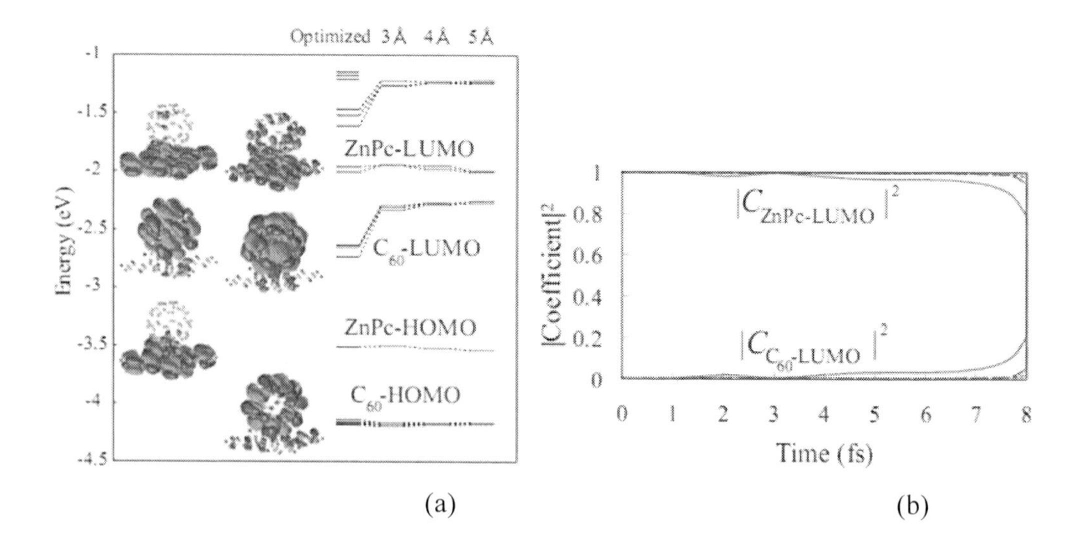

Figure 5 (a) Energy eigenvalues and the corresponding wave functions of a system composed of C_{60} and ZnPc for the optimized geometry and the geometry with intermoleculer distances of d=3, 4, and 5 Å; (b) Time evolution of the absolute squared coefficients for the ZnPc and C_{60} eigenstates in the wave packet of ZnPc-LUMO level. Solid, dash, and dotted curves correspond the systems with d=3, 4, and 5 Å. Adopted from [27]

It can be seen that when the intermolecular distance increases, the C_{60} or ZnPc character is almost preserved in each level. As a result, the mixing ratio between eigenstates of ZnPc-LUMO and C_{60}-LUMO in the system of d=3 Å is much higher than that in the systems of d=4 and 5 Å, which implies that the net probability of electron transfer from the ZnPc-LUMO to C_{60}-LUMO is the largest at d=3 Å. On the contrary, the eigenstates of ZnPc-HOMO and C_{60}-HOMO levels are hardly mixed and stay at the pure eigenstate of ZnPc-HOMO regardless to the intermolecular distances. This suggests the electron transfer process from ZnPc to C_{60} and hole immobility at ZnPc, and thus, charge separation achieves. In addition, the charge separation becomes the most prominent when the intermolecular distance between ZnPc and C_{60} is d=3 Å. These dynamics simulations can be applied in other OPVs with D/A heterojunction and by attaching side branches to donor material, e.g., widened photo-absorption range.

2.4. Charge Collection

Electrons and holes generated from excitons dissociation at the D/A interface will transfer in the acceptor and donor layer and then get collected by anode and cathode, respectively, as shown in Fig. 6. As for charge carrier transportation in active films with a planar HJ, total current in the donor layer equals the hole current[28]

$$J_D = q\mu_D[pF(V,x) - \frac{kT}{q}\frac{dp}{dx}] \tag{7}$$

where, p is hole density, $F(V, x)$ is electric field heterogeneously distributed in active films, and V is applied bias. Various solutions of $\eta_{CC}(V)$ can be deduced from Eq. (7). As pointed out in almost works with high *PCE* performance, $\eta_{CC}(V=0)$ can be simply assumed as ~100% for planar HJ OPVs.[19][29]

When the device structure is both hybrid planar-mixed heterojunction (PM-HJ) for small molecule and bulk heterojunction (BHJ) for polymers, percolated pathways for donor and acceptor should firstly form by the means of complete phase separation, as shown in Fig. 6(b). For the complex mixed film, the determination of $\eta_{CC}(V)$ is more difficult than that for planar OPVs. Herein, the analyzing result of $\eta_{CC}(V)$ proposed by Sokel and Hughes is adopted as[30][31]

$$\eta_{CC}(V) = [\frac{\exp[e(V_0 - V)/kT]+1}{\exp[e(V_0 - V)/kT]-1} - \frac{2kT}{e(V_0 - V)}] \tag{8}$$

where, V_0 is compensation voltage determined by solving J_{light}-J_{dark}=0. J_{dark} and J_{light} is current density under dark condition and light illumination, respectively.

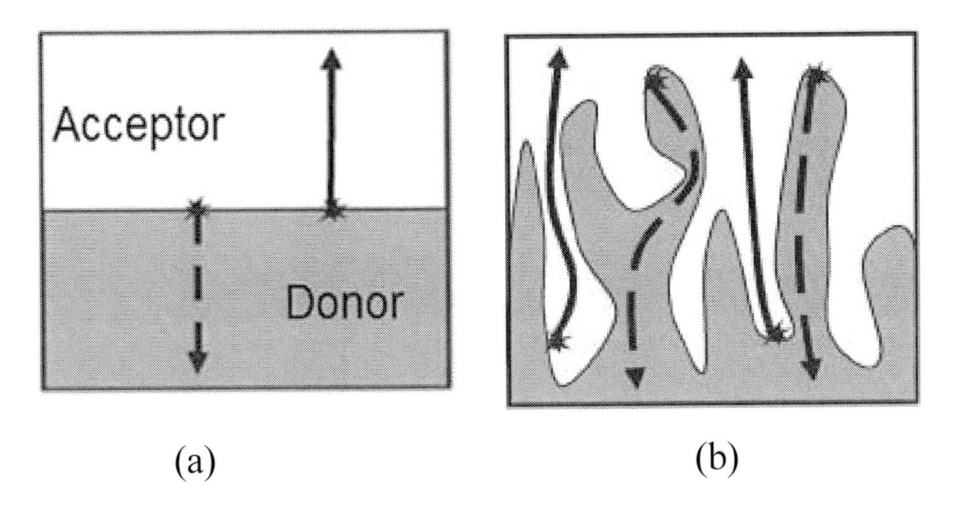

Figure 6. (a) Charge transportation in active films with a planar HJ and (b) an optimal planar-mixed heterojunction (PM-HJ) or polymer bulk heterojunction (BHJ).

2.5. Equivalent Circuit Model

Based on the discussion above, the detailed simulation for physical photovoltaic process in the active layers is too complex. As an alternative, a conventional equivalent circuit model from inorganic solar cell is usually introduced to simulate photovoltaic performance of OPVs, e.g., *J-V* characteristics, as shown in Fig. 7(a). [25][32] Then the *J-V* characteristics can then be expressed by the generalized Shockley equation:

$$J = \frac{R_p}{R_S + R_p}\left[Js\left\{ \exp(\frac{q(V - JR_S)}{nK_B T}) - 1 \right\} + \frac{V}{R_p} - J_{ph}(V) \right] \tag{9}$$

Therefore, crucial parameters judging the qualities of an OPV, e.g., series resistances R_S, parallel resistances R_P, the diode ideality factor n, the reverse saturation J_S and photocurrent densities J_{ph} can be extracted based on Eq. (9).

However, instinctive difference between inorganic silicon solar cell and OPVs is discovered that intermediate states, known as coulombically bound polaron pairs, would firstly form in OPVs before exciton dissociation.[26] Evidence for such intermediate states was observed from the long wavelength emission of polymer blends.[26][33][34] According to this photovoltaic process, it's more appropriate to consider polaron-pairs generation rate as a constant for a given light intensity rather than current. Thus, an improved equivalent circuit is proposed by Mazhari[35] and revised by Huang et al. [21][36], as shown in Fig. 7(b), which can be used to all kinds of organic solar cells. J_p is polaron density, assumed as constant, while J_{ph}(V) is dependent on the extra bias. D_{rec} represents the loss due to polaron-pair recombination, and D_{ext} model polaron-pair dissociation.

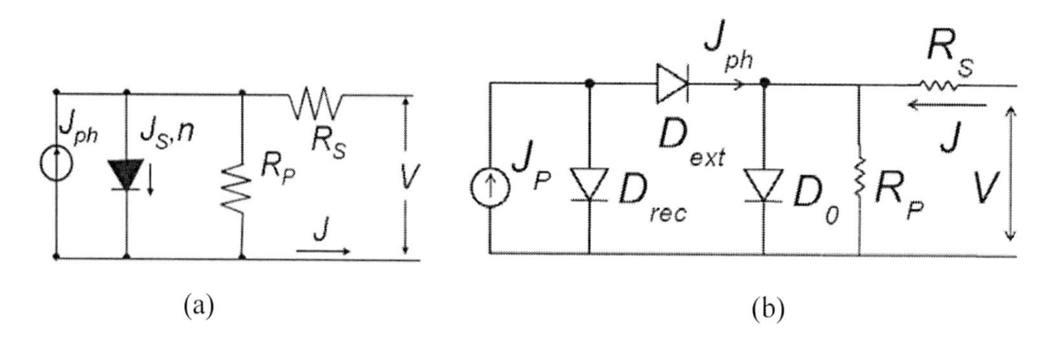

Figure 7 (a) Normal equivalent circuit for an organic photovoltaic cell; (b) an improved equivalent circuit for an organic photovoltaic cell. Adopted from [21]

3. Organic Photovoltaic Semiconductors

Organic electronic or optoelectronic materials are conjugated semiconductors where both optical absorption in the ultraviolet (UV)-visible part of the solar spectrum and electric current transportation are due to partly delocalized π and π* orbitals. By introducing novel materials and improved material engineering during the past 30 years especially in the last decade, a rapid enhancement in power conversion efficiency has been achieved. Generally, there are two major classes of organic semiconductors, i.e., the small molecular semiconductors with molecules weight about several hundred such as pentacene, C_{60}, and the π-conjugated polymers such as P3HT and MDMO-PPV (poly[2-methoxy-5-(3,7-dimethyloctyloxy)]- 1,4-phenylenevinylene). Herein, to review and specify the typical properties of organic photovoltaic materials to indicate future material engineering, a brief summarization of some intensely studied electronic donor and acceptor materials, and also other functional materials will be proposed.

3.1. Donor Materials

3.1.1. Small Molecular Donor Materials

Porphyrins e.g., MTPP and MOEP where M refers different metal center were firstly studied as photoactive materials to fabricate OPVs, partly as the molecular structure of porphyrin is a synthetically tractable analogue of chlorophyll, which is similar to the molecules used in plants to collect photons and generate energy via photosynthesis.[37] With extensively conjugated π systems, porphyrins are able to quickly transfer electron to an acceptor, they also possess high absorption coefficients in blue light region and moderately in the green regions of the visible spectrum. Moreover, via synthetic modification or metal insertion into the cavity, the redox properties can be tunable.[38] These attributes make porphyrins attractive semiconductors for the active layer of OPV devices. However, the short exciton diffusion length L_D (<10nm) of porphyrins and the relatively low charge carrier mobility limit their power conversion efficiency *PCE.* [39]

Compared with porphyrins, materials with functional group phthalocyanines (PC) offer various advantages as potential active layer in OPVs, such as higher hole mobility,[40] generally longer exciton diffusion length,[41] and an absorption over a wider spectral range.[42] Chemical structures of several PC derivatives with different metallic centers are presented in Fig. 8.

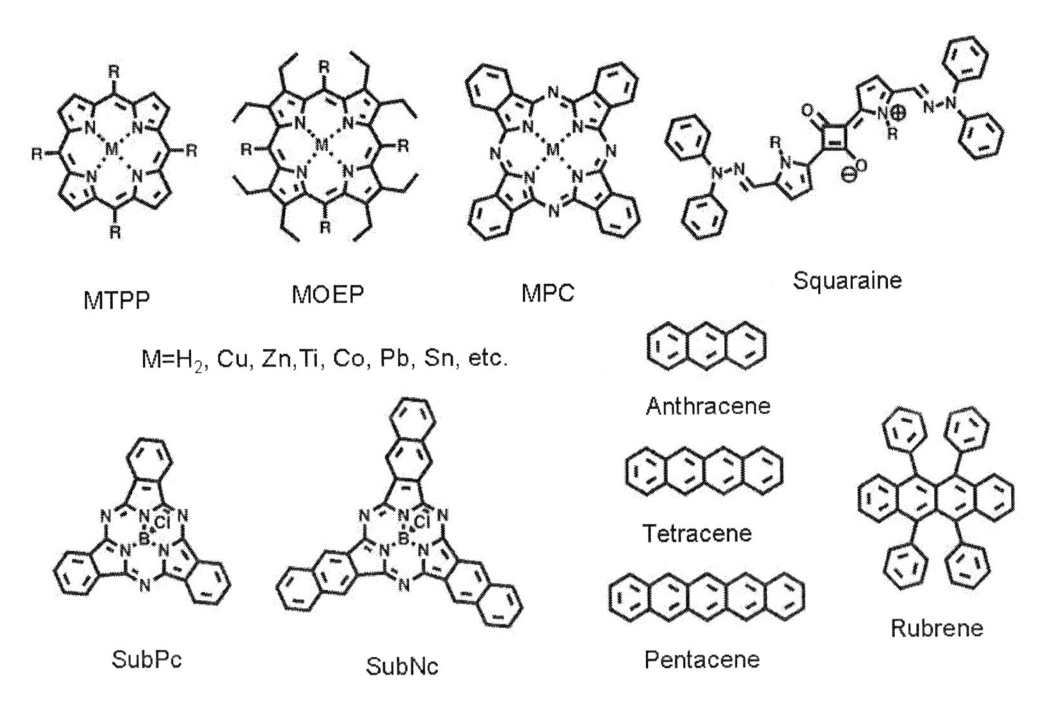

Figure 8. Chemical structures of small molecular electronic-donor materials in OPVs. MTPP= metallotetraphenylporphyrin; R- is usually a phenyl derivative; MOEP= metallooctaethylporphyrin; MPC= metallophthalocyanine; SubPc= chloro[subphthalocyanine] boron(III); Rubrene= 5,6,11,12-tetraphenyltetracene.

Based on the seminal work of Tang in 1986, CuPc and 3,4,9,10-perylenetetracarboxylic-bis-benzimidazole (PTCBI) were vacuum-evaporated on an ITO-coated glass substrate forming a planar HJ OPVs.[1] In addition to perylene diimide acceptors, fullerenes such as C_{60} are frequently paired with Pcs in OPVs as an active layer.[4] Notably, the D/A combination based on CuPc/C_{60} is among the most commonly studied for small molecule OPVs at present, and high *PCEs* were obtained via device structure optimization.[4][43][44][45] However, relatively low $V_{OC}\approx0.55$ V limits further improvement *PCE*>6% due to its relatively high HOMO level 5.2 eV, as summarized in Table 1.

Table 1. Characteristics of small molecular donor materials.

Material	HOMO (eV)	LUMO (eV)	L_D (nm)	μ_h (cm^2/Vs)	Ref.
pentacene	5.1	3	65±16 [a]	1.5×10^{-1} [b]	[29][46][47]
CuPc	5.2	3.2	10±3 [a]	7.4×10^{-4} [b]	[19][46][48]
ZnPc	4.8	3.3	30±10 [a]	0.022 [c]	[39][49][50]
SnPc	5.8	3.2	-	$2\pm1\times10^{-10}$ [b]	[51]
TiOPc	5.4	4	-	9×10^{-6} [c]	[52][53]
SubPc	5.6	3.6	12±2 [a]	-	[8][54]
SubNc	5.4	3.6	23±5 [a]	-	[8][55]
ClAlPc	5.3	4	5±1 [a]	-	[8][56]

[a] Determined by fitting EQE of single planar heterojunction cells.
[b] Measured by the space charge limited current method.
[c] Measured by the organic field effect transistor measurement.

Similar limitations have also been pointed out based on D/A pairs of pentacene, tetracene, zinc phthalocyanine (ZnPc) or tin-phthalocyanine (SnPc) with C_{60},[57][49] although pentacene has long exciton diffusion length $L_D\approx65$ nm and high hole mobility $\mu_h=1.5\times10^{-1}$ cm^2/Vs in organic field effect transistor measurement and SnPc has high extinction coefficient covering 650 nm to 900 nm,[51] as shown in Fig. 9.

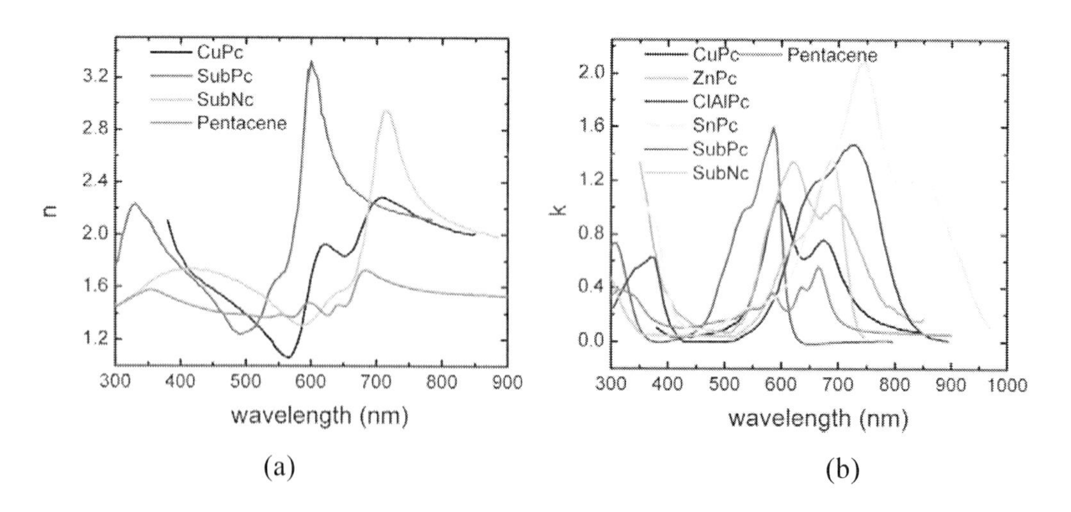

Figure 9. (a) Refraction $n(\lambda)$ and (b) extinction coefficient $k(\lambda)$ of small molecule donor materials in OPVs. Adopted from Refs. [20][46][58][59][60].

To further optimize *PCE* by improving V_{OC} parameter, decreased HOMO level about 5.4 eV has obtained from modified material TiOPc by simply replacing metal center of CuPc.[53]

Subsequently, cholro [subphthalocyanine]boron(III) (SubPc) with low HOMO level 5.6 eV is employed in a bilayer OPV with C_{60} to enhance V_{OC}≈1 V. [7][54][58] Meanwhile, SubPc has high extinction coefficient k=1.6 around 580 nm,[54] which can be adopted in tandem cells to balance the absorption amount of two subcells.[8] Moreover, due to its nonplanar geometry contrasts with classical Pc rings and imparts, SubPc exhibits a high solubility and a low tendency to aggregate.[55] Moreover, choro- [subnaphthalocyanine]boron(III) (SubNc) was fabricated via the judicious modification of fused benzene rings to the trio of isoindole units.[55] Because of the greater conjugation imparted by the additional benzene rings, SubNc exhibits a reduced tendency to aggregate and a broader red-shifted absorption spectrum, but a higher HOMO energy comparing to SubPc.[61] Polyacenes such as anthacene,[62][63] tetracene[64][65][66] and pentacene [29][67][68][69][70] as well as phenyl-substituted 5,6,11,12-tetraphenyltettures, rubrene,[68][71][72] were also employed in OPV as donors. They have also been used as dopant in common systems, such as CuPc/C_{60} and ZnPc/C_{60}.[73][74] For example, the photovoltaic performance could be improved by doping pentacene, as the high mobility of pentacene provides percolation paths that can effectively improve hole mobility in CuPc mixture layer and result in a better mobility balance between the CuPc mixture layer and the C_{60} layer.

Squaraines have recently been reinvestigated [75][76] and effectively applied to small molecule OPV devices, after initial appraisals[77][78] produced device efficiencies of only ~0.02%. By varying the amine donors or aryl groups, it is possible to design a number of symmetrical or unsymmetrical squaraines whose thin film absorptions extend into the near infrared about 1000 nm, and high extinction coefficients and acceptable hole mobility $\mu_h \approx 10^{-4}$-10^{-5}cm^2/Vs in OTFT measurement for electro-optic devices, making them excellent candidate for OPV application.[75]

3.1.2. Polymer Donor Materials

In the first report of solution processed BHJ solar cells, Yu et al. used poly [2-methoxy,5-(2′-ethyl-hexyloxy)-p-phenylene vinylene] (MEH-PPV) as an electronic donating material, [79] which were later substituted by easier processable MDMO-PPV. Due to the relatively large bandgap energy and low mobility of PPV-type polymers, highest *PCE* remained unimproved,[80][81][82] and the general interest in this material type gradually faded. During the last five years, research efforts have focused on poly(alkyl-thiophenes), in particular on P3HT, for its high mobility in pure film e.g., $\mu_h=1\times10^{-1}$ cm^2/Vs in the OTFT measurement,[83] good solubility, nano-crystalline property in blend film with fullerene with a $\mu_h=2\times10^{-4}$ cm^2/Vs measured by the means of photogeneration and charge extraction by linearly increasing voltage (photo-CELIV) measurements[84] and relatively low bandgap 1.7 eV,[85] as shown in Table 2.

By carefully controlling the morphology of the BHJ active layer P3HT : PCBM combined with polymer nanospheres and fullerene clusters, *PCE*> 5% have been achieved.[92][93] However, the limitations of P3HT based solar cells such as high band gap with relatively high HOMO level, which limits the absorption wavelength from 420 to 560 nm and high exciton binding energy, as a result, relatively low V_{OC}≈0.6 V performance becomes the bottle neck to hamper P3HT : PCBM solar cells from achieving *PCE*>6%.[94] Thus, to further improve V_{OC} and *PCE* of OPVs without decreasing J_{SC} and *FF*, several novel polymer donor materials have been synthesized, as shown in Fig. 10.

Table 2. Characteristics of various polymer donor materials

Material	HOMO (eV)	LUMO (eV)	hole mobility (cm^2/eV)	Ref.
MDMO-PPV	5.3	2.8	5×10^{-7} [b]	[85][86]
P3HT	5.2	3.5	1×10^{-1} [a] 2×10^{-4} [d]	[85][83][84]
poly(thiophene-thienylenevinylene)	4.93	2.96	-	[87]
APFO-Green5	5	3.4	8×10^{-4} [a]	[88]
PF10TBT	5.4	3.4	-	[100]
PCDTBT	5.5	3.6	-	[22]
PCPDTBT	5.3	3.57	2×10^{-2} [a c]	[89]
PTB1	4.9	3.2	4.5×10^{-4} [b]	[90]
PTB4	5.12	3.31	7.7×10^{-4} [b]	[90]
PBnDT-4DTBT	5.26	2.94	3.8×10^{-5} [b]	[91]
PBnDT-DTBT	5.33	3.17	1.6×10^{-5} [b]	[91]
PBDTTT-E	5.01	3.24	4×10^{-4} [b]	[14]
PBDTTT-C	5.12	3.35	2×10^{-4} [b]	[14]
PBDTTT-CF	5.22	3.45	7×10^{-4} [b]	[14]

[a] Measured by organic field effect transistor measurement;
[b] Measured by space charge limited current;
[c] Not been corrected for serial resistance of the OFET;
[d] measured by photo-CELIV measurements.

One approach to improve the photocurrent is to strengthen the absorption ability of P3HT through the UV and visible regions, e.g., poly(thiophene- thienylenevinylene) in the type of poly(3-vinylthiophenes).[87] The incorporation of chromospheres conjugated to the backbone through the 3-vinyl linkage leads to a broadened absorption spectrum from 400 to 500 nm, and thus, higher J_{SC} and *PCE* in poly (thiophene-thienylenevinylene) : PCBM blend system can be obtained comparing with that in P3HT : PCEM.

The second optimizing method is to decrease the HOMO level of polymers by material engineering. One of the commonly used strategies is to adjust HOMO, LUMO levels and bandgap energy of materials by the means of repeating electron-rich group (donor, D) with deep HOMO levels e.g., thiophene, cyclopentadithiophene, fluorene or carbazole [17] and electron-poor units (acceptor, A) with high LUMO energy such as 2,1,3- benzothiadiazole [95][96] to form the polymer backbone, namely "donor-acceptor approach" or $(-D^+-A^--)_n$ structure.[17] The main reason is that inter-charge transfer effect can increase the double-bond character rather than the single bond in the polymer. As a result, HOMO and LUMO energy levels can be effectively modified based on the reduced bond-length alternation effect.[17] Accordingly, several promising materials with different polyfluorene derivatives such as APFO, APFO-Green5, PF10TBT have been synthesized by Andersons and co-workers, which exhibited prominent photovoltaic performances.[97]-[100] For instance, the PF10TBT : PCBM blend system had a *PCE*= 4.2% (AM1.5 corrected for the spectral mismatch) due to the enhanced V_{OC} =1 V for decreased HOMO level ~5.4 eV and maintained high J_{SC} for high absorption coefficient, as shown in Fig. 11. Also, Park et al. fabricated a novel PCDTBT comprising carbazole and 2,1,3-benzothiadiazole as repeating $(-D^+-A^--)_n$ structure.[22] And BHJ OPVs based on PCDTBT : $PC_{70}BM$ blend (1:4 ratio) achieved an efficiency *PCE* as high as 6.1% due to improved V_{OC}=0.88 V by decreased HOMO about 5.5 eV.

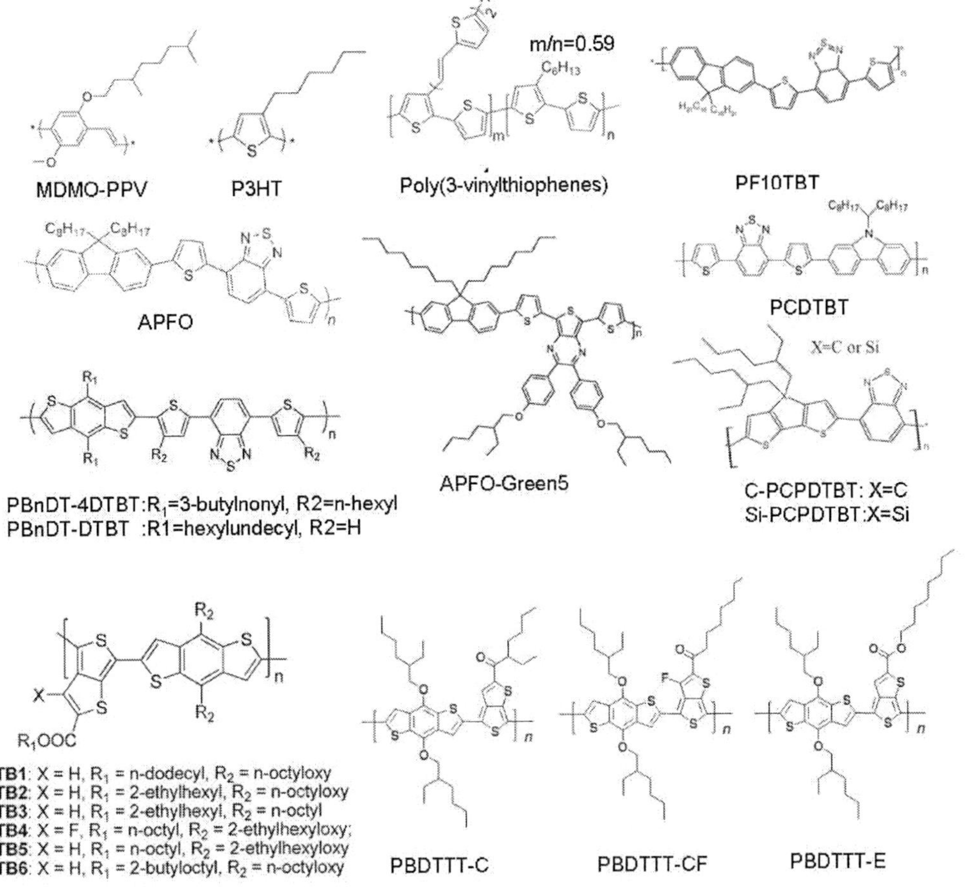

Figure 10. Commonly used and alternative polymers as donor materials: PF10TBT=poly[9,9-didecanefluorene-alt-(bis-thienylene) benzolthiazole]; APFO= poly[{2,7-(9,9-dialkylfluorene)}-alt-{5,5-(4,7-di-2′-thienyl-2,1,3-benzothiadiazole)}]; PCDTBT= poly[N-9”-hepta-decanyl-2,7- carbazole-alt-5,5-(4′-7′-di-2-thienyl-2′,1′,3′-benzothiadiazole); PCPDTBT= poly[2,6-(4,4-bis- (2-ethylhexyl)-4H-cyclopenta[2,1-b;3,4-b2]-dithiophene)-alt-4,7-(2,1,3-benzothiadiazole)]; PBnDT- 4DTBT=poly[4,8-dialkylbenzo[1,2-b:4,5-b′]dithiophene-alt-4,7-di(4-hexyl-2-thienyl)-2,1,3- benzothiadiazole]; PBDTTT= poly[4,8-bis-substituted-benzo[1,2-b:4,5-b′]dithiophene-2,6-diyl-alt-4- substituted-thieno[3, 4-b]thiophene-2,6-diyl].

Thirdly, to achieve a higher photocurrent, much attention was paid to lower bandgap systems. The cyclopentadithiophene-based polymers were the first low bandgap series presenting high OPV performance.[101][102][103] PCPDTBT as the most prominent candidate of this class of polymers has a measured optical bandgap of about 1.45 eV, as well as high hole mobility μ_h= 2×10^{-2} cm^2/Vs measured by OTFT device.[89] Due to a broad absorption spectrum, PCPDTBT : PCBM blend with a ratio 1:1 by weight shows photocurrent production at wavelength longer than 900 nm.

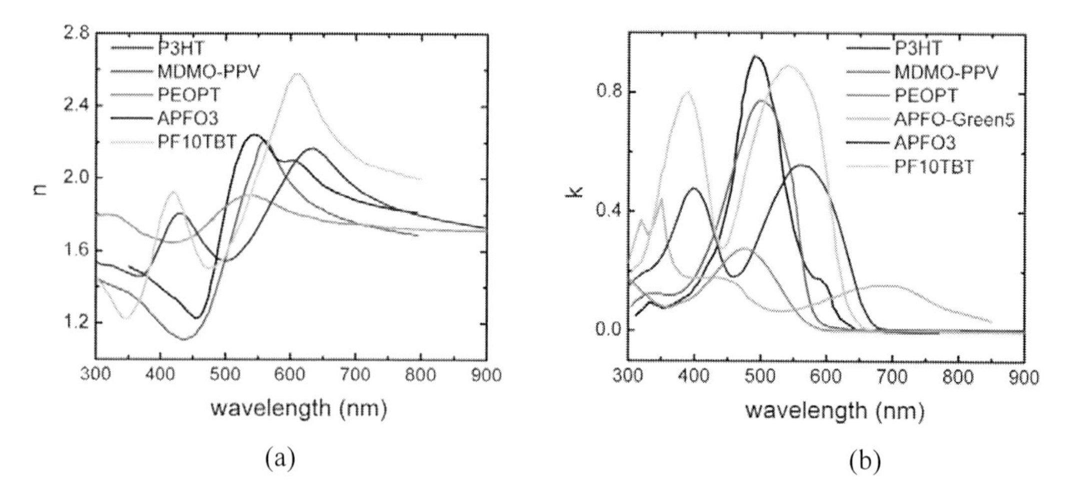

Figure 11. (a) Refraction $n(\lambda)$ and (b) extinction coefficient $k(\lambda)$ of small molecule donor materials in OPVs. Adopted from Refs. [85][104][105][106][107][108][109].

Moreover, Yu et al. developed a series of polymers namely PTB1 to PTB6 based on alternating donor type benzodithiophene and acceptor type thieno[3,4-b]thiophene units with different HOMO levels and almost decreased optical bandgap about 1.6 eV.[90] Among them, PTB4 : PCBM blend system had the highest performance PCE=5.9% due to the lowest HOMO of 5.12 eV and high J_{SC}=13.9 mA/cm^2. Chen and Yang et al. developed three novel materials as PBDTTT-C, PBDTTT-E and PBDTTT-CF with synchronously decreased HOMO and LUMO levels. Again, PBDTTT-CF : PCBM system had the highest PCE=7.73% due to highest V_{OC}=0.76 V for the lowest HOMO 5.22 eV and the highest J_{SC}=15.2 mA/cm^2 for hole mobility 7×10^{-4} cm^2/Vs of blend films.[14] The excellent performance of thieno[3,4-b]- thiophene-based polymers can be attributed to the high oxidation potential offered by fusing an electron deficient benzene with two flanking thiophene units, as well as the entirely planar and symmetrical structure, which is an important prerequisite in order to achieve high mobility.[91]

As a conclusion, an ideal donor polymer for a BHJ solar cell should have a low bandgap to maximize light absorption and a high hole mobility. The low HOMO energy level of donor is needed to achieve a high V_{OC}. Meanwhile, high molecular weight and good solubility are also required to achieve the optimal morphology, which is important to maximize J_{SC} and FF.[91]

3.1.3. Acceptor Materials

In 1986, Tang fabricated the first heterojunction organic solar cell using PTCBI as electron acceptor, and CuPc as electron-donor. Then, perylene diimides such as PTCDA and PTCDI have emerged as small-molecular acceptor materials mainly attributed to their characteristics such as unique liquid crystalline behavior, a large absorption coefficient, high electronic affinity and mobility, and excellent photochemical and thermal stability. [114][115] For instance, PTCDI with higher absorption ability was used instead of C_{60} as acceptor and coevaporated with pentacene to form PM-HJ structure. Thus, high J_{SC}=8.6 mA/cm^2 and PCE=2.0% were achieved under an incident light power of 80 mW/cm^2.[105]

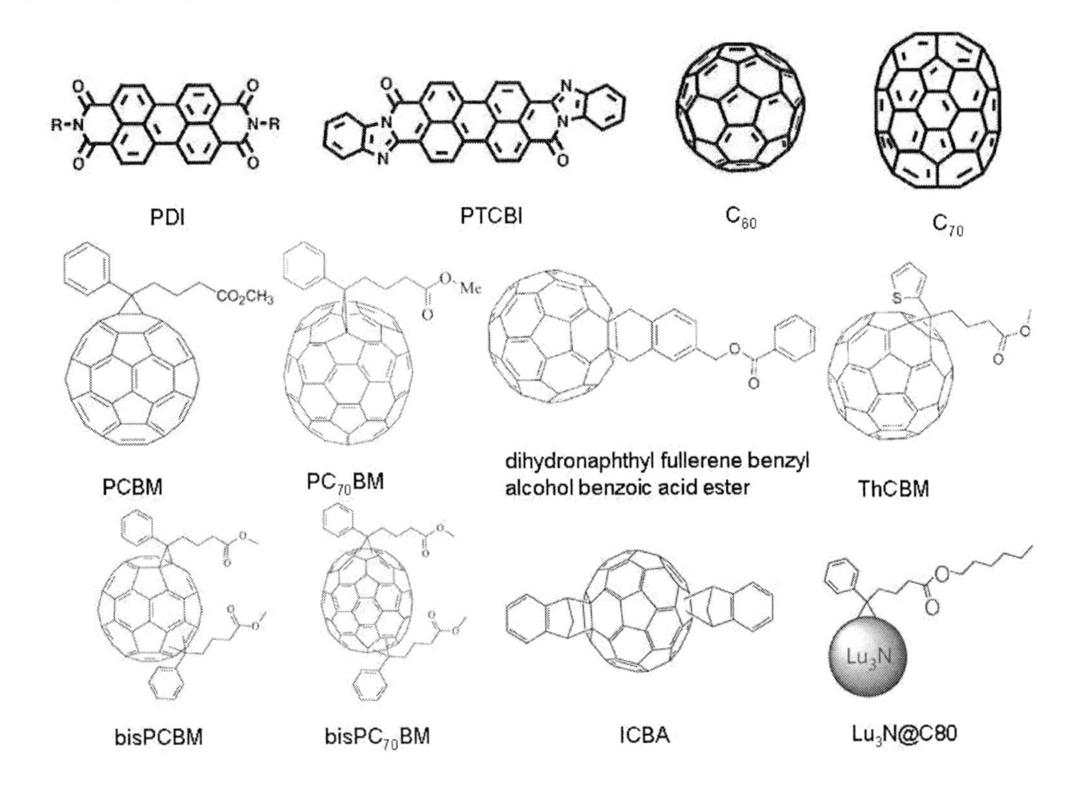

Figure 12. Commonly used and alternative acceptor materials: PDI= perylene diimide; PTCBI = 3,4,9,10-perylenetetracarboxylic bis-benzimidazole; PCBM = [6,6]-phenyl-C_{61}-butyric acid methyl ester; $PC_{70}BM$= [6,6]-phenyl-C_{71}-butyric acid methyl ester; ThCBM = 1-(3-methoxycarbonyl) propyl-1-thienyl-[6,6]-methanofullerene; ICBA = indene-C_{60} bisadduct.

As one of the most intensely studied materials, fullerenes are considered as ideal acceptor materials for both small molecular and polymer OPVs, as they have a set of truly unique characteristics such as high electron affinity, long exciton diffusion length as well as high electron mobility. The triply-degenerated LUMO of fullerenes allows the molecule to be reversibly reduced with up to six electrons, thus illustrating its capability to stabilize negative charge.[116] Importantly, the photo-induced charge transfer from excited donors is several orders of magnitude faster than back-transfer.[129]

For polymer OPVs, although several acceptor molecule have been tested, such as conjugated polymers, carbon nanotubes, perylenes, and inorganic semiconducting nanoparticles,[130] only derivatives of C_{60} and C_{70} have provided high efficient devices. As polymer OPVs are almost fabricated from solution process, fullerene derivatives with organic solubilizing groups are synthesized to prevent C_{60} from aggregating. PCBM was firstly used in OPV devices in 1995,[131] and since then, no significant better acceptors have been found despite the fact that the position of the HOMO and LUMO levels and the optical absorption are not ideal for most polymer donor materials. It is worth to mention that $PC_{70}BM$ often exhibits higher absorption, and consequently, enhances photocurrent due to its spherical unsymmetry, as shown in Fig. 13.

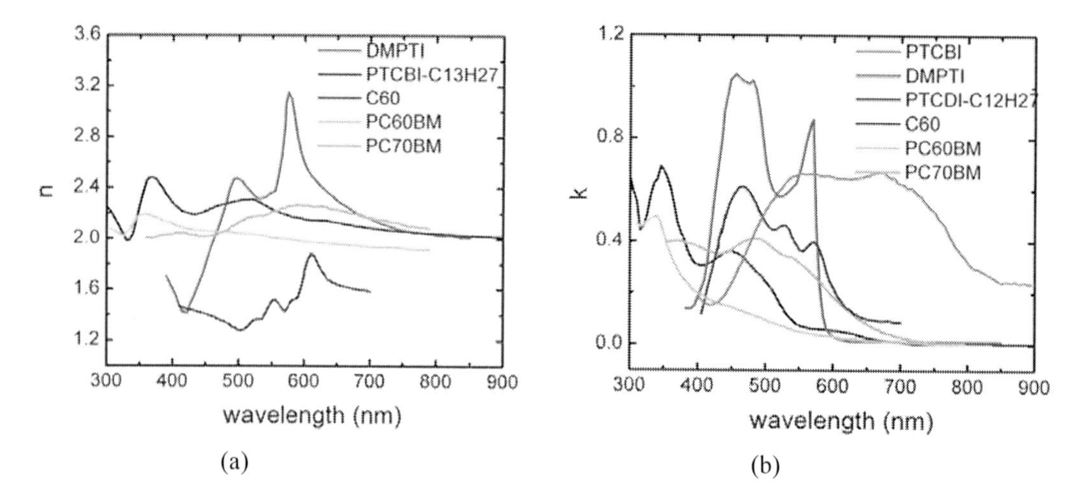

Figure 13. (a) Refraction and (b) extinction coefficient of common acceptor materials in OPV devices. Adopted from Refs. [19][20][60][107][108]

A variety of other soluble C_{60} and C_{70} derivatives [132][133][134][135][136] have been synthesized to improve the miscibility, charge carrier mobility, or other aspects of performance that are influenced by the structure of soluble fullerene, which are summarized in Table 3. However, the substitution of fullerene with a various solubilizing groups has only led to small variation in electronic structure, which can give the highest boost in efficiency.[17]

Table 3. Characteristics of different acceptor materials

Material	HOMO (eV)	LUMO (eV)	Mobility (cm^2/eV)	Ref.
PTCBI	6.1	4.4	2.4×10^{-6} b	[19][48]
PTCDA	6.8	4.6		[19]
C_{60}	6.2	4.0	5.1×10^{-2} b	[48][118][119]
C_{70}	6.4	4.3	1.3×10^{-3} c	[120][121]
PCBM	6.1	4.3	2×10^{-7} c	[122][123]
$PC_{70}BM$	6	4.3	1×10^{-3} c	[22][124]
bisPCBM	-	4.2 a	7×10^{-8} c	[125]
ICBA	-	4.1 a	-	[126]
$IC_{70}BA$	5.8	4.1 a	-	[127]
$Lu_3N@C_{80}$-PCBH	-	4.0 a	4×10^{-4} b	[128]

a Estimated from the shift of the first reduction potential with respect to PCBM or $PC_{70}BM$

b Measured in blend films by means of space charge limited current

c Measured by organic field effect transistor measurement

As mentioned above, raising the LUMO of the acceptor will directly result in a higher V_{OC} without affecting the absorption of the cell. Hummelen et al. introduced bisPCBM and bis$PC_{70}BM$, which is the bisadduct analogue of PCBM and $PC_{70}BM$, respectively. LUMO level of bisPCBM was increased by 0.1 eV from PCBM,[125] as shown in Table 3. Recently, Zhao et al. reported a novel indene-C_{60} bisadduct (ICBA) with an increased LUMO level about 1.17 eV higher than that of PCBM,[17] and P3HT : ICBA blend system has a high

PCE=6.48% with V_{OC}=0.84 V.[137] Recently, Zhao et al. also synthesized indene-C_{70} bisadduct ($IC_{70}BA$), with the LUMO energy level 0.19 eV higher than that of $PC_{70}BM$ and a higher V_{OC} of 0.84 V in blend system with P3HT.[127] If more multiadducts added onto PCBM e.g., tris-PCBM, improving the V_{OC} by more than 200 mV can be obtained, but trapping of electrons in the tris-fullerenes has so far prevented an overall efficiency optimization.[138] Drees et al. proposed a method of modifying the LUMO level by introducing novel trimetallic nitride endohedral into fullerenes as an acceptor for OPV applications.[128] As a result, the reduced energy offset of $Lu_3N@C_{80}$-PCBH to P3HT in charge transfer process and increases V_{OC} to 260 mV above to reference devices made with PCBM acceptor.

After intensive investigation of numerous materials in the past decade, it has been demonstrated that several characteristics of acceptor materials should possess for a BHJ solar cell: strong absorption complementary to the absorption profile of the donor material to maxmize the photon absorption; modified LUMO level to ensure an optimized LUMO-level offset, and subsequently efficient charge transfer and a higher V_{OC} at the same time; sufficient electron mobility in composites with the donor.[17] Promise has been held that by using the improved acceptor and combining it with previously reported low-bandgap donor polymers, OPV efficiencies greater than 10% would finally be achieved.

3.1.4. Functional Materials

Charge leakage between two electrodes and exciton quenching at the organic/ electrode interface represent two main loss mechanisms in many devices, which would commonly induce low shunt resistance R_P, low *FF*, low J_{SC}, and thus, lower *PCE*. Thereby, various functional materials serving as interfacial layers (IFL) have been introduced into device designation on both sides of the active layer to prevent these energy loss processes, among which organic materials are summarized in Fig. 14.

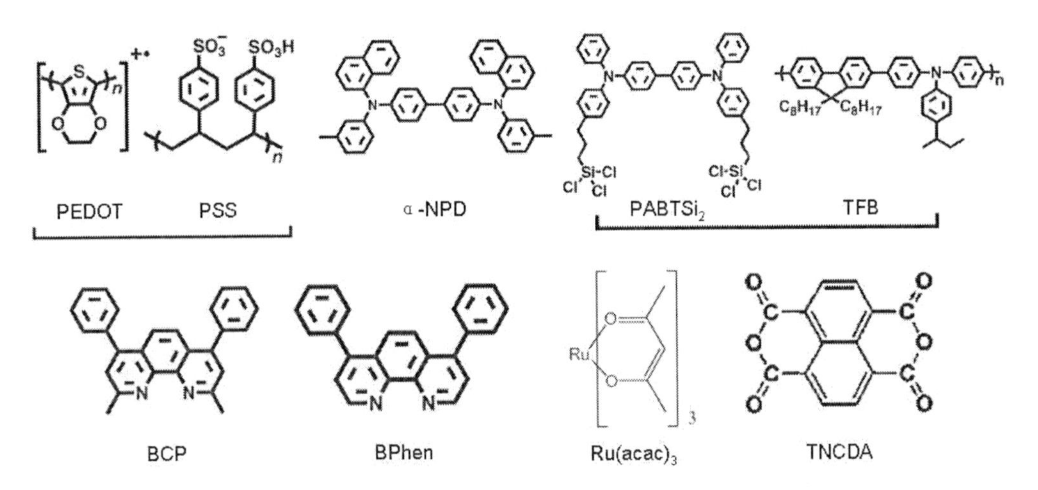

Figure 14. Commonly used functional materials as an interfacial layer: PEDOT = poly(3,4-ethylenedio- xythiophene); PSS=poly(styrene sulfonate); α-NPD=N,N′-di-1-naphthyl-N,N′- diphenyl-benzidene. TFB=poly[9,9-dioctylfluorene-co-N-[4-(3-methylpropyl)]-diphenylamine; TPDSi2=4,4′-bis [(p- trichlorosilylpropylphenyl)-phenylamino]biphenyl; BCP=2,9-dimethyl-4,7-diphenyl-1,10-phenan-throline; BPhen = 4,7-diphenyl-1,10-phenanthroline; Ru(acac)3=tris(acetylacetonato)ruthenium (III); TNCDA=naphthalene tetracarboxylic anhydride.

On the anode side of the OPVs, PEDOT:PSS is conventionally deposited between the ITO and active layer by spin-coating a blended aqueous dispersion of PEDOT and PSS onto clean ITO. The PEDOT:PSS film exhibits the following functions: (1) preventing surface spikes from shorting out the device by planarizing the ITO;[43][139] (2) enhancing the work function of anode e.g., ~4.7 eV of ITO to a consistent and reliable ~5.1 eV and forming ohmic contacts with almost donor materials;[43][82] (3) creating a more uniform surface conductivity than that of ITO,[140] and thus improving charge collection at the ITO surface; (4) blocking electrons and increasing the V_{OC} as compared to devices without an IFL.[141][142][143] However, there are still some problems with PEDOT:PSS. First, the electron-blocking character in PEDOT:PSS is unexceptional and incomplete. [141][142] Second, it can lead to undesirable cross-talk between multiple devices on a common substrate and is itself rather inhomogeneous. [144][145] Finally and most importantly it is highly acidic with pH ≈ 1, which is sufficient to corrode the underlying ITO substrate.[146] To overcome the aforementioned problems, efforts have been paid to replace PEDOT:PSS, and several organic materials e.g., NPD,[75] PABTSi$_2$:TFB[51], and inorganic materials e.g., WO_3[147] V_2O_5[148] MoO_3 [51] have been tested and produced similar or enhanced performance, their energy levels are summarized in Table 4 to expediently design device structure.

Table 4. HOMO (or valence band) and LUMO (or conduction band) levels of functional materials

Material	HOMO (eV) (or VB)	LUMO (eV) (or CB)	Ref.
PEDOT:PSS	5.1	2.2	[12]
NPD	5.5	1.7	[46]
MoO_3	5.3	2.3	[148]
V_2O_5	4.7	2.4	[148]
BCP	6.2	1.6	[19]
BPhen	6.4	2.9	[150]
$Ru(acac)_3$	4.9	2.8	[151]
ZnO	7.8	4.4	[152]
TiO_2	7.4	4.2	[20]
TiOx	8	4.3	[22]

Meanwhile, interfacial modification at the cathode is no less important. An interfacial layer between the active layer and the cathode needs to fulfill many roles. Firstly, it needs to display hole blocking characteristics and good electron transport properties, and prevent excitons from quenching at the active layer/cathode interface namely "exciton blocking layer (EBL)". Secondly, it should protect the active layer from being damaged by hot metal atoms during cathode deposition. Finally, it should fulfill the function of an optical spacer to control the light intensity distribution within the active layer. To date, a wide range of organic materials e.g., BCP[149], Bphen[150] $Ru(acac)_3$[151] and inorganic materials e.g., ZnO[153] TiO_x,[13] have been employed for this purpose. Consequently, interfacial modification at the cathode has been considered utmost importance in further optimizing the maximum efficiency from an optimized solar cell.

In tandem OPV devices, which consists of two or more cells, the intermediate layer serving as the recombination center is crucial and should possess a number of properties including: (1) charge transportation to ensure a low series resistance; (2) high optical transparent across the solar spectrum; (3) low energy barriers for both electron and hole extractions; (4) easy-fabrication process; (5) protection for prior-deposited active layer during fabricating solution-based active layer of back cell.[154] Generally, metal thin films e.g., Ag[45] or Al/Au,[155] can be used as the semitransparent intermediate layer for tandem cells. However, as the film thickness is not high enough to protect the prior-deposited polymer layer and light loss is usually high, thus, such intermediate layer alone is not suitable for solution-processed tandem cells. The innovation came by using combination of an n-type semiconductor layer e.g., TiOx,[13] ZnO[156] in contact with the front cell and a p-type semiconductor e.g., PEDOT:PSS layer in contact with the back cell, which creates a barrier for Ohmic transport and enforce effective recombination of electron and hole at the interface with equal rates.

4. Small Molecular OPV Cells

4.1. Planar Heterojunction

For small-molecular solar cells, CuPc and ZnPc were frequently used as donor materials [157][158] in combination with C_{60} as acceptor material. [54][159] However, this material pair only leads to V_{OC} ≈0.5 V at 100 mV cm^{-2} incident light. Therefore, in order to further improve J_{SC} and V_{OC}, new low bandgap donor materials with lower HOMO level broadening the light absorption spectrum are needed to introduce and discuss. The optical and structural properties of chloroboron (III) subnaphthalocyanine (SubNc) was characterized by Vereet et al.,[58] and then they fabricated planar heterojunction (PHJ) OPVs using SubNc as donor material with following stacks: ITO/SubNc (13 nm)/ C_{60} (40 nm)/BCP (10 nm)/Al with an active area 3 mm^2.

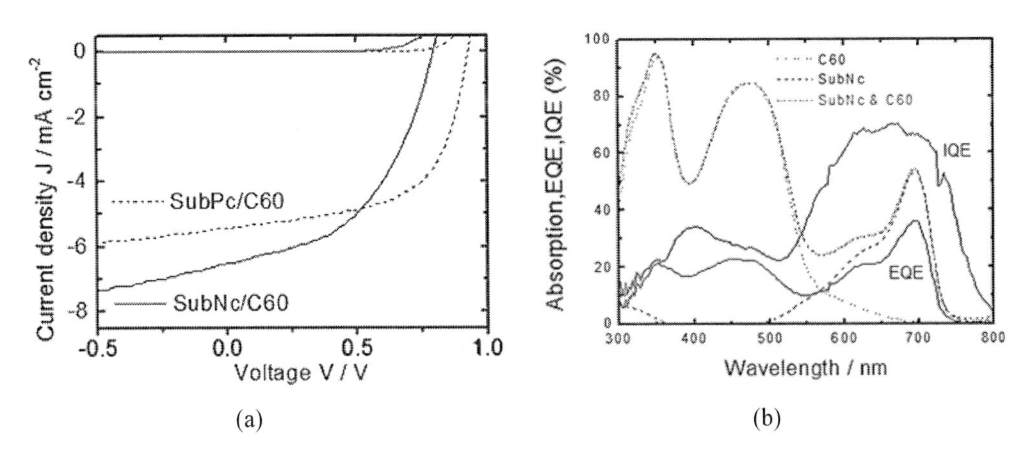

Figure 15. (a) Current density-voltage (*J-V*) characteristics of SubNc based OPVs; (b) absorption, EQE and IQE of C_{60}, SubNc and SubNc/ C_{60} cell. Adapted from [58].

The current density-voltage (*J-V*) characteristics are shown in Fig. 15 (a). The device based on SubNc exhibited a high V_{OC} of 0.79 V, producing a power conversion efficiency of

2.5 %. As shown in Fig. 15 (b), the strong and narrow red absorption of SubNc can broaden the absorption spectrum of solar cells and leaded to a high photocurrent. Especially, utilizing the complementary absorption of SubNc and SubPc in tandem cell, the overall coverage of the solar spectrum would be helpful.

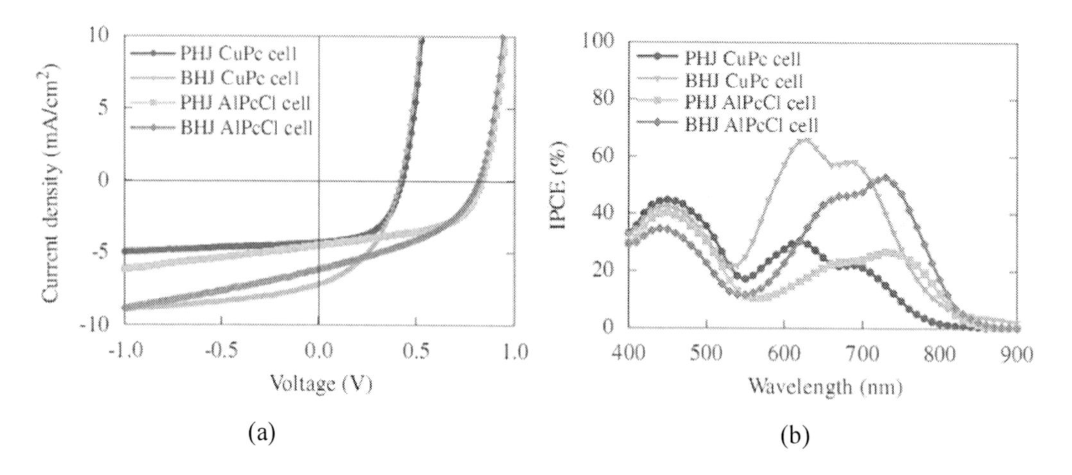

Figure 16. (a) Current density–voltage (*J-V*) characteristics and (b) incident photon-to- current collection efficiency (IPCE) of CuPc and ClAlPc cells. Adapted from [56].

Kim et al. [56] fabricated small molecule OPVs with PHJ and PM-HJ (or BHJ in Ref [56]) using chloride aluminum phthalocyanine (ClAlPc) or copper phthalocyanine (CuPc) as donor materials. The maximum power conversion efficiency of 2.0 % was obtained in the device with a structure of ITO/ClAlPc (10 nm)/ ClAlPc:C_{60} (20 nm)/ C_{60} (30 nm)/BCP (7.5 nm)/Al. Fig. 16 (a) shows the *J-V* characteristics of CuPc, and ClAlPc cells and the device performances of these OPVs are also summarized in Table 5.

Table 5. Photovoltaic performance of CuPc- and ClAlPc-based OPVs.

Devices	J_{SC} (mA/cm^2)	V_{OC} (V)	*FF*	*PCE* (%)	R_SA (Ω cm^2)	R_PA (Ω cm^2)
PHJ CuPc cell	4.32	0.44	0.54	1.02	10.30	1019.10
BHJ CuPc cell	7.19	0.43	0.42	1.29	10.40	262.70
PHJ ClAlPc cell	4.52	0.84	0.52	1.86	14.00	839.40
BHJ ClAlPc cell	6.18	0.82	0.42	2.00	15.60	232.10

From Table 5, it can be seen that both ClAlPc and CuPc cells have J_{SC}. Notably, J_{SC} of BHJ cells are higher than that of PHJ cells, resulting from the increased interfacial area in BHJ cells. The enhancement in J_{SC} was also consistent with the incident photon-to-current collection efficiency (IPCE) spectra shown in Fig. 16 (b).The fill factor (*FF*) of BHJ cell is lower than that of PHJ cell in both ClAlPc and CuPc cells, which due to the low shunt resistant (R_PA) of BHJ cells. The V_{OC} of ClAlPc cells are compared with CuPc, leads to the higher power conversion efficiency. The V_{OC}s are 0.84 and 0.82 V for PHJ and BHJ cells, respectively, which is about twice than that of CuPc cells. The increased V_{OC}, due to the lower HOMO level of ClAlPc of organic solar cell can be further improved by selecting donor material with low HOMO level and high charge carrier mobility.

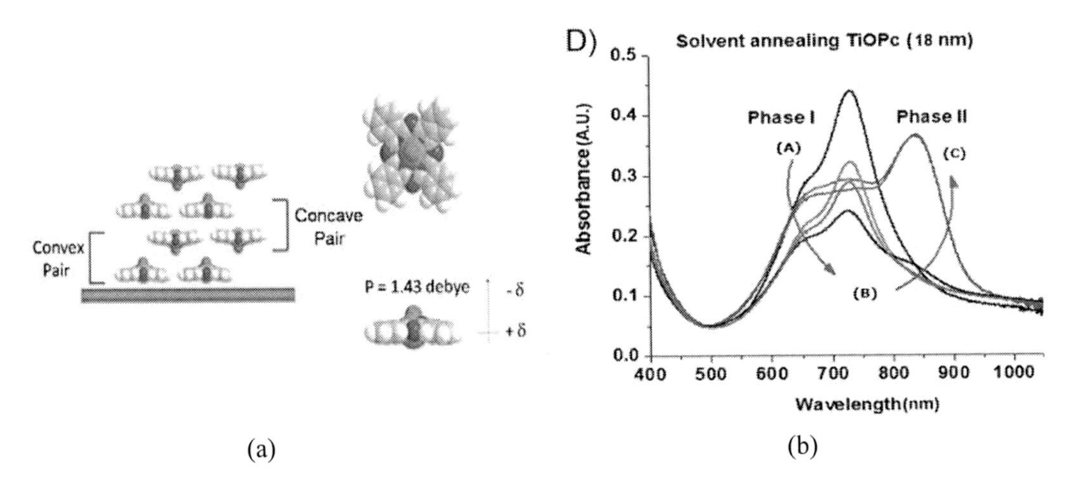

Figure 17. (a) Molecular models of TiOPc showing out-of-plane oxygen perpendicular to the base of the structure and the definition of concave and convex pairs in the triclinic unit cell for the Phase II polymorph; (b) UV-Vis spectra of an 18 nm TiOPc film, showing the change in light absorbance as a function of solvent annealing time from all-Phase I (λ_{max}=720 nm) to mixtures of Phase I and Phase II polymorphs, and finally TiOPc films fully transformed to the Phase II polymorph (λ_{max}=850 nm). Adopted from [160]

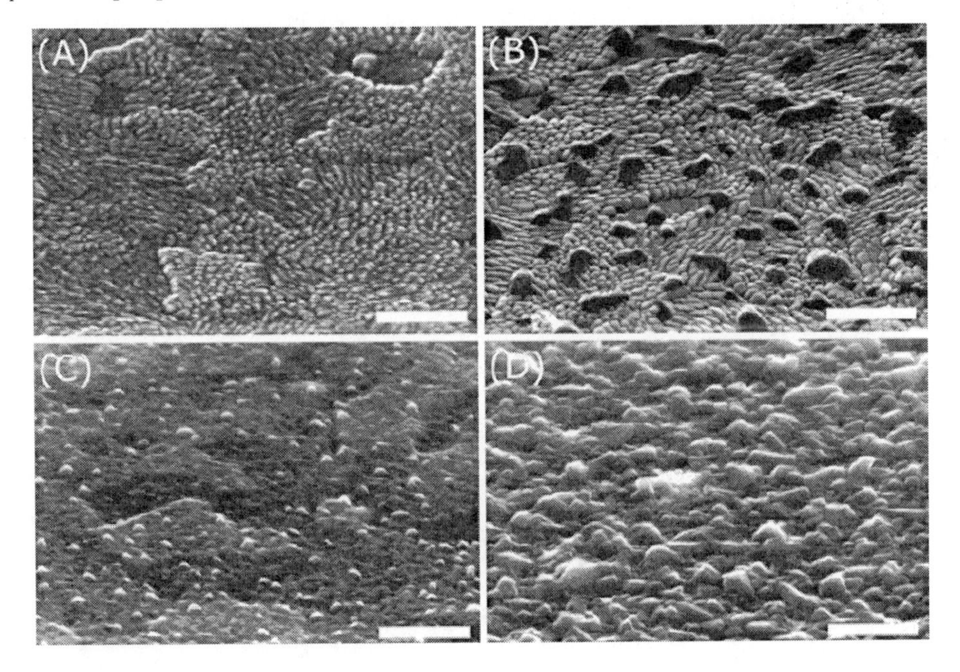

Figure 18. FE-SEM images of 2 nm TiOPc (A, B) and 20nm TiOPc (C, D) films on ITO. A) as deposited, Phase I polymorph; B) after solvent annealing to achieve the Phase II polymorph; C) as deposited, Phase I polymorph; D) after solvent annealing. As-deposited TiOPc films are conformal to oxygen plasma-etched ITO substrates.

As well known, J_{SC} can be increased by extending the absorption spectra of organic solar cells into the near-IR without significant loss of V_{OC}. Placencia et al. [160] provided OPVs incorporating highly textured titanyl phthalocyanine (TiOPc)/C_{60} heterojunction with good near-IR photoactivity. TiOPc films can transform from Phase I to Phase II by solvent annealing, as shown in Fig. 17. Compared with Phase I, the broader visible to near-IR spectrum in Phase II polymorph of TiOPc results from a combination of exciton splitting due to cofacial aggregation of Pc, and from charge transfer character in the absorbance band of this polymorph.[161]-[167] The differences on film morphologies of Phase I and II polymorph are discovered by FE-SEM measurement, showing in Fig. 18.

Interestingly, they investigated three types of TiOPc/C_{60} OPVs. The first type was based on Phase I- only TiOPc film with a thickness 18 nm or 20 nm. The second type used 20 nm TiOPc films completely converted to the Phase II polymorph. The third type used 2 nm layer of the TiOPc Phase II polymorph overcoated with thicker layers of Phase I polymorph. The performances of three types OPVs are summarized in Table 6.

Table 6. Performance of three types of OPVS based on TiOPc/C_{60} heterojunction.

Devices	V_{OC} (V)	J_{SC} (mA/cm^2)	*FF*	*PCE* (%)
1 : Phase I (20 nm)	0.59	9.0	0.48	2.6
2 : Phase II (20 nm)	0.57	15.10	0.53	4.2
3 : Phase I + Phase II (16 + 2 nm)	0.59	11.2	0.50	3.3

It can be seen that J_{SC} was significantly increased, leading to the increase of *PCE*. In addition, even though of the film is the Phase I polymorph, the texturing introduced by solvent annealing 2 nm TiOPc films is sufficient to enhance photocurrent in OPVs. With the processing conditions introduced here, J_{SC} values apparently as high as the best nanotextured CuPc/C_{60} heterojunction reported to date, [168],[169] and the best pentacene/C_{60} heterojunctions [170],[32],[69] have been achieved. However, short exciton diffusion lengths appear to still be the limits for such Pc/C_{60} heterojunctions. In order to further increase power conversion efficiency, enhanced nano-texturing to form stable mixed heterojunctions will be needed.

4.2. Planar-Mixed Heterojunction

Since the properties of D/A interface have great influences on the performance of OPVs, the concept of blend film structure based on polymer donor and acceptor materials namely bulk heterojunction (BHJ) has been adopted to fabricate a small-molecular OPVs with a planar-mixed heterojunction (PM-HJ) structure. Thus, PM-HJ device can increase the D/A interfacial area for exciton dissociation and also increase the thickness of active layer for light absorption [171].

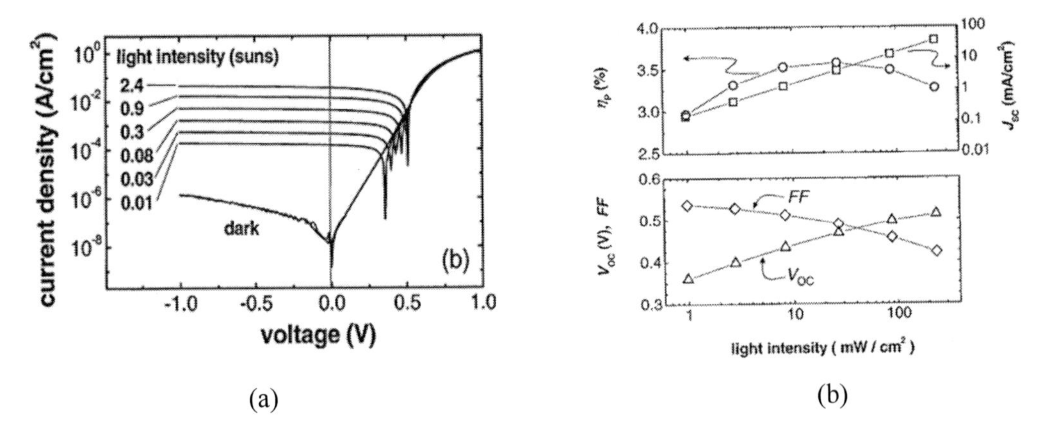

Figure 19. (a) Current density-voltage (J-V) characteristics of the device in dark and under various illumination intensities; (b) photovoltaic performance characteristics of the device under various illumination intensities. Adapted from [172]

Uchida et al. [172] reported an efficient OPVs with a structure of ITO/CuPc:C_{60} (1:1 33 nm)/C_{60} (10 nm)/BCP (7.5 nm)/Ag. The device has a high power conversion efficiency of 3.5 % under 1 sun (AM 1.5G) illumination. It was attributed to the high J_{SC} due to low series resistance of only 0.25 $\Omega{\cdot}cm^2$. The additional C_{60} layer results in an increased optical field at the D/A interface by displacing the active region farther from the reflective metal cathode.[19] They also discussed the relationship between the performance of organic solar cell and light intensity, as shown in Fig. 19. In addition, it was found that η_p was reduced after annealing, suggesting that the CuPc : C_{60} film undergo phase separation during the deposition process.

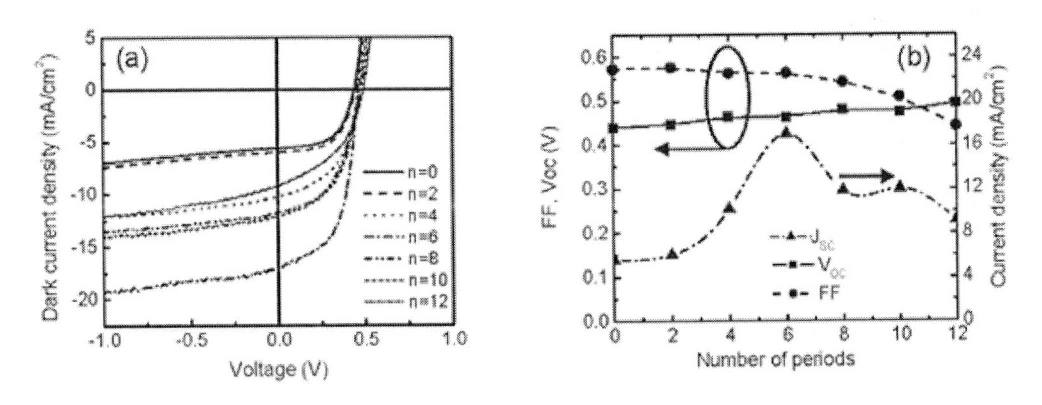

Figure 20. The performance of PM solar cells based on CuPc and C_{60}. (a) Current density–voltage (J-V) characteristics; (b) photovoltaic parameters as a function of the number of periods (n). Adapted from [173].

Yang et al. fabricated an OPV with the structure of CuPc (14.5 nm)/[C_{60} (3.2 nm)/CuPc (3.2 nm)]$_n$/C_{60} (50.0 nm)/BCP (10 nm)/Ag, grown by the process of organic vapor phase deposition (OVPD).[173] Their photovoltaic performances incorporating an all-organic nanocrystalline network are shown in Fig. 20. The PCE= 4.4 % has achieved at number of mixed layer n=6, corresponding to a total film thickness of 38 nm, which is almost two folds more than the optimum thickness of a mixed CuPc:C_{60} solar cell active region. They reported the solar cell efficiency showed minor (~5 %) degradation when exposed to ambient for 24 hours. Furthermore, even higher efficiencies may be obtained by changing the nanoxrystalline size,[169][174] or varying the DA ratio,[175] incorporating more than two molecular components to broaden the coverage of solar spectrum, and employing multiple cells in a tandem structure [149].

4.3. Tandem OPVs

Due to the limited absorption width of organic materials, only a small fraction of solar flux can be harvested by a single-layer bulk heterojunction organic solar cell. In addition, the thickness of active layer is limited by the low charge carrier mobility of most organic materials. A tandem or multi-junction solar cell, consisting of multiple layers each with their specific absorption maximum and width, can overcome these limitations and can cover a large part of the solar flux.

Yakimov and Forrest [177] presented the first multiple-heterojunction solar cells based on small-molecule organic materials. All single heterojunction cells were made of CuPc as a donor and 3,4,9,10-perylenetetracarboxylic bis-benzimidazole (PTCBI) as an acceptor.[149] They stacked two, three, or five sub-cells in series. The structure of the tandem solar cell was shown in Fig. 21. The ultrathin Ag separation layer provides recombination sites for the free charge carriers arriving from the bottom to top cell. When the thickness of Ag interlayer was larger than 0.5 nm, the photocurrent was decreased while the V_{OC} kept constant. This was due to the high absorption coefficient of Ag, leading to a reduction of light intensity at top cell. Since the sub-cells were connected in series, meaning that the current of tandem cell is limited by the lower current of two sub-cells, which is the top cell. Recently, the metallic interlayer has attracted much attention as its properties have great influence on the performance of tandem solar cells [178].

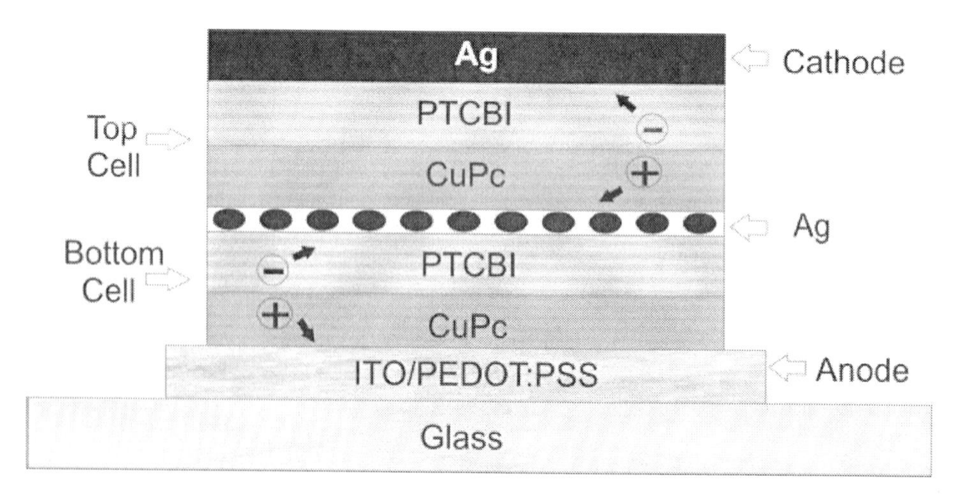

Figure 21. Structure of an organic tandem solar cell based on CuPc and PTCBI. Adapted from [177].

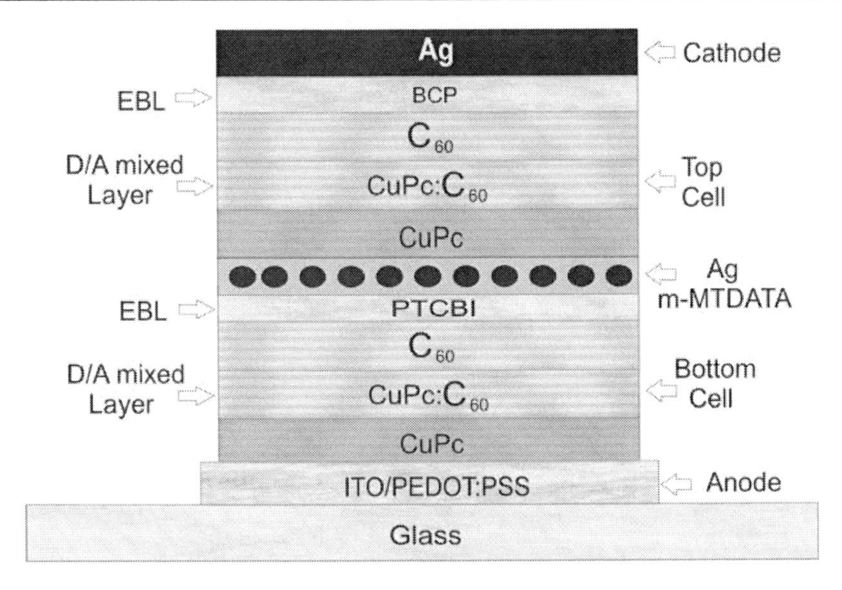

Figure 22. Structure of PM-HJ organic tandem solar cell based on CuPc and C_{60}. Adapted from [45].

Xue and Forrest et al. reported a stacked solar cell based CuPc and C_{60}, which was called hybrid planar-mixed heterojunction (PM-HJ) device [44][45]. The structure of the whole device was depicted in Fig. 22. Ag nanoclusters with a typical thickness of 0.5 nm buried in a 5 nm thick m-MTDATA, which was p-doped with 5 mol % F_4-TCNQ, was used as recombination sites instead of the ultrathin Ag interlayer. As a result, Ag particles can serve as scattering centers for incident light in the middle of the device, leading to the improvement in optical absorption of the active layer of the subcells. Then the film thicknesses of the homogeneous layers of CuPc and C_{60} and mixed layers have to be optimized in order to realize current matching. The *PCE* of PM-HJ tandem cells using different thicknesses for active regions are summarized in Table 7. It can be seen that cell 2 has the highest performance as the thicknesses used here are the best tradeoff between the electrical and optical properties.

Table 7. Layer thickness in nanometers and *PCE* of various PM-HJ tandem organic solar cells

cell	CuPc	CuPc:C_{60}	C_{60}	PTCBI	CuPc	CuPc:C_{60}	C_{60}	BCP	*PCE*(%)
1	10	18	2	5	2	13	25	7.5	5.4
2	7.5	12.5	8	5	6	13	16	7.5	5.7
3	9	11	0	5	5	10	21	10	5.0

Series connected tandem organic solar cells with peak power conversion efficiencies of 5.15 % were reported by Rand and Cheyns et al. [8]. The structure of the OSC is shown in Fig. 23. They used various small molecule phthalocyanine-based donor materials with highly complementary absorption. It is noted that efficient recombination between subcells has been shown in literature using thin metal layers [181]-[183], highly doped organic layers [184], or metal-oxides [185]. Similarly, they used an optimized recombination zone (RZ), consisting of 5 nm PTCBI, thin 0.1 nm Ag [186] and 2 nm MoO_3. Consequently, from Figure 23 we can see that due to the higher complementary absorption between SubNc and SubPc than between ClAlPc and SubPc, the tandem cell using SubNc as donor material in front cell has a higher short circuit current of 4.3 mA/cm^2, and finally a higher power conversion efficiency of 5.15 %. In addition, a large open circuit voltage about 1 V was obtained in single cells with C_{60} as accepter material and SubPc, ClAlPc, SubNc as donor materials, respectively, leading to a V_{OC} of 1.92 V in tandem cells.

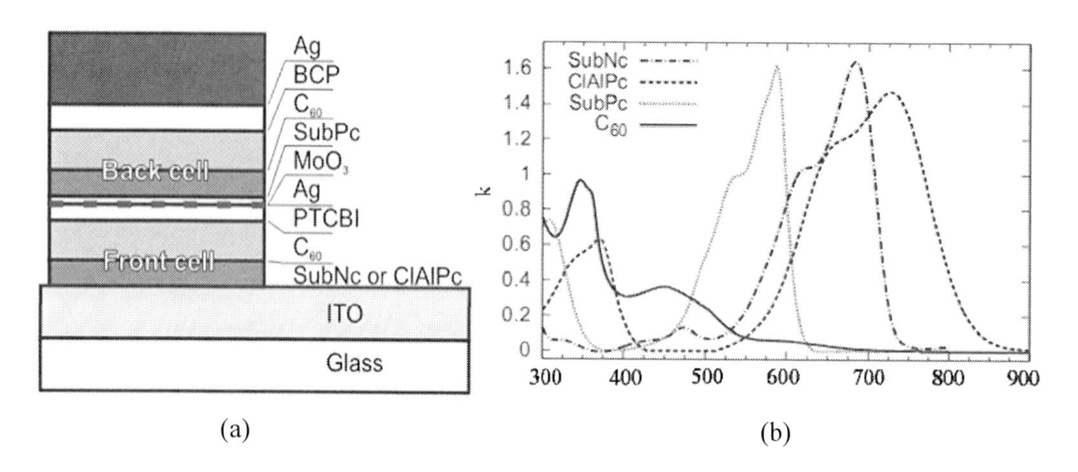

Figure 23. (a) Structure of organic tandem solar cells with phthalocyanine-based donor materials; (b) extinction coefficient (k) of SubNc, ClAlPc, SubPc and C_{60}. Adapted from [8].

4.4 Inverted OPVs

Conventional organic solar cells are fabricated on transparent substrates such as glass or plastic, coated with a transparent anode such as ITO or fluorine doped tin oxide (FTO). While in an inverted structure, the donor material is placed adjacent to the transparent cathode, which is fabricated on the substrates. This architecture eliminates the reliance for comparatively high cost transparent substrates. In addition, it allows for use in applications such as semitransparent power-generating coatings, or for growth on flexible and inexpensive opaque substrates.

J. Meiss et al. [187] reported the zinc phthalocyanine (ZnPc):C_{60} based PM-HJ inverted OPVs using ultrathin metal layer (Ag or Al/Ag) as transparent top contacts, and then an organic capping layer aluminum tris(quinolin-8-olate) (Alq_3) was introduced. The current density-voltage (*J-V*) characteristics and EQE of the cells with different thickness of capping layer are shown in Fig. 24. The device with 50 nm Alq_3 capping layer showed a highest power conversion efficiency of 1.06 %. Simulations showed that an Alq_3 layer can drastically improve light absorption within active layers of solar cell and influence the light reflection, which can be depicted by considering a microcavity effect between semitransparent top contact and the reflecting back contact. It is obvious that ITO-free organic solar cells with transparent metal contacts are a feasible, cost-efficient method.

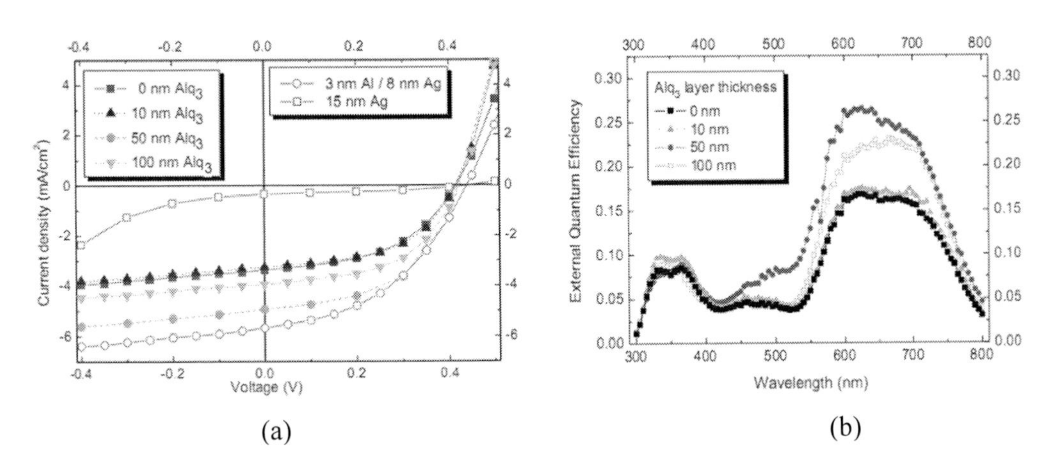

Figure 24. Current density-voltage (*J-V*) characteristics of the cells without and with capping layer; (b) external quantum efficiency of the devices without and with capping layer. Adapted from [187].

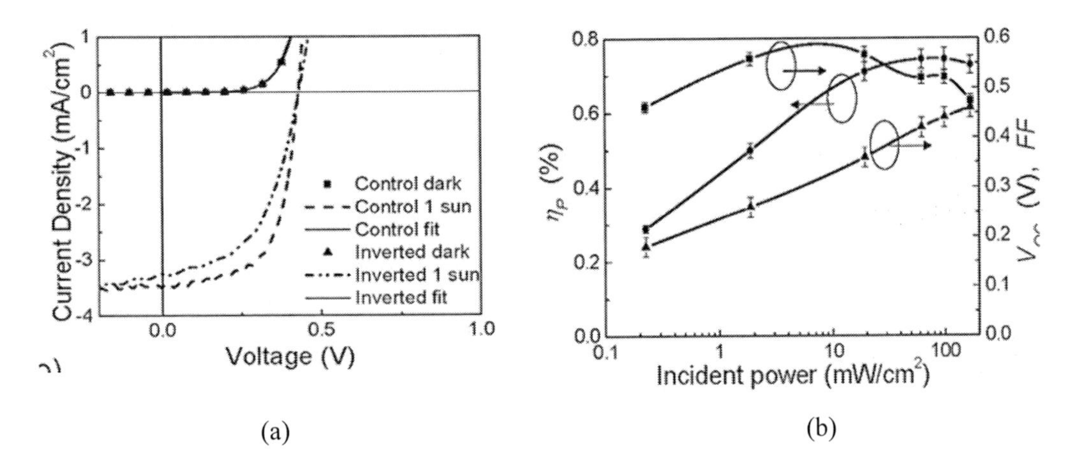

Figure 25. (a) Current density-voltage (*J-V*) characteristics for a conventional and optimized inverted cell; (b) photovoltaic parameters versus incident power density. Adapted from [188].

Tong et al. [188] fabricated top-illuminated, inverted, small molecule solar cells on reflective substrates employing CuPc as a donor and PTCBI as an acceptor, with an ITO cathode and a Ni anode. This architecture can lower the costs, allowing for use in applications such as semitransparent power-generating coatings, or for growth on flexible and inexpensive opaque substrates. Fig. 25 shows the photovoltaic performance of the CuPc/PTCBI heterojunction based inverted solar cells.

4.4. *P-I-N* Architecture

Falkenberg et al. [189] reported *p-i-n* type OPVs with the device structure of ITO/NDP2 (1 nm)/MeO-TPD:NDP2 (30 nm, 30:1) (*p*)/C_{60}:ZnPc (37.7 nm, 1:1) (*i*)/C_{60} (26 nm)/electron transporting layer (ETL):NDN1 (~40 nm)/Al. The photovoltaic performances of the devices are shown in Fig. 26. A power conversion efficiency of 2.83 % was achieved by introducing

NTCDA as an ETL. It was attributed to the advantageous position of the HOMO and LUMO levels of NTCDA, which guarantee a loss free charge extraction from the active layers and exciton blocking simultaneously. Therefore, by selecting proper transport materials with high dupability, high thermal and morphological stability, high charge carrier mobility, as well as high transparency, the performance of *p-i-n* OPVs can be further improved.

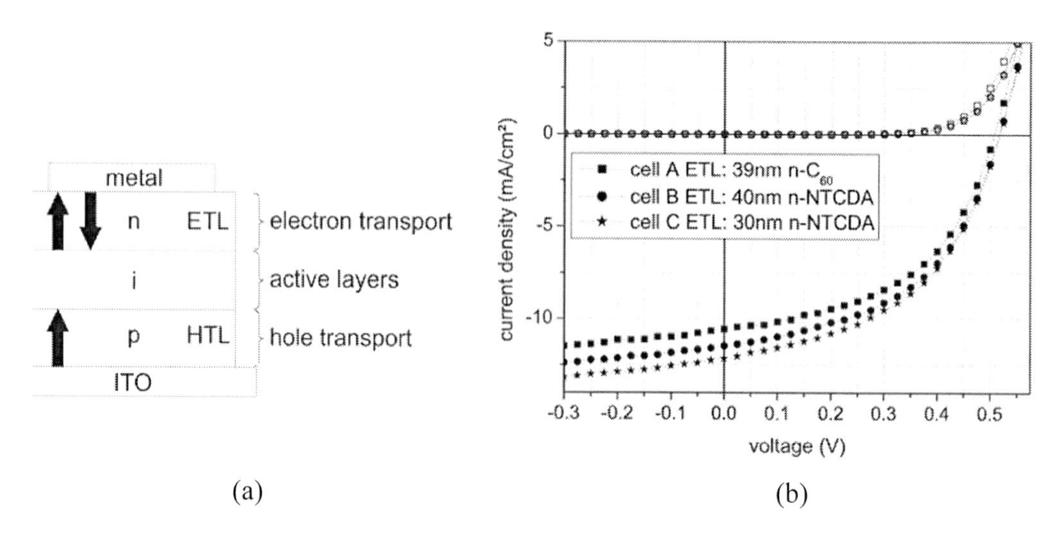

Figure 26. (a) Schematic structure of an OPV according to the *p-i-n* concept; (b) current density–voltage (*J-V*) characteristics of OPVs with different electron transport layer. Adapted from [189].

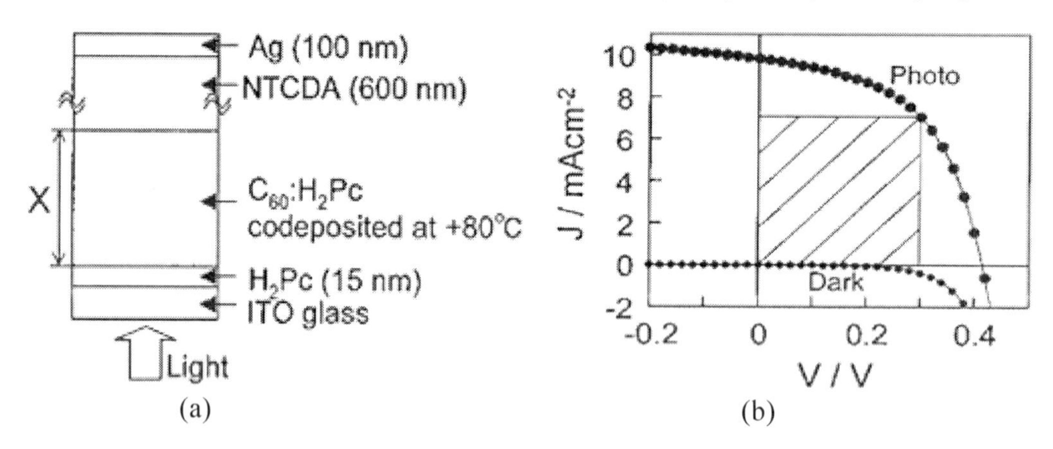

Figure 27. (a) Structure of three-layered *p-i-n* cell; (b) Current density–voltage (*J-V*) characteristics of three-layered cell incorporating a 130-nm-thick C_{60} : H_2Pc (5:6) interlayer codeposited at +80 °C. Adapted from [190].

Suemori et al. [190] reported a three-layered OPV device with a *PCE*=2.5%. Fig. 27 shows the device structure and *J-V* characteristics of the devices. These three-layered cells have a *p-i-n* like energetic structure. It is believed that the nanostructure of codeposited films influences the photocurrent generation process. Therefore they controlled the nanostructure of codeposited films by varying the substrate temperature during vacuum co-evaporation [191][192]. The result showed that the optimized substrate temperature for C_{60}:H_2Pc

interlayer was +80 °C. Meanwhile, the highest performance was obtained when the thickness and C_{60} : H_2Pc ratio of the codeposited layer was optimized to 130 nm and 5:6, respectively.

A morphological illustration of cross section is depicted in Fig. 28(a). The film has a nanocomposite structure somewhere between crystalline H_2Pc and amorphous C_{60}. Therefore, the efficient photocurrent generation is due to the formation of spatially separated routs for electrons and holes. Consequently, for lower and higher substrate temperature, molecular mixtures and crystalline-crystalline composites were confirmed as being formed [193][194]. From Fig. 28(b), it can be seen that the devices with different thickness of C_{60}:H_2Pc interlayer almost has the same value of open circuit voltage (V_{OC}). This suggests that V_{OC} was determined by built-in electric field, created by the difference in the Fermi levels (E_F) of NTCDA and H_2Pc. As a result, the *p-i-n* energetic structure was supported by the dependence of J_{SC} and V_{OC} of the three-layered cells on C_{60} : H_2Pc thickness.

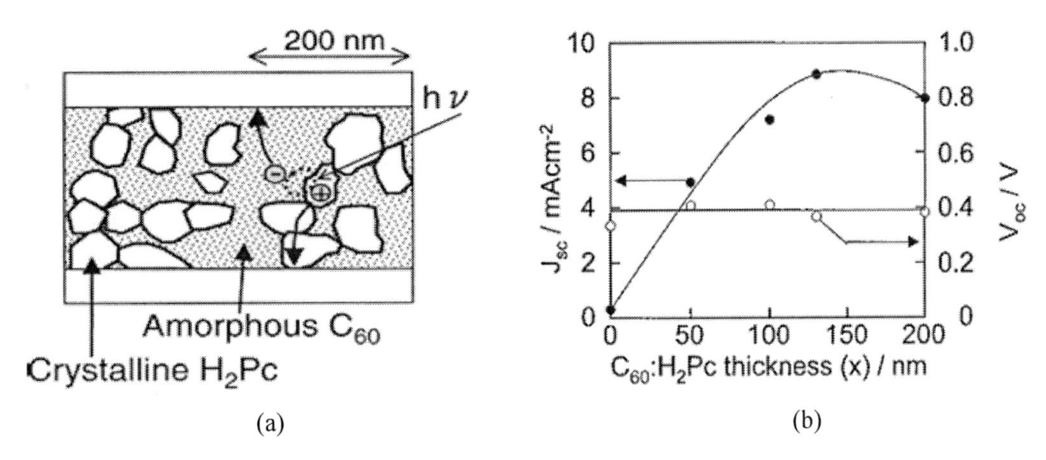

Figure 28. (a) Morphological illustration of a cross section; (b) dependence of J_{SC} and V_{OC} of the three-layered cells on C_{60}:H_2Pc (1:1) thickness (x). Adopted from [194].

4.5. OPVs with Multi-Charge Separation (MCS) Structure

It is well known that V_{OC} is determined by energy difference between the highest occupied molecular orbital (HOMO) of donor material (D) and the lowest unoccupied molecular orbital (LUMO) of acceptor material (A) [195]. Based on this assumption, inserting fluorescent materials (D1) with low HOMO level and high light absorption between donor (D2) and acceptor (A) is an effective method to obtain large V_{OC} [196][196][197]. When an ultrathin interlayer was used, there were multicharge separation (MCS) interfaces, D1/A and D2/A. Due to the lower HOMO of D1, V_{OC} can be enhanced. [68]

M. Y. Chan et al. [198] reported by introducing rubrene as a dopant in standard CuPc/C_{60} OPVs, a power conversion efficiency of 5.58 % was achieved. It was attributed to a high V_{OC} due to a low HOMO level in rubrene. In addition, the complementary absorption of rubrene broadens the absorption spectrum in the 500-550 nm, thus, higher J_{SC} was obtained. Also, an ultrathin layer can be used to substitute doping in order to make the fabrication process more controllable and reproducible. Murata et al. developed OPVs with an ultrathin layer of CuPc inserted between pentacene and fullerene (C_{60}) to form multi-charge separation (MCS) interfaces. [68] The device structure, the model of MCS interface and atomic force microscope (AFM) images are shown in Fig. 29.

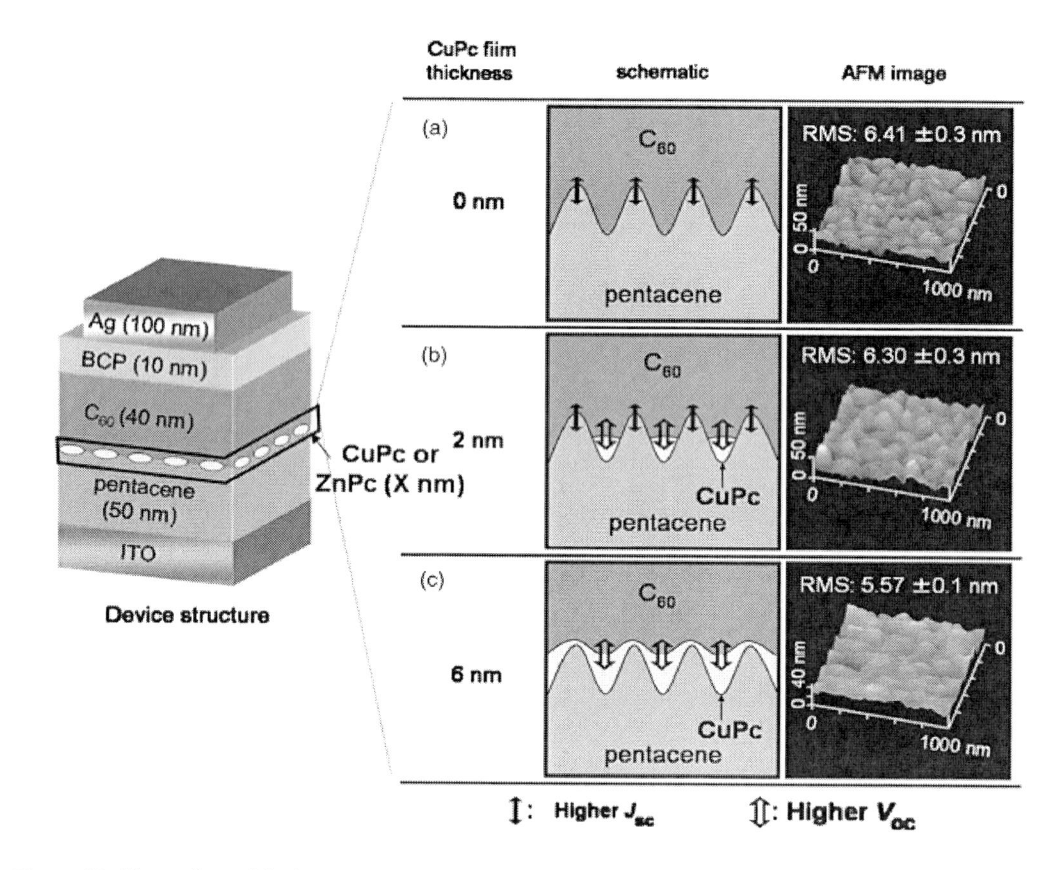

Figure 29. Illustration of device structure with MCS interface and atomic force micrOPVope (AFM) images of the surface of (a) pentacene (50 nm), (b) CuPc (2nm)/pentacene (50 nm) and (c) CuPc (6nm)/pentacene (50 nm) on ITO substrate. Adapted from [68].

It was found that V_{OC} was enhanced as increasing the thickness of CuPc, which corresponded to the lower HOMO level of CuPc. Whereas, the J_{SC} exhibited plateau in the 0-2 nm range and then decreased. This can be ascribed to the shorter lifetime of singlet excited state of CuPc compared with pentacene. In order to adjust a balance between V_{OC} and J_{SC}, the thickness of ultrathin interlayer should be optimized.

Based on the previous discussion, it revealed that an effective MCS structure should consist of subject donor material with high charge carrier mobility or long exciton diffusion length L_D, ultrathin inserted donor material with strong light absorption and lower HOMO level or ultrathin acceptor material with higher LUMO level. [196]

4.6. OPVs with Other Architecture

Another important and popular approach to improve the performance of organic photovoltaic cell is to use anode or cathode buffer layer, which has been widely studied

[202]-[205][21]. One role of cathode buffer layer is to form an interlayer between organic material and metallic cathode, which prevents metal atoms from diffusing into active layer and improves charge carrier transport and power conversion efficiency. On the other hand, the buffer layer can block the oxygen molecules from permeation, which enhances the device stability. In addition, buffer layer adjacent anode can control the interface morphology between ITO and donor material, reducing the leakage current. Meanwhile, the work function of anode can be increased by covering a buffer layer, which is helpful for charge collection.

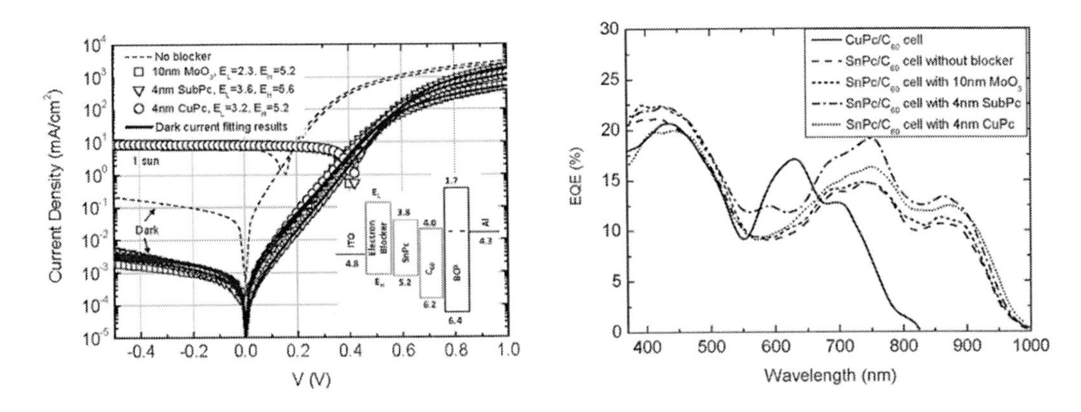

Figure 30. Current density-voltage (*J-V*) characteristics of the cells without and with several blocking layer; (b) external quantum efficiency (EQE) of the devices. Adapted from [51].

N. Li et al. [51] also reported that by employing several different organic and inorganic electron blocking layers (CuPc, SubPc [54] and MoO_3 [148]) between donor layer and anode contact in SnPc/C_{60} heterojunction solar cells, the open circuit voltage was dramatically increased. The performance of devices with a structure of ITO/electron blocking layer/SnPc (10 nm)/C_{60} (40 nm)/BCP (10 nm)/Al is shown in Fig. 30. The power conversion efficiency was increased from 0.45 % for a SnPc device without the blocking layer to a maximum 2.1 % after inserting a SubPc layer. It was attributed to the high V_{OC} due to the low electron leakage current, which was decreased by two orders of magnitude compared to cells lacking the layers. Herein, the introduction of SubPc electron blocking layer resulted in a twofold increase in V_{OC} from 0.16 V to 0.40 V. This work is helpful for deeper understanding the origins of V_{OC} and the importance of reducing dark current to improve *PCE*.

4.7. Brief Summary

It is clear that for organic solar cells with single junction architecture, the donor materials should possess the properties as follows: strong optical absorption over a broad wavelength rang extending to near infrared region (NIR), high hole mobility of 10^{-2} cm^2/Vs, and long exciton diffusion length (L_D). These properties allow coarse BHJ cells and thick films. In contrast, for tandem solar cell with multijunction, thin and smooth films absorbing over a limited spectral window are preferred. Therefore, by introducing new materials, controlling the evaporation process and optimizing film morphology, the performance of organic solar cells based on small molecular materials can be further improved.

5. Polymer OPVs

Polymer solar cells have evolved as a promising cost-effective alternative to inorganic solar cells based on mono-crystalline Si[206] due to their potential to be low cost, light weight, and flexible.[207] Comparing with small-molecular OPVs, polymer solar cells have special properties e.g., simple and time-saving fabrication process via spin-coating,[208] brush painting[207],[209], inkjet printing,[210],[211] doctor blading[133] or gravure,[213] et.al., and therefore large-area fabrication. [207]-[213] Intense theory analysis has proved that bulk heterojunction (BHJ) organic solar cells (OPVs) based on polymer and fullerene are competent to achieve high power conversion efficiency PCE >10%[16]. Thus, this section would dwell on the fabrication, characterization, photovoltaic properties and optimization strategies for BHJ OPVs.

5.1. Morphology Controlling

It has been found that efficient charge transfer (or exciton dissociation) can take place at large area interface between mixed electron donor polymer and electron acceptor fullerene molecular when their LUMO level offsets are suitable, thus, to overcome the limited exciton diffusion length ~10 nm in polymer.[207] To design high efficiency OPVs with a bulk heterojunction, one of the most significant targets is to form continuous pathways for the transportation of electrons in acceptor and holes in donor. Then, the thickness of organic bulk films could be increased to maximize light absorption at a low level of recombination loss.[214] The importance of generating continuous crystalline pathways in bulk heterojunction OPVs based on poly(3-hexylthiophene) (P3HT): 6,6-phenyl C_{61} butyric acid methylester (PCBM) system has been realized by observing low photovoltaic performance of devices without post-treatment due to unoptimized morphology.[215] Two extensively studied strategies to effectively produce percolated pathways combined with polymer nanospheres and fullerene clusters are "thermal annealing" [17][122][216]-[221] and "solvent annealing"[222]-[227]. Then, the concept of adopting solvent mixture to enhance the vertical phase separation of bulk film will be addressed.[14] [228]-[236]

5.1.1 Thermal Annealing

Based on previous research, the importance of thermal annealing process has been commonly realized to optimize the photovoltaic performance of OPVs with bulky heterojunction.[216]-[218] Thermal annealing can be adopted either after fabricating polymer film (pre-annealing) or after making the final device (post- annealing). To obtain film morphology with well-organized polymer crystallinity, the annealing temperature and time needed to be optimized. The effect of thermal annealing of the P3HT : PCBM system on device performance was revealed by X. Yang et al. that P3HT crystals grow continuously into longer fibrils distinguished from TEM image as shown Fig. 31.[218]

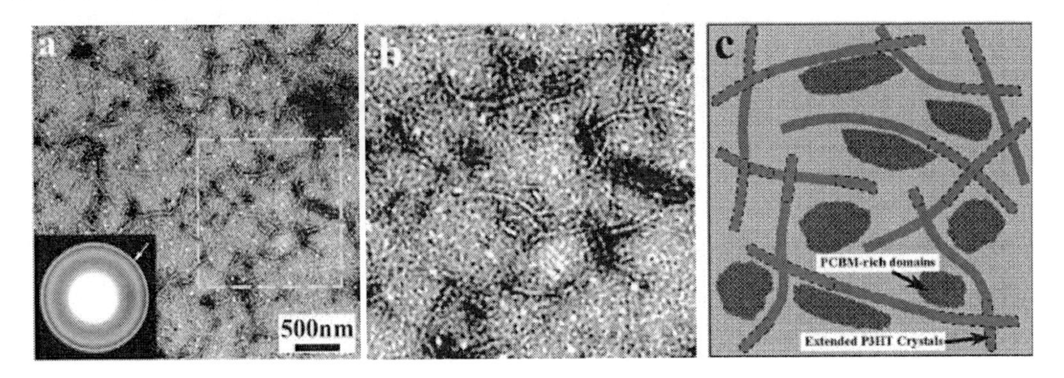

Figure 31. BF TEM images of the overview (a) and zoom in (b), and the corresponding schematic representation (c) of the thermal annealed RR-P3HT : PCBM active layer. Adopted from [218]

Meanwhile, the segregated PCBM crystalline structure (or clusters) after thermal annealing could be also detected. X. Yang postulated that P3HT crystallites form boundaries to restrict the extensive diffusion of PCBM molecules and stabilize large-scale crystallization. Thus, long term stability ~1000 hr under 70 °C of the photovoltaic properties of these devices was achieved upon thermal annealing.[218] Similarly, by choosing solvent chlorobenzene (CB) rather than toluene to dissolve bulk film poly2-methoxy-5-(3,7-dimethyloctyloxy)-1,4-phenylenevinylene (MDMO-PPV) : PCBM, apparent phase separation of polymer as long fibrils and fullerene as clusters can be distinguished in Fig. 32.[219]

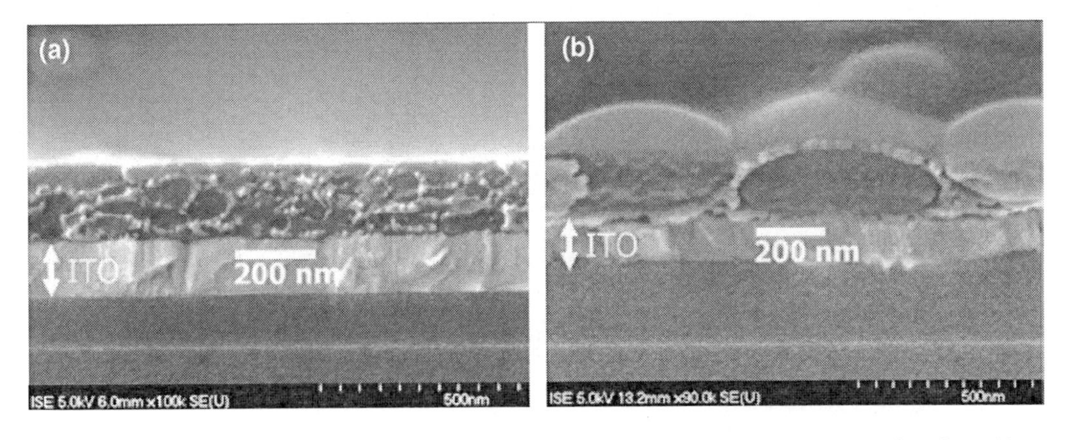

Figure 32. SEM cross-section images of MDMO-PPV:PCBM blend films cast on ITO-glass from (a) chlorobenzene and (b) toluene solution. The brighter little objects are polymer nanospheres, whereas the large inclusion in (b) is a PCBM cluster. Adopted from [219].

Moreover, H. Hoppe et al. also presented the effect of different annealing temperature on film morphologies. While up to 150 °C the polymer nanospheres can be detected, for temperatures higher than 165 °C these spheres are not visible anymore and the polymer has been molten. Meanwhile, by studying the AFM images of P3HT : PCBM bulk film under different annealing temperatures, the surface of annealed film under 110 °C shows a much coarser texture with broad hill-like features compared to other films annealed at 75 °C and 150 °C, especially, the as-cast film with very smooth roughness RMS 0.377 nm.[216] Additionally, the optimized nano-crystalline of bulk film would also enhance the long-wavelength (near-infrared) absorption of polymers, which could also be observed from the changing film color from orange to purple. Thus, higher photocurrent would be expected due to improved light absorption. Moreover, the improved crystallinity would enhance hole mobility by a factor of 10 due to improved cross-linked percolation pathway of polymer.[216]-[218]

There is no doubt that these fundamental researches upon optimized film morphology motivate improved device performance of OPVs based on bulk structure. By carefully optimizing post-annealing and annealing of RR-P3HT:PCBM blend with an external bias, Padinger's work achieving 3.5% PCE attracted tremendous attention in this aspect.[220] Further extensive studies on the thermal-annealing approach followed, and PCE values up to 5% were reported by A. J. Heeger group in 2005.[122] Other variation of the thermal annealing such as microwave annealing approach was reported by Chen et al.[221] More device performance based on various composition of P3HT:PCBM bulk heterojunction OPVs at different thermal annealing process are summarized in Ref [17].

5.1.2. Solvent Annealing

To optimize photovoltaic performance of OPVs with bulk heterojunction, proper host solvent to dissolve both electrical donor and acceptor materials should first be selected to control film morphology due to different solvent properties, such as boiling point, vapor pressure, solubility, and polarity. As shown in Fig. 32, MDMO-PPV:PCBM blend films casted from solvent chlorobenzene had a better phase separation including polymer nanospheres and PCBM cluster than that from toluene solution.[219]

Recently, intense research by Y. Yang group have pointed out that the solvent-removal process "solvent annealing" after fabricating blend films can effectively optimize the film morphology before thermal annealing.[222]-[226] Typical process is that the solution is spin-coated at low speed, and then, the wet polymer film were left to dry slowly in petri dishes until the color of the polymer films changed from orange to purple.

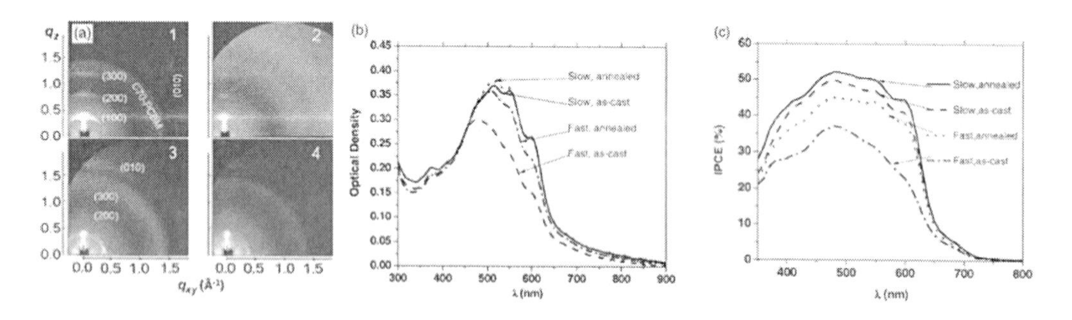

Figure 33 (a) 2D GIXRD patterns of RR-P3HT:PC70BM (1:1 ratio) films(1) DCB (1, 2 dichlorobenzene), 1000 rpm, 30 s; 2) DCB, 1000 rpm, 90 s; 3) CB (chlorobenzene), 1000 rpm, 90 s; 4) CB, 3000 rpm, 90 s). (b) UV-vis spectra for RR-P3HT:PC70BM (1:1 ratio) films, for fast- and slow-grown films from 1,2,4-trichlorobenzene (TCB), before (dash line) and after (solid line) annealing. The films were spun cast at 3000 rpm for 50 s (film thickness ~50 nm) and the thermal annealing was done at 110 °C for 15 min inside glove box. (c) IPCE of RR-P3HT:PC70BM solar cells with fast-grown and slow-grown active layers: before (dash line) and after (solid line) annealing. Adapted from [227]

Chu et al. have revealed that controlling the solvent-evaporation rate with various solvent boiling point and drying time can improve the crystalline and absorption of P3HT chains,

[227] as shown in Fig. 33(a) by grazing-incidence x-ray diffraction (GIXRD) measurement. From Fig. 33(a), the crystal lattice of spin-coated blend films can be detected in sample 1 due to remained small amounts of 1,2 dichlorobenzene (DCB) residue before solvent annealing. On the contrary, slight crystal lattice signal of sample 2 is due to little solvent residue. Samples 3 and 4 have similar situation in solvent CB. The improved crystalline of polymer RR-P3HT in blend film with solvent 1,2,4-trichlorobenzene (TCB) can also be confirmed by enhanced absorption coefficient as shown in Fig. 33(b). Thus, OPVs with slow growth (or solvent annealing) blend films have the highest incident photon conversion efficiency (IPCE) in Fig. 33(c).

5.1.3. Solvent Mixture

Besides thermal annealing and solvent annealing, solvent mixture represents a new promising method to improve solar cell morphology and enhance device performance. F. Zhang et al. found that by introducing a small amount of chlorobenzene into chloroform host solvent, significant improvement in photocurrent density and IPCE were obtained in poly (2,7-(9,9-dioctyl-fluorene)-alt-5,5-(40,70- di-2-thienyl-20,10,3- benzothiadiazole)):PCBM blend system.[14] Also, J. Peet et al. reported that by adding alkanethiol into host solvent toluene device performance based on P3HT : PCBM blend can be enhanced due to longer carrier lifetime with ordered structure in morphology.[228] Alkanethiol is also found effective to achieve the highest efficiency record for low band gap polymer solar cells.[229]

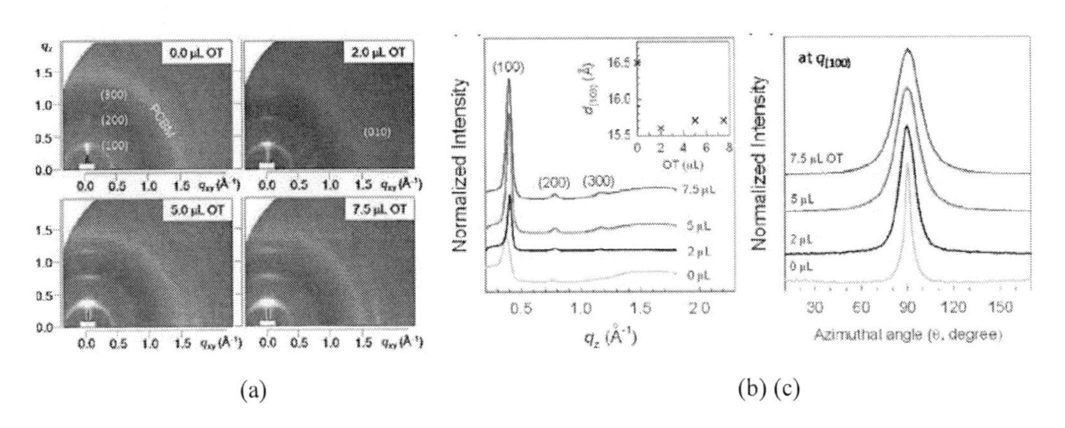

Figure 34. (a) 2D GIXRD patterns of films with different amounts of OT. (b) 1D out-of-plane X-ray and (c) azimuthal scan (at *q*(100)) profiles extracted from part a. Inset of b: calculated interlayer spacing in (100) direction with different amounts of OT.

Recent researches have proved that solvent mixture combined with host solvent such as CB or DCB and small amount of additive solvent such as 1,8-octanedithiol (OT) have a significant effect on nano-crystalline morphology and device performance based on P3HT:PCBM blend system.[230], [231] The fabricating process of blend films in Ref [231] is that all the solutions were spin-coated at 3000 rpm for 70 s until the films were dried and neither thermal annealing nor solvent annealing were carried out to exclude their effect, thus, this process is known as "fast grown". Notably, the host solvent CB is evaporated at first during spin-coating due to higher boiling point than OT. Then, the remaining OT will continuous dissolve PCBM instead of polymer, thus, induce phase separation and crystallization of polymer. As shown in Fig. 34(a), 2D GIXRD measurements of "fast grown"

P3HT:PCBM blend films present that by adding more and more volume of additive solvent OT, the densities of crystalline signals become stronger. Meanwhile, the 1D XRD experiments of these films also reflect the improved crystallization of blend films with more volume of OT, as shown in Fig. 34(b) and (c).

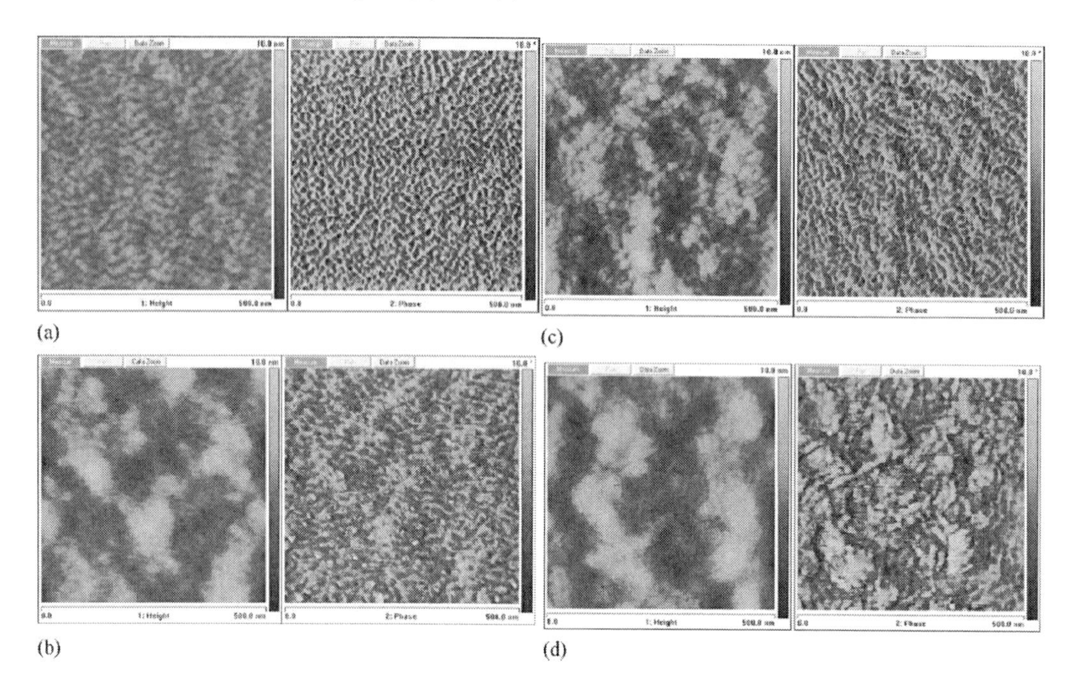

Figure 35. Tapping mode AFM images of films with different amounts of OT in 500 nm × 500 nm. Left: topography. Right: phase images. (a) 0 μL, (b) 7.5 μL, (c) 20 μL, and (d) 40 μL of OT. The scale bars are 10.0 nm in the height images and 10.0° in the phase images.

Table 8. Boiling points, vapor pressure of solvent for organic polymers and fullerene.

(Number) Solvent	Boiling points (°C)	Vapor pressure at 30 °C (Pa)	PCBM solubility (mg/ml)	Mixture Host/ Additive	Ref.
(1) alkylthiols	283	-	-	-	-
(2) alkanethiol	160	1.77 k	-	-	-
(3) chloroform	61	2.12 k	25	-	-
(4) chlorobenzene (CB)	132	1600	25	-	-
(5) 1,2-dichlorobenzene (DCB)	198	200	100	-	-
(6) 1,2,4-trichlorobenzene (TCB)	213	38.6	-	-	-
(7) o-dichlorobenzene (ODCB)	180	181	30		
(8) toluene	110	4.89 k	10	-	-
(9) 1-chloronaphthalene (Cl-naph)	259	2.26	-	(5)/(9)	[232]
(10) 1,8-octanedithiol (OT)	270	1	19	(4)/(10)	[230][231]
(11) di(ethylene glycol)diethyl ether	189	100	0.3	(5)/(11)	[230]
(12) hexane	68	2.47 k	-	(7)/(12)	[233][234]
(13) nitrobenzene (NtB)	211	47	-	(4)/(13)	[235][236]
(14) N-methyl-2-pyrrolidone	229	10	18	(5)/(14)	[230]
(15) 1,8-Diiodooctane				(4)/(15)	[14]

Moreover, the tapping-mode AFM images in Fig. 35 show that the surface roughness of blend films with more volume of additive solvent OT become much higher, and highly ordered fibrillar crystalline domains of P3HT are clearly visible.

As a conclusion, to optimize the morphology of blend polymer films by solvent mixture, two crucial principles of choosing addictive are proposed by Y. Yao et al.: (1) selective solubility of the fullerene component; (2) a higher boiling point (lower vapor pressure) than the host solvent.[230] Herein, chemical and physical properties of usual organic solvents are summarized in Table 8 to select designing effective solvent mixture for bulk heterojunction OPVs. Moreover, strategies combined with above three methods could also be useful to optimize film morphology and device performance.

5.2. High Efficiency Regular BHJ OPVs

During the last several years, the power conversion efficiency PCE of regular bulk heterojunction OPVs based on the most intensely studied materials P3HT and $PC_{61}BM$ blend are reaching the bottle neck values about 5 % under the standard solar spectrum, AM 1.5G. [12][93] Specifically, after the short-circuit current voltage J_{SC} is optimized to a maximum value of 12 mA/cm^2, the limited open-circuit voltage of V_{OC} about 0.65 V has be further improved to break through *PCE*>10%.[12][16] Herein, this subsection would dwell on some promising OPV devices with high photovoltaic performance and mention the fabrication process at the same time.

As pointed out that maximum V_{OC}≈(1/e)(|E_{Donor}HOMO|-|$E_{Acceptor}$LUMO|)-0.3V at given donor and acceptor pairs,[16] thus, improved OPVs with high V_{OC}>0.65 V and PCE >4.2 % have emerged by enlarging the energy offset between the HOMO of donor and the LUMO of acceptor by modifying functional group of organic materials.[14][22] [91][110][117] [125][237]-[239] Once the V_{OC} has been improved, preserving other parameters such as J_{SC} >9 mA/cm^2 and *FF* >50% is crucial to obtain high power efficiency. From the summarized photovoltaic parameters with high *PCE* >4.5% in Table 9, high J_{SC} and *FF* performances insure their successful strategies to enhance V_{OC} except for Ref [240] with relatively low J_{SC} =4.4 mA/cm^2 and *FF*=40%. Thus, the detailed analyses of their strategies will be expatiated at full steam to be beneficial for the future improvement of OPVs performance.

Chen et al. have reported BHJ OPVs with high *PCE* >7% based on polymer materials PBDTTT and fullerene PCBM, as shown in Fig. 36.[14] Comparing with PBDTTT-E and PBDTTT-C, the alkyl group not only results in a lower HOMO but also a lower LUMO.

Meanwhile, with the addition of an alkyl group, the introduction fluorine of PBDTTT-CF further lowered both HOMO and LUMO levels. H.-Y. Chen pointed out that a change in both the HOMO and LUMO levels with just one synthetic modification to the polymer backbone has not been reported in any polymer system.[14] As a result, enhanced high *PCE* of PBDTTT-CF : PCBM system is attributed to the both enhanced V_{OC}=0.76 V and J_{SC}=15.2 mA/cm^2 confirmed by the highest internal quantum efficiency (IQE) approaching 100% in Fig. 36 (d).

Table 9. High photovoltaic performances based on regular BHJ OPVs.

Device Structure	V_{OC} (V)	J_{SC} (mA/cm^2)	*FF* (%)	*PCE* (%)	Treatment (annealing)	Ref.
[a] P3HT/PCBM (1:0.8)/TiOx/Al	0.61	11.1	66	[b]5.0	150°C, 10min	[12]
[a] PBDTTT-E/PCBM(1:1)/Ca/Al	0.62	13.2	63	5.15	1,8-	
[a] PBDTTT-C/PCBM(1:1)/Ca/Al	0.70	14.7	64	6.58	Diiodooctane	[14]
[a] PBDTTT-CF/PCBM(1:1)/Ca/Al	0.76	15.2	67	7.73	3% ratio in CB	
[a] PSBTBT:PCBM(1:1)/Ca/Al	0.68	12.7	55	5.1	140°C, 15min	[237]
[a] PTB1:$PC_{71}BM$(1:1)/Ca/Al	0.56	15.6	63	5.6	-	[117]
[a] PCDTBT:$PC_{70}BM$(1:4)/TiOx/Al	0.88	10.6	66	6.1	70°C, 60min	[22]
[a] BisDMO-PFDTBT:PC71BM(1:3)/-	0.97	9.10	51	4.5	110°C, 10min	[238]
[a] PSiF-DBT:PCBM(1:2)/Al	0.90	9.50	51	[c]5.4	-	[110]
[a] PBnDT-4DTBT/PCBM(1:1)/Ca/Al	0.81	9.70	56	4.3	Solvent	[91]
[a] P3HT:bisPCBM(1:1.2)/Sm/Al	0.73	9.14	62	4.5	110°C, 5 min	[125]
[a] P3HT:ICBA(1:1)/Ca/Al	0.84	9.67	67	5.44	Solvent	[126]
[a] P3HT:Lu3N@C_{80}-PCBH(1:1)/LiF/Al	0.81	8.64	61	4.2	110°C, 10 m	[239]
[a] oligothiophene:PCBM(1:2)/LiF/Al	0.94	4.4	40	1.65	-	[240]

[a] Referring to the anode ITO/PEDOT:PSS; [b] incident light density Pin=90 mW/cm^2; [c] Pin=80 mW/cm^2

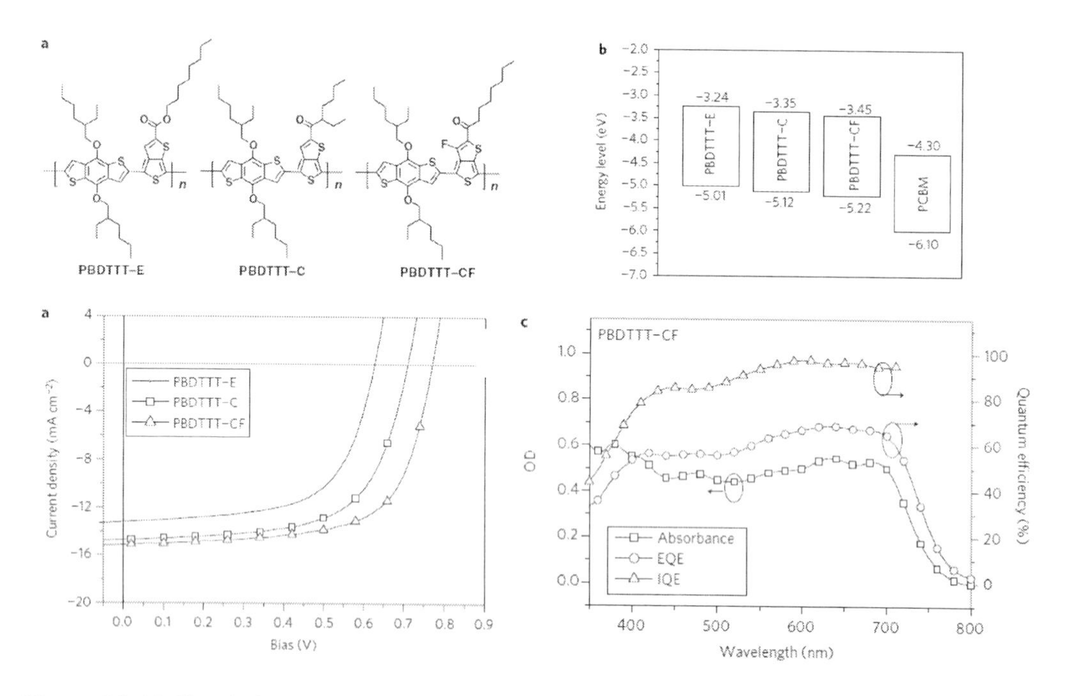

Fig.ure 36. (a) Chemical structure and (b) energy levels of PBDTTT-E, PBDTTT-C and PBDTTT-CF. (c) Current density versus voltage (J-V) curves and (d) external quantum efficiency (EQE), absorption (OD, optical density), and internal quantum efficiency (IQE) of BHJ OPVs with structure ITO/PEDOT:PSS/Donor : PCBM/Ca/Al. Adapted from [14].

On the other hand, only solvent mixture treatment (with 3% volume ratio of 1,8-Diiodooctane in host solvent CB) of bulk films were adopted to control film morphology. Consequently, well organized phase separation and high hole mobility ~7×10^{-4} cm^2/Vs (determined by SCLC method [225]) of PBDTTT-CF : PCBM film were achieved. Moreover, small optical absorption band ~1.45 eV polymer PSBTBT has been synthesized by Hou et al. to enhance more absorption of sunlight.[237] Thus, V_{OC} = 0.68 V, high J_{SC} =12.7 mA/cm^2 and *PCE*=5.1% were obtained by thermal annealing active films at 140 °C for 15 min. Similarly, in spite of relatively low V_{OC}=0.56 V, high J_{SC}=15.6 mA/cm^2 was obtained by mixing acceptor $PC_{71}BM$ with PTB1 rather than $PC_{61}BM$. Thus, high *PCE*=5.6% was achieved.[117]

Besides above slightly improved V_{OC} less than 0.8 V, great efforts have been applied to modify donor materials to decrease their HOMO energy levels.[22], [238], [110], [91] By introducing PCDTBT polymer with ~5.5 eV low HOMO forming an energy offset 1.2 eV with $PC_{70}BM$, Park et al. have obtained a high V_{OC}=0.88 V as shown in Fig. 37.[22] Meanwhile, high J_{SC} =10.6 mA/cm^2 was optimized by adopting inorganic oxide TiOx serving as optical spacer and hole blocking layer and by carefully controlling film morphology using solvent DCB. As a result, the PCE=6.1% was achieved mainly attributed to high IQE approaching 100%.

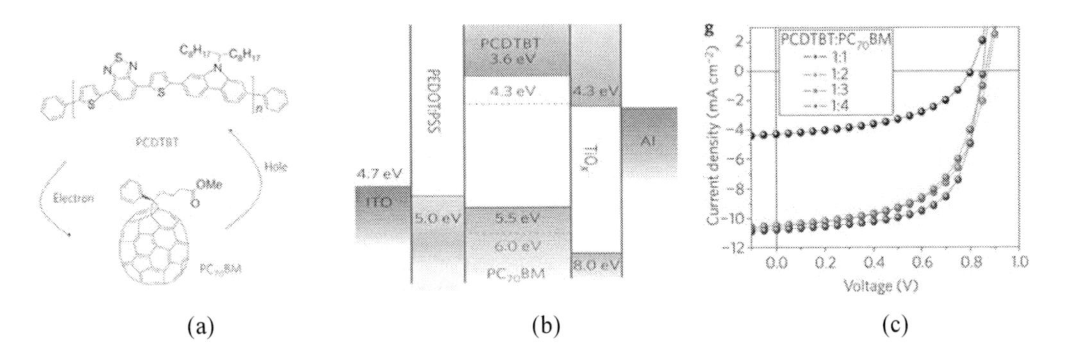

Figure 37. (a) Chemical structure and (b) energy levels of PCDTBT and $PC_{70}BM$. (c) J-V curves of OPVs with an active layer PCDTBT:$PC_{70}BM$ (1:1 to 1:4). Adapted from [22].

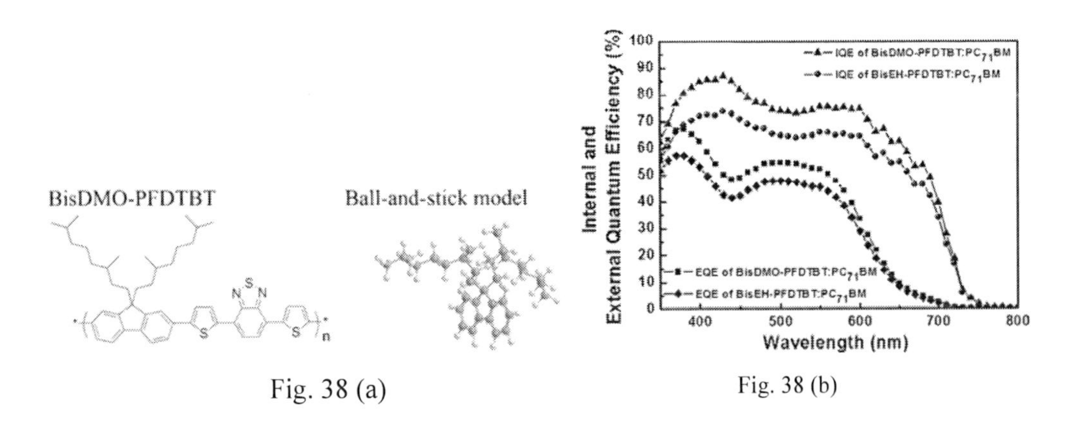

Figure 38. (a) Chemical structure of BisDMO- PFDTBT, (b) EQE and IQE properties of OPVs with bulk layer BisEH-PFDTBT or BisDMO- PFDTBT:$PC_{71}BM$ (1:3). Adopted from [238].

M.-H Chen et al. pointed out that two unbulky methyl groups located on the third and seventh carbon of the BisDMO-PFDTBT fluorene derivative, which not only decreases the steric effect, resulting in more compact π-π stacking, but also increases the solubility.[238] As a result, BisDMO-PFDTBT has a low HOMO level 5.5 eV with a band gap of 1.9 eV. Thus, a high energy offset of ~1.2 eV with $PC_{71}BM$ leads to a high V_{OC} of 0.97 V with a thin active-layer thickness of only 47nm as shown in Fig. 38. Also, Wang and Cao et al. developed a polymer PSiF-DBT with low HOMO level 5.39 eV as shown in Fig. 39. Thus, ~1.1 eV high energy offset with $PC_{60}BM$ results in a high V_{OC} of 0.90 V. [110].

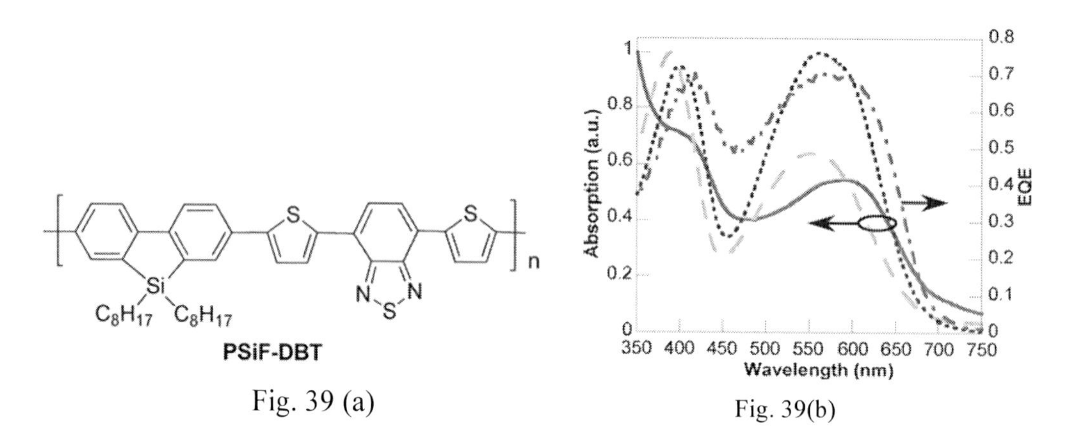

Figure 39. (a) Chemical structure of PSiF-DBT, (b) Absorption and EQE properties of OPVs with bulk layer PSiF-DBT:$PC_{60}BM$ (1:2). Adopted from [110].

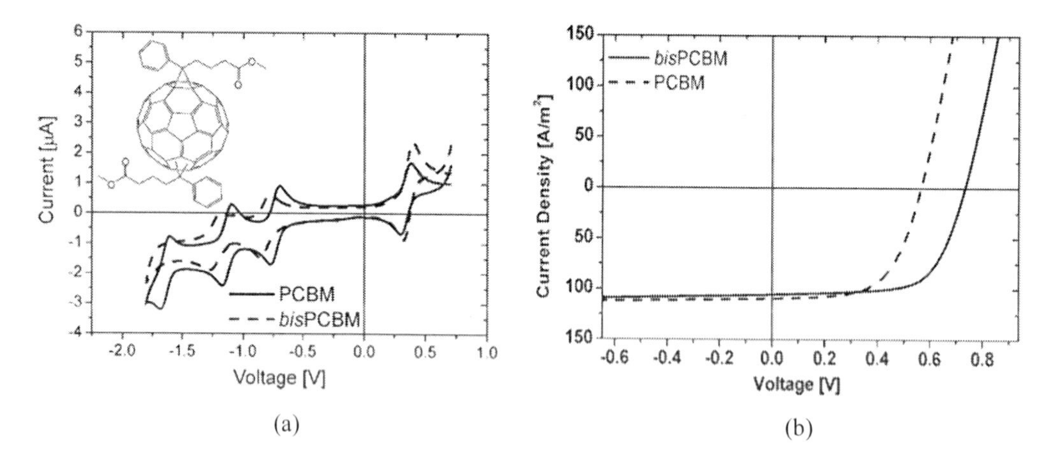

Figure 40. (a) Cyclic voltammetry measurements performed on PCBM (solid line) and bisPCBM (dashed line). (b) Current-density versus voltage curves of P3HT:PCBM and P3HT:bisPCBM solar cells. Adopted from [125].

To enlarge the energy offset between HOMO of polymers and LUMO level of acceptor materials to upgrade the LUMO energy of acceptors is also feasible. By introducing the same functional group to PCBM, bisPCBM has a high LUMO level 3.7 eV and an unchanging HOMO 6.1 eV.[125] Then, the OPV devices based on bulk film P3HT: bisPCBM has a higher V_{OC}=0.73 V than that 0.58 V based on P3HT: PCBM blends, as shown in Fig. 40. Also, novel fullerene C_{60} derivatives ICMA and ICBA with high LUMO levels 3.86 eV and 3.74 eV than PCBM were synthesized by Y. J. He and Y. F. Li et al. As a result, both high

V_{OC} of 0.84 V and PCE of 5.44% were obtained, as shown in Fig. 12.[126] Moreover, novel trimetallic nitride endohedral fullerenes named Lu3N@C80-PCBH with high LUMO has been synthesized by Ross et al. Therefore, V_{OC} was improved to 0.81 V.[239]

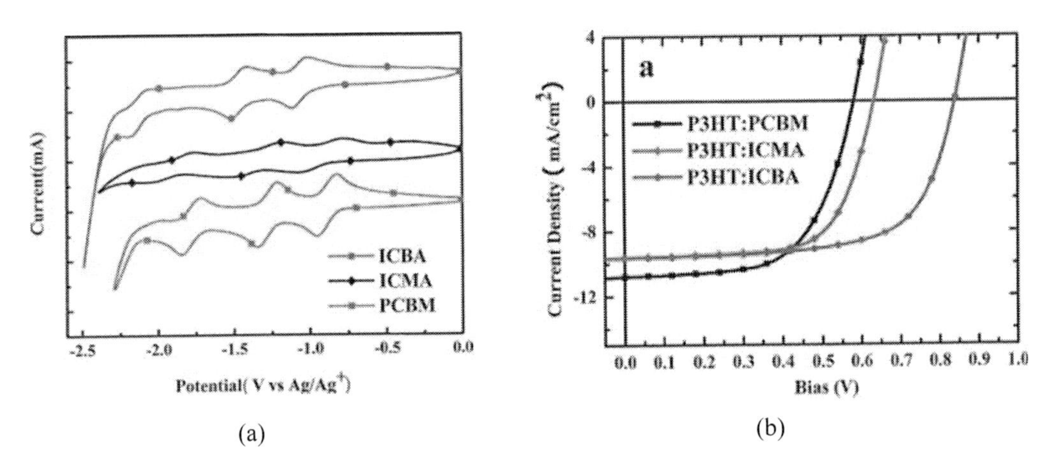

(a) (b)

Figure 41. (a) Cyclic voltammetry measurements performed on PCBM, ICMA and ICBA. (b) Current-density versus voltage curves of P3HT:PCBM, P3HT:ICMA and P3HT:ICBA solar cells. Adopted from [126].

In summary, instinctive maximum V_{OC} can only be optimized by modifying functional group to decrease HOMO level of donor polymers and enhance LUMO energy of acceptors such as fullerene C_{60} derivatives. Meanwhile, the fabrication process of BHJ OPVs should be carefully optimized to insure high J_{SC} and FF, for instance, proper host solvent, thermal annealing, solvent annealing, solvent mixture and also optical spacer, hole blocking layer, and so on.

5.3. Inverted BHJ OPVs

In the regular BHJ devices with a structure of ITO/PEDOT:PSS/blend film/Ca (or LiF)/Al, the PEDOT:PSS is commonly used as anode buffer layer not only to smooth the ITO surface, but also to enhance hole injection due to the decreased work function of ITO.[240] However, due to the acidic nature of PEDOT:PSS, it's not helpful to the lifetime of OPVs for interface instability due to etched ITO and affected blend films.[241],[242] Meanwhile, Ca and Al cathode would be easily oxidized in air even with a delicate encapsulation to decrease lifetime.[243],[244] Therefore, one of most intensely studied strategy is adopting inverted structure, typically, ITO/inorganic oxides (or thin metal layer with high work function)/ P3HT: PCBM/PEDOT:PSS (or low work function inorganic oxides)/cathode.

The capability of ITO serving as electron collecting cathode has been proved when modified by inorganic oxides (or metals) with high work function. Another crucial reason of adopting inverted structure is that vertical phase separation of polymer blend films are currently observed by M. C. Quiles et al.[245] and Y. Yang group.[254] They pointed out that conventional fabricated polymer blend films based on thermal, solvent annealing and solvent mixture treatments would result in PCBM rich at anode and polymer donors rich at film surface (near cathode), namely "vertical phase separation". Then, inverted OPVs can commendably utilize this film property, thus, inverted structure is considered as another improving method for BHJ OPVs.[245] Herein, representational inverted BHJ devices are summarized in Table 10 to aim at the performance optimization of OPVs.

Table 10. Device characteristics of inverted BHJ OPV cells.

Device Structure	V_{OC} (V)	J_{SC} (mA/cm^2)	FF (%)	PCE (%)	Ref.
ITO/ZnO/P3HT:PCBM/Ag	0.57	11.22	47.5	2.58	[246]
ITO/ZnO NP/P3HT:PCBM/PEDOT:PSS/Ag	0.62	11.17	54.3	3.3	[247]
Ag-grid/PEDOT:PSS/ZnO/C_{60}/P3HT:PCBM/ PEDOT:PSS/Ag	0.60	9.39	57	3.2	[248]
ITO/TiOx/P3HT:PCBM/PEDOT:PSS/Au	0.56	9.0	62	3.1	[249]
ITO/PTE/TiOx/P3HT:PCBM/PEDOT:PSS/Ag	0.56	10.2	64	3.6	[250]
FTO/TiO_2/P3HT:PCBM/PEDOT:PSS/Au	0.64	12.4	51.1	4.07	[251]
FTO/TiO_2/P3HT:PCBM/PEDOT:PSS/Ag	0.61	11.27	62.8	4.3	[252]
ITO/Cs_2CO_3/RR-P3HT:PCBM/V_2O_5/Al	0.56	8.42	62	2.25	[253]
ITO/Cs_2CO_3/P3HT:PCBM/V_2O_5/Al	0.59	11.13	63	4.2	[254]
ITO/Cs_2CO_3/P3HT:FPCBM/V_2O_5/Al	0.56	8.64	55.7	2.7	[255]
ITO/Cs_2CO_3/P3HT:PCBM/PEDOT:PSS/Ca/Al	0.60	7.03	48	2.0	[256]
ITO/Al/P3HT:PCBM/MoO_3/Al	0.58	10.95	61.4	3.9	[257]
ITO/Ca/P3HT:PCBM/MoO_3/Ag	0.65	8.28	65.9	3.55	[258]

Hau et al. fabricated unencapsulated inverted flexible BHJ OPVs based on P3HT : PCBM blend with zinc oxide (ZnO) nanoparticles serving as cathode buffer layer, both high J_{SC}=11.17 mA/cm^2 and PCE=3.3% were obtained due to high electron mobility ~0.066 cm^2/Vs of ZnO nanoparticles.[246] On the other hand, device stability was obviously improved comparing with conventional devices due to good ability at obstructing from diffusion of H_2O and O_2 into active layers of inverted anode PEDOT:PSS/Ag.[246] Another interesting work based on metal grid/conducting polymer hybrid transparent electrode has been proposed by Zou et al. to replace ITO for the fabrication of inverted OPVs.[248] The OPV with a structure of Ag-grid (5 μm/50 μm)/PEDOT:PSS/ ZnO/C_{60}/P3HT:PCBM/ PEDOT:PSS/Ag presented a highest performance of J_{SC}=9.39 mA/cm^2 and PCE=3.2% due to the reduced lateral resistance, as shown in Fig. 42, where Ag-grid (5 μm/50 μm) (with or without PEDOT:PSS)/ZnO acted as transparent anode.

As well known, inorganic oxide TiOx can be used as optical spacer, hole blocking layer between active blend film and cathode due to its high electron transporting mobility and proper energy levels with 4.3 eV conduction band and 8.0 eV low valence band.[12],[22] Thus, when TiOx is used in inverted devices to modify ITO and fluorinedoped tin oxide (FTO) as a cathode, high photovoltaic performances with PCE=3.1% in Ref [249] and 3.6% in Ref [250] based on regular P3HT:PCBM OPVs were obtained. The J-V curves from R. Steim et al.'s work [250] of various inverted OPVs with contrast devices are shown in Fig. 43 Another metal dioxide titanium (TiO_2) commonly used in dye-sensitized cells can also be adopted to form electron selective contact at the ITO interface in inverted OPVs.[251],[252] Both G. K. Mor et al. [251] and W.-H. Baek et al. [252] obtained high J_{SC}>11 mA/cm^2 and PCE>4% based on P3HT:PCBM OPVs, which proving TiO_2 as one of promising cathode modifying materials in inverted OPVs. Meanwhile, the lifetime study of device

FTO/TiO_2/P3HT: PCBM/PEDOT:PSS/Ag shows good stability >500 hours in air, as presented in Fig. 44.[252]

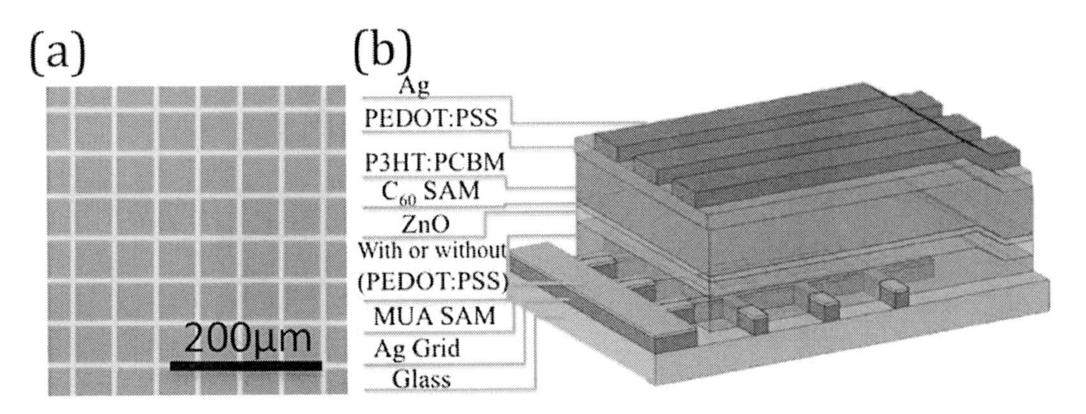

Figure 42. (a) Optical microscopic image of silver grid with 5 µm width separated by a distance of 50 µm. (b) Device configuration of the polymer solar cell using Ag grid as the transparent electrode with or without conductive PEDOT:PSS layer. Adopted from [248].

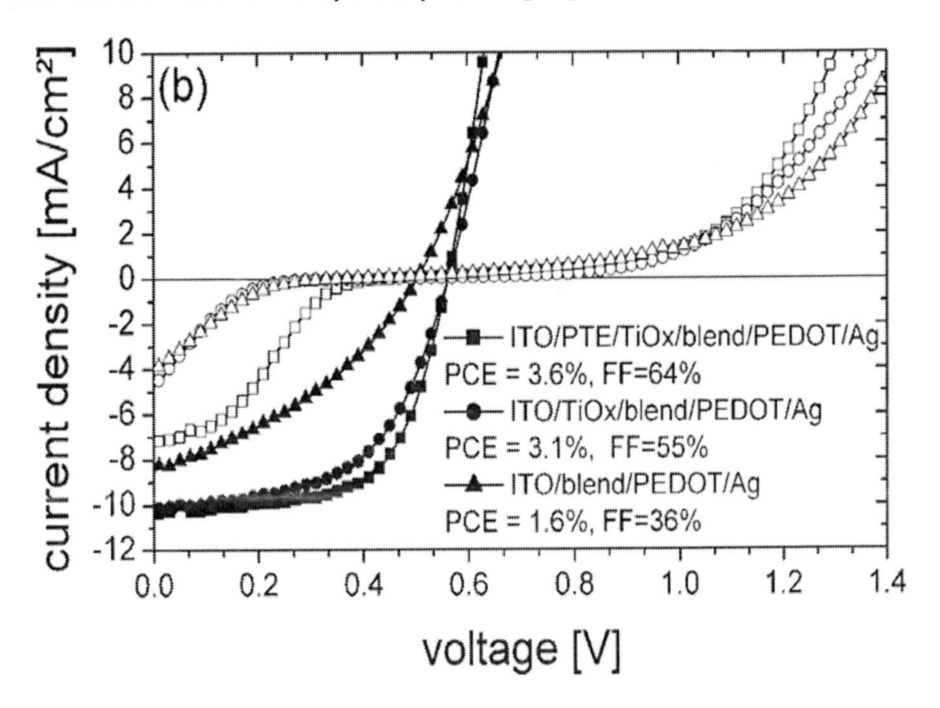

Figure 43. J-V characteristics of the solar cells studied under illumination with UV light illumination (filled symbols) and without UV illumination, where PTE is polyoxyethylene tridecyl ether. Adopted from [250].

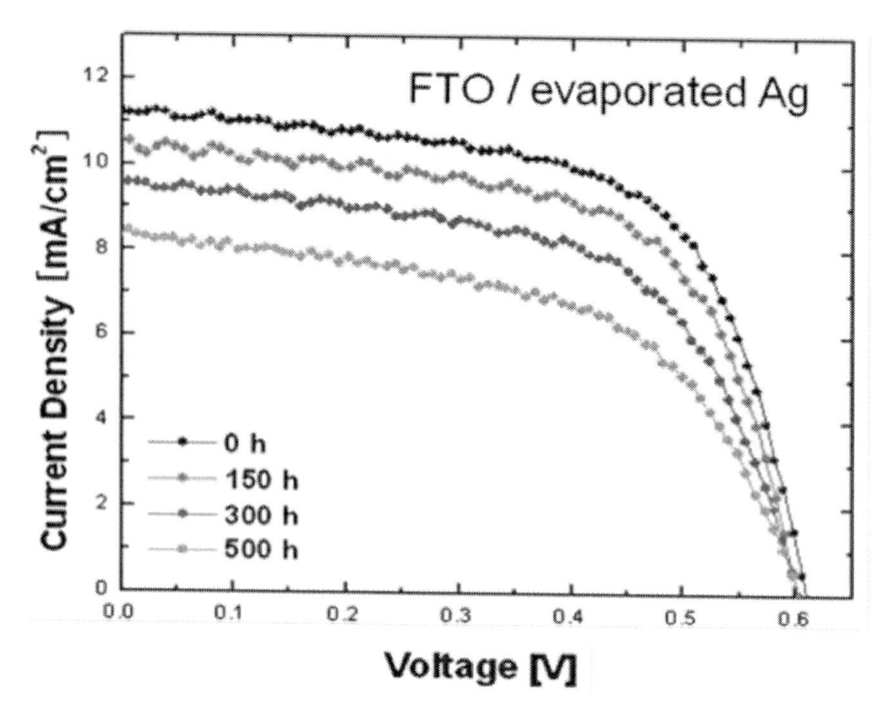

Figure 44. J-V curves of inverted solar cells FTO/TiO_2/P3HT:PCBM/PEDOT:PSS/Ag (thermal evaporated) at various time in air, showing good device stability. Adopted from Ref [252].

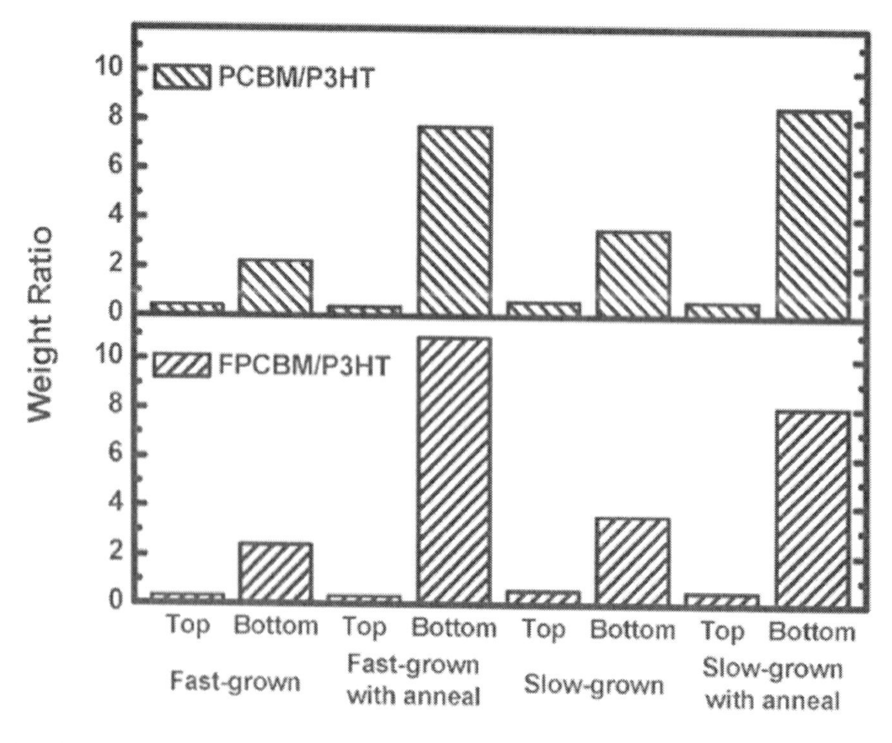

Figure 45. The compositions at the top and bottom surfaces of the blend films either PCBM/P3HT or FPCBM/P3HT spin-coated on Cs2CO3 coated glass. Adopted from [255].

Moreover, another inorganic cesium carbonate (Cs_2CO_3) has been introduced in inverted OPVs to modify ITO as a cathode. Y. Yang group has presented fundamental researches based on Cs_2CO_3 and vanadium oxide (V_2O_5).[253]-[255] First of all, the vertical phase separation morphology of P3HT:PCBM films has been observed that the weight ratios of PCBM/P3HT and FPCBM/P3HT on Cs_2CO_3 (thermal annealed at 170 °C for 20 min) coated substrates are higher than 7 when blend films fabricated by two different treatments namely fast-grown with annealing, and slow-grown with annealing rather than that without annealing, as shown in Fig. 45.[222][255] Fig. 46 shows the height and phase AFM images of the exposed top and bottom surfaces of fast grown films. Thus, because of this vertical phase separation, inverted BHJ OPVs based on solution processed Cs_2CO_3 have high J_{SC}=11.13 mA/cm^2, FF=63% and PCE=4.2%.[254] Furthermore, formation of interface dipole layers at the ITO surface has been discovered, as shown in Fig. 47(a). The direction of the dipole moments points from the ITO surface to vacuum, and hence reduce the work function of ITO surface, as shown in Fig. 47(b).[259]

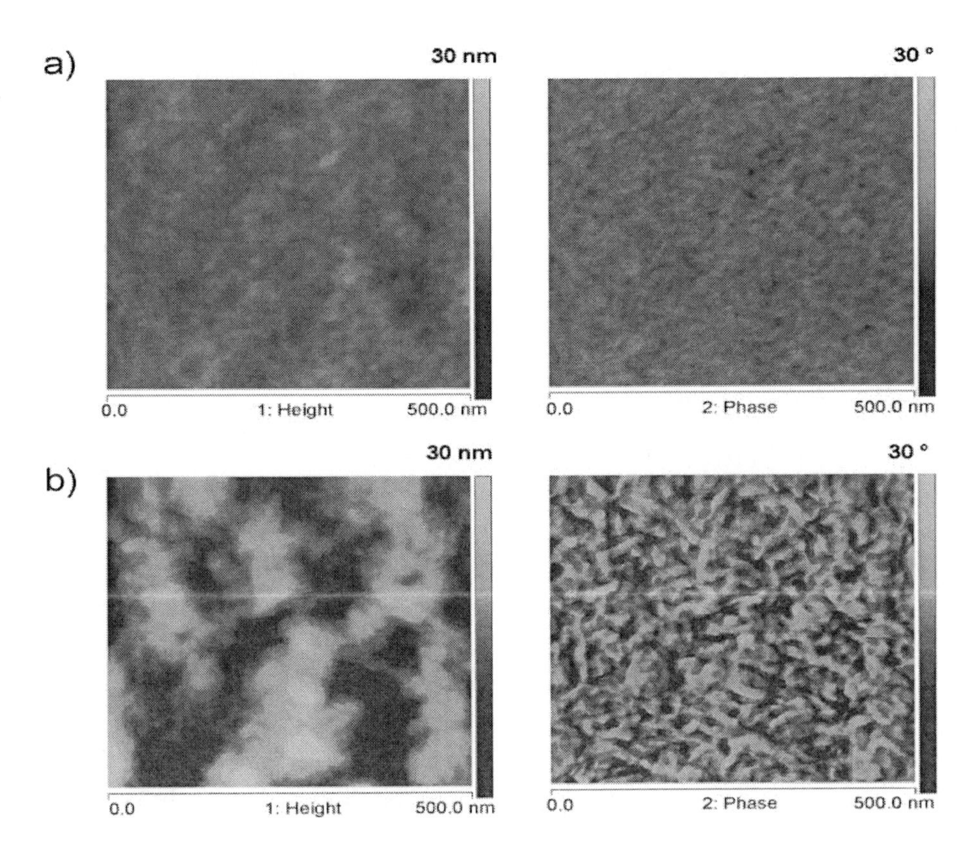

Figure 46. AFM topography (left) and phase (right) images of a) top and b) bottom surfaces of the exposed P3HT networks. Adopted from [255].

Last but not least, specific metals with high work functions can also be adopted to modify ITO as a cathode. For instance, by depositing ultrathin layer Al (1.2 nm) [257] or Ca (1 nm) layer [258] on ITO surface and inserting 3-5 nm MoO_3 between active layer and anode, both the high J_{SC}>8 mA/cm^2 and PCE>3.5% can be obtained.

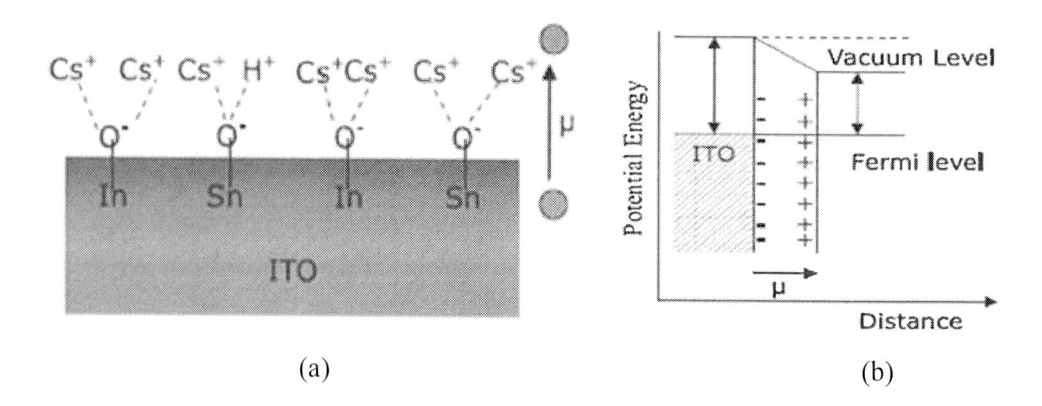

Figure 47. (a) Scheme for the formation of dipole layer on ITO and its effect on reducing the work function of ITO. (b) Schematic of the semi-transparent laminated device. Adopted from [259].

It's an interesting to investigate that the commonly used blend polymer films can be used as not only forward charge transporting layer in regular OPVs but also reverse transporting layer in inverted BHJ OPVs. The crucial advantages of inverted structure on device lifetime and application in blend films with vertical phase separation have also been investigated.[247],[252] Furthermore, another motivation for adopting organic oxides is to provide design flexibility for tandem cells.[13] As well known, for solution-processed tandem BHJ OPVs, it is difficult to realize a multilayer structure without dissolution of the layers underneath.[260] As commonly used in inverted OPVs, inorganic oxides would provide protection to the underlying polymer layer against the subsequent solution coating.[261] Thus, the following discussion would dwell on tandem BHJ OPVs including inorganic oxides as intermediate layers.

5.4. Tandem BHJ OPVs

As pointed out in previous researches, two major energy losses in solar cells with single active layer are the sub-band-gap transmission and the thermalization of hot charge carriers.[261] Thus, the first reason for adopting tandem structure is that series-connected subcells with different absorption band can achieve a maximum theoretical efficiency under nonconcentrated sunlight, e.g., M. A. Green reported high efficiency 33.8% inorganic solar cells based on three subcells GaInP/GaInAs /GaInAs structure.[262] Secondly, the lowcharge transporting mobility of polymer blend films e g. ~10^{-4} cm^2/Vs for hole and electron and short exciton diffusion length, e.g., ~10 nm for polymer donors and ~40 nm for fullerene acceptors (as summarized in section 3) hinder the realization of a thick active layer in single device from achieving a maximum absorption of sunlight.[263] Thus, combining several subcells with various materials can help to more efficiently cover the spectrum of sunlight.

Accordingly, G. Dennler et al. proposed design rules for BHJ tandem solar cells towards PCE=15 % at the first time,[17],[263] which are revealed as very useful to theoretically design and experimentally optimize high J_{SC} and *PCE* tandem devices.[13] [154] [185][260][264]-[266] From the ideal band diagram of tandem cells and simulated efficiencies in Fig. 48, the optimized band gap of donor material in bottom (or front cell) and top cells (or back cell) are around 1.55 to 1.7 eV and 1.1 to 1.55 eV, respectively. Hence, low band gap polymers less than 1.55 eV are exigently needed to enhance the absorption of near infrared sunlight ranging from 800 to 1100 nm.[267] Hence, this subsection would dwell on some successful designations of tandem cells to instruct obtaining high J_{SC} and PCE. Meanwhile, crucial issues on intermediate layer between two subcells would also be analyzed.

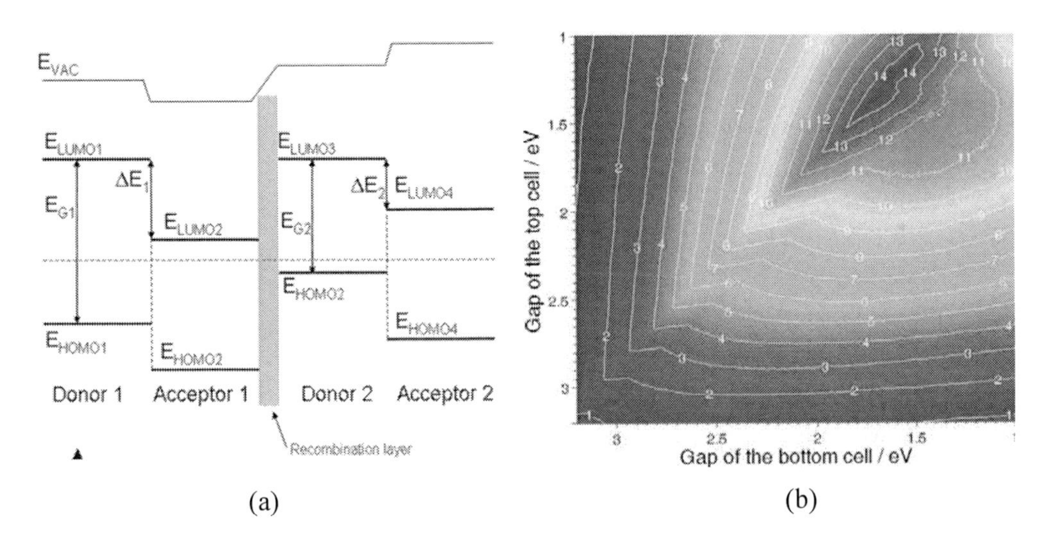

Figure 48. (a) Simplified band diagram of tandem cells composed of two subcells connected in series by a recombination layer. (b) Efficiency of an OPV tandem device versus the band gap of both donors. We assumed that the difference between the LUMO of the donor and the acceptor is 0.3 eV, that the maximum EQE of the subcells are 0.65, and that the IQE of the bottom device is 85%. Adopted from [17],[263].

To find out compatible techniques in semiconductor field to optimize tandem cells, some successful researches with various device structures are summarized in Table 11. Typically, the designation of tandem cells is related with three aspects. Firstly, selecting two subcells with almost equal J_{SC} at different absorption range due to J_{SCtan}=Min (J_{SC1}, J_{SC2}). Secondly, V_{OCtan} of tandem cell, the only parameter to improve PCE, should be enhanced as high as possible due to $V_{OCtan} \cong V_{OC1}+V_{OC2}$. Thirdly, semitransparent intermediate layer connecting two subcells should be proper selected to serve as an efficient recombination centers of holes and electrons. Meanwhile, intermediate layer should be fabricated without damaging active layer of front cell, and protect front cell from dissolution during fabricating button cell in BHJ OPVs. Thus, the following discussions are mainly focused on the above three points.

The first type of tandem BHJ OPVs is a combination of front BHJ cell and back small molecular cell. Major motivation of adopting this structure is that the back cell is fabricated by vacuum deposition process, thus, the active blend layer of front cell is not described. G. Dennler et al. presented a tandem structure with front cell P3HT:PCBM and a back cell ZnPc/ZnPc:C_{60}/C_{60}.[264] V_{OC} equals the sum of two subcells, however, J_{SC} is obviously twice decreased. Main reason would be 1 nm Au does not serve as an effective recombination layer. Another designation with a compound intermediate layer LiF/Al /MoO_3 is proposed by D. W. Zhao et al. [154] LiF/Al can effectively collect electrons from front cell and MoO_3 serve as a hole collector, as a result, effective recombination interface of holes and electrons forms, as shown in Fig. 49. Also, the absorption amount of front cell and back cell were equally optimized which resulted in high J_{SC}=6.05 mA/cm^2.

Table 11. Device characteristics of tandem (TD) BHJ OPVs with parameters of front cell (FC, or button cell), back cell (BC, or top cell)

Device Structure		V_{OC} (V)	J_{SC} (mA/cm^2)	FF (%)	PCE (%)	Ref.
[a)]/P3HT:PCBM (front cell) /Au (1 nm) /ZnPc -ZnPc:C_{60}-C_{60} (back cell)/Al	BC	0.47	9.3	50	2.2	[264]
	FC	0.55	8.5	55	2.6	
	TD	1.02	4.8	45	2.3	
[a)]/P3HT:PCBM /LiF/Al /MoO_3/CuPc/ CuPc:C_{60}/C_{60}/BPhen/Ag	BC	0.45	7.83	47.6	1.68	[154]
	FC	0.63	6.54	51.3	2.11	
	TD	1.01	6.05	46.2	2.82 [b)]3.88	
[a)]/MDMO-PPV:PCBM/ITO (20 nm) +PEDOT:PSS/MDMO-PPV:PCBM/Al	BC	0.84	4.6	58	2.3	[265]
	FC	-	-	-	-	
	TD	1.34	4.1	56	3.1	
[a)]/PCPDTBT:PCBM/TiOx (8 nm) + PEDOT:PSS/P3HT:PC_{70}BM/TiOx/Al	BC	0.63	10.8	69	4.7	[13]
	FC	0.66	9.2	50	3.0	
	TD	1.24	7.8	67	6.5	
[a)]/P2:PC_{71}BM/TiOx+PEDOT:PSS/P3HT: PC_{71}BM/TiOx/Al	BC	0.62	8.47	59	3.1	[260]
	FC	0.79	8.16	52	3.3	
	TD	1.33	6.17	56	4.6	
[a)]/PFTBT:PCBM/ZnO + pH neutral PEDOT/pBBTDPP2:PC_{70}BM/LiF/Al	BC	0.61	6.2	57	2.2	[185]
	FC	0.98	5.5	52	2.8	
	TD	1.58	6.0	52	4.9	
[a)]/P3HT:PC_{70}BM/Al+TiO_2/PEDOT:PSS/P SBTBT:PC_{70}BM/TiO_2:Cs/Al	BC	0.67	10.7	56	3.94	[266]
	FC	0.60	9.27	67	3.77	
	TD	1.25	7.44	63	5.84	

[a)] Referring to the anode ITO/PEDOT:PSS; [b)] Referring to incident light density Pin=300 mW/cm^2.

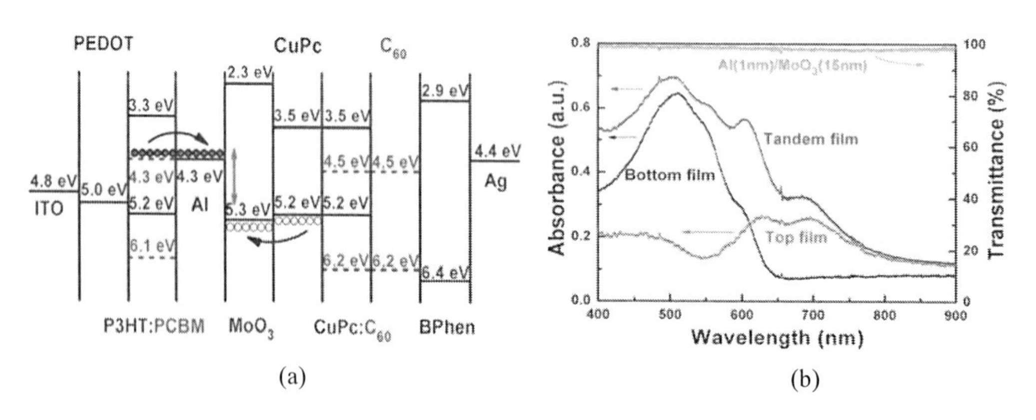

Figure 49. Energy level diagram of polymer-small molecule tandem OPV. (b) Absorption spectra of the bottom film P3HT:PCBM (120 nm), top film CuPc (7.5 nm)/CuPc:C_{60} (12.5:12.5 nm)/C_{60} (27.5 nm) and tandem film P3HT:PCBM/LiF (0.5 nm)/Al (1 nm)/MoO_3 (15 nm)/CuPc /CuPc:C_{60}/C_{60} films. Transmittance spectrum of Al(1 nm) /MoO_3(15 nm) intermediate layer. Adopted from [154].

The first high efficiency tandem polymer OPVs fabricated by all-solution processing was reported by J. Y. Kim and A. J. Heeger et al. with a structure ITO/ PEDO :PSS/PCPDTBT : PCBM/TiOx + PEDOT:PSS/P3HT : $PC_{70}BM$/TiOx/Al. [13] Their devices presented doubled V_{OC}, both high J_{SC} and PCE>6.5%, not only because of almost equally optimized absorption amount of two subcells, but also due to the effective recombination interface formed by TiOx and PEDOT:PSS, as shown in Fig. 50. The TiOx intermediate layer was deposited from solution (by means of sol-gel chemistry) with no substantial interlayer mixing. Also, the spin-coated PEDOT:PSS layer protected the active layer of front cell from dissolution during fabricating P3HT:$PC_{70}BM$ film. Moreover, the photovoltaic properties of two subcells reached their highest performances with both J_{SC}>9 mA/cm^2 and PCE>3% at relatively low thickness ~130 nm. This work inspired various further applications with different active layers. For instance, Y. C. Chen et al. obtained high PCE=4.6% with a tandem OPVs combined with front layer P2:$PC_{71}BM$ and back layer P3HT:$PC_{71}BM$ based on TiOx/PEDOT:PSS intermediate layer.[260].

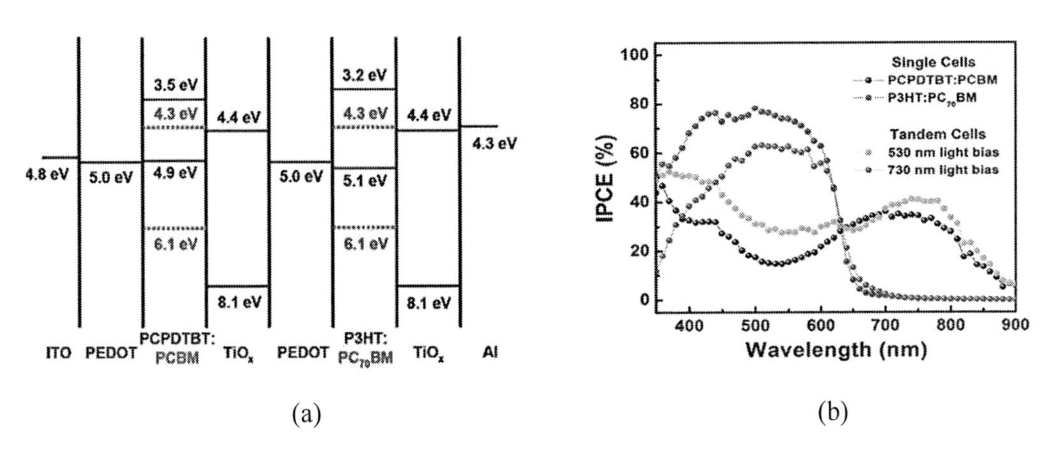

Figure 50. (a) Energy-level diagram of each of the component materials. (b) IPCE spectra of single cells and a tandem cell with bias light. Adopted from [13].

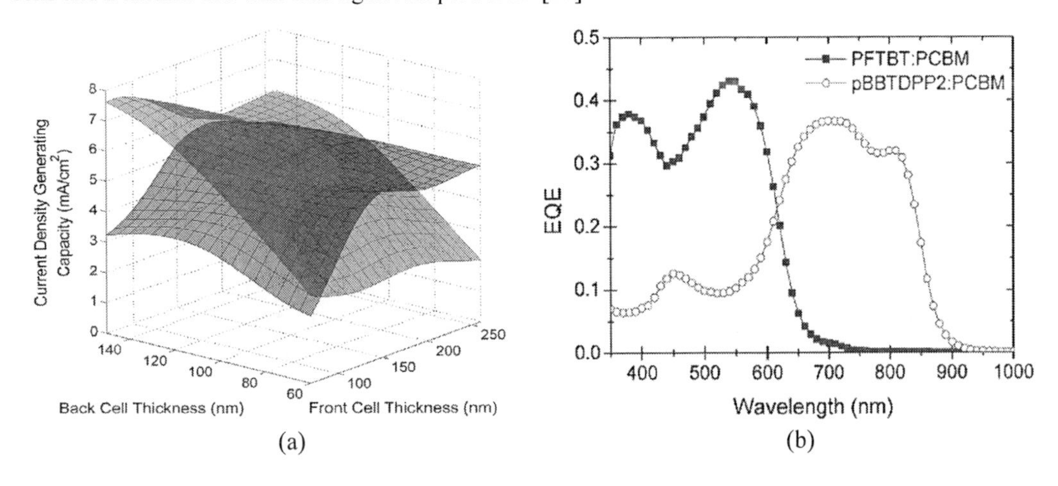

Figure 51. (a) 3D plot of current-density-generating capacities of the front (wine) and back cell (green) versus the front and back cell layer thicknesses; (b) the spectral response of a tandem cell with a 180-nm-thick front cell and a 125-nm-thick back cell. Adopted from [185].

Other solution processed inorganic oxides like ZnO and TiO_2 were also applied as an effective intermediate layer to collect electrons. [185][266] J. Gilot et al. fabricated a 30 nm ZnO layer by spin-cast from a 10 mg/mL solution of ZnO nanoparticles in acetone followed by a 15 nm pH-neutral PEDOT:PSS layer. The final device is ITO/PEDOT: PSS/PFTBT:PCBM/ZnO+pH neutral PEDOT/pBBTDPP2:PC_{70}BM/LiF/Al. [185]

The most concerning point in this research is that J_{SC} can be improved to a maximum value based on the thickness optimization of two active films by the means of optical transfer matrix theory, as shown in Fig. 51. Almost crucial parameters V_{OC}=1.58 V, and PCE=4.9% were all doubly enhanced and J_{SC} of two subcells were equally optimized. In S. Sista et al.'s work, Al (0.5 nm)/TiO_2/PEDOT4038 acts as an effective recombination layer to connect front cell of P3HT:PC_{70}BM film and PSBTBT: PC_{70}BM blend, although conducting band of TiO_2 seems higher than LUMO of PC_{70}BM, as shown in Fig. 52.[266] The thin layer of n-type nanocrystalline TiO_2 film was spin-casted from 0.25 wt% of TiO_2 solution in a 1:1 volume ratio of 2-ethoxyethanol and ethanol at 1000 rpm, followed by thermal annealing at 120 °C for 10 min. Moreover, a thin layer TiO_2:Cs was inserted between active film and Al cathode. This device also presented high J_{SC} =7.44 mA/cm^2 and PCE=5.84%.

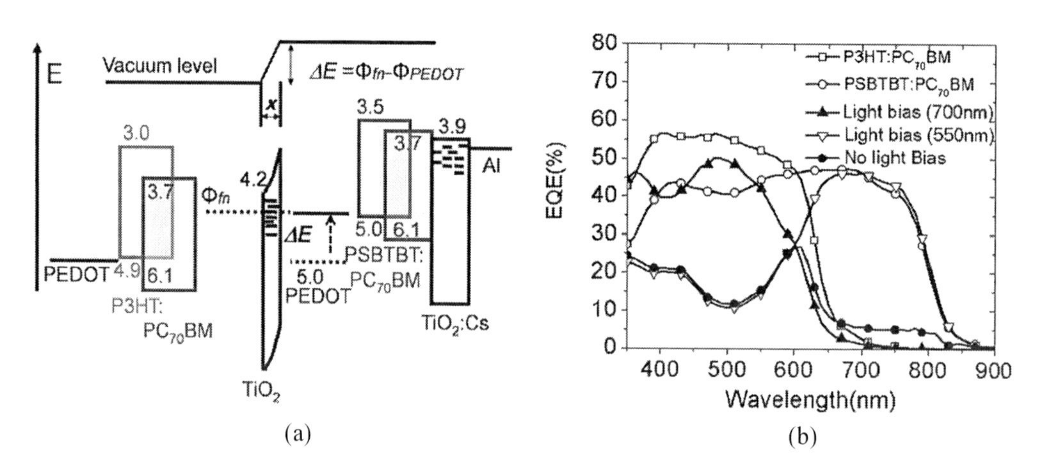

Figure 52. (a) Energy level diagram of tandem solar cells containing two BHJs as subcells and TiO_2/PEDOT4083 as an interconnection layer; (b) EQE of subcells in a tandem structure (with and without monocolor light bias) and reference single cells. Adopted from [266].

6. Summary and Outlook

During last two decades researching on organic solar cells, the OPVs power conversion efficiencies have reached 5.5% for small molecular tandem cells and ~8% for bulk heterojunction OPVs. However, remarkable improvements are urgent required to reach PCE>15% with a tandem structure, not only by synthesizing novel donor and acceptor materials, but also by optimizing device structures both for single and tandem OPVs. Based on the presentation of high PCE OPVs in above sections, combination of novel donor and acceptor materials would rapidly lead to PCE commonly beyond 7%. Some optimization strategies related to crucial photovoltaic parameters are summarized as follows.

Firstly, the maximum V_{OC} of OPVs is refrained by energy offset between the HOMO level of donor and LUMO level of acceptor. Thus, to enhance V_{OC} as much as possible, novel low band gap donor materials with decreased HOMO energy level, long exciton diffusion length and high carrier mobility would be expected. Meanwhile, novel multi-adduct fullerenes with increased LUMO level and dissolving abilities instead of single substituted fullerenes would also become one of future researching targets. However, accompanying with the enlarged offset between low band gap donor HOMO and acceptor LUMO, the decreasing energy offset between the donor and acceptor LUMO levels should not be less than 0.3 eV required for an efficient electron transfer and high J_{SC}. The successful strategies that steadily enhance V_{OC} with improved J_{SC} and *FF* (or no decreasing at least) should be referred and further improved.

Secondly, to obtain high J_{SC} and FF for both planar-mixed small molecular cells and BHJ cells, scrupulous controlling film morphology may be the most important issue. For planar-mixed OPVs, percolating paths are formed across the film consisting of a mixture of a planar stacking e.g., CuPc, and SubPc and spherically symmetric molecule e.g., C_{60}. The nano-structured, spatially distributed donor acceptor interface is responsible for efficient exciton diffusion in the mixture. In BHJ OPVs, an ideal morphology consists of nano-scale phase separation, with an interpenetrating network for efficient exciton dissociation and carrier extraction. The BHJ phase separation should combine aggregated fibrils of polymer donor and clusters of acceptor domains, in which their nano-sizes are comparable to short exciton diffusion length e.g., 10 nm. Several effective treatments including thermal annealing, solvent annealing, and solvent mixture are discussed in detail in above section. Usually, once adopting one of these post-treatments, doubly improved photovoltaic performances can be obtained in recent researches. Recently, the detailed vertical composition profiles of polymer donor and acceptor in blend films can be deduced using ellipsometry and manually controlled by modifying substrates, which would also benefit for further optimization of OPVs such as inverted device structure.

Finally, the instinctive mechanism of energy loss in tandem OPVs needs further study to achieve doubled V_{OC} and high *PCE* without decreasing J_{SC}. According to the discussion in Section 5, two common places to optimize tandem OPVs will be addressed here. Firstly, the optimized device structure of two subcells with best performance in single cell should be carefully redesigned in tandem structure to balance absorption amount of two subcells, in other words, to optimize equally high J_{SC}. Detailed designing process can be found in previous researches.[268]-[271] Secondly, intermediate interface formed by semitransparent electron collector and hole transportation layer e.g., TiOx and PEDOT:PSS possesses much higher recombination efficiency than ultra-thin (or discrete) metal dots, which may be another future trend.

ACKNOWLEDGMENT

This work was partially supported by the National Science Foundation of China (NSFC) via Grant no. 60736005 and 60425101-1, the Foundation for Innovative Research Groups of the NSFC via Grant no. 60721001, Provincial project via grant no. 9140A02060609DZ0208, Program for New Century Excellent Talents in University via Grant no. NCET-06-0812, SRF for ROCS, SEM via Grant no. GGRYJJ08-05, and Young Excellent Project of Sichuan Province via Grant no. 09ZQ026-074.

References

[1] Tang, C. W. *Appl. Phys. Lett.* 1986, *48,* 183.

[2] Brabec, C. J. ; Sariciftci, N. S.; Hummelen, J. C. *Adv. Funct. Mater.* 2001, *11,* 15.

[3] Poullikkas, A. *Energy Policy* 2009, *37,* 3673.

[4] Xue, J. ; Uchida, S.; Rand, B. P.; Forrest, S. R. *Appl. Phys. Lett.* 2004, *84,* 3013.

[5] Xue, J.; Rand, B. P.; Uchida, S.; Forrest, S. R. *J. Appl. Phys.* 2005, *98,* 124903.

[6] Gommans, H.; Aernouts, T.; Verreet, B.; Heremans, P.; Medina, A.; Claessens, C. G.; Torres, T. *Adv. Funct. Mater.* 2009, *19,* 3435.

[7] Gommans, H.; Cheyns, D.; Aernouts, T.; Girotto, C.; Poortmans, J.; Heremans, P. *Adv. Funct. Mater.* 2007, *17,* 2653.

[8] Cheyns, D.; Rand, B. P.; Heremans, P. *Appl. Phys. Lett.* 2010, *97,* 033301.

[9] Marcel, D. *Handbook of Conducting Polymers, Vol. 1 (Ed: J. Kanicki ,T. A. Skotheim), New York 1985* .

[10] Yu, G.; Gao, J.; Hummelen, J. C.; Wudl, F.; Heeger , A. J. *Science* 1995, *270,* 1789.

[11] Halls, J. J. M.; Walsh, C. A.; Greenham, N. C.; Marseglia, E. A.; Friend, R. H.; Holmes, A. B. *Nature* 1995, *376,* 498.

[12] Kim, J. Y.; Kim, S. H.; Lee, H. –Ho.; Lee, K.; Ma, W.; Gong, X.; Heeger, A. J. *Adv. Mater.* 2006, *18,* 572.

[13] Kim, J. Y.; Lee, K.; Coates, N. E.; Moses, D.; Nguyen, T.-Q.; Dante, M. A. J. Heeger, *Science* 2007, *317,* 222.

[14] Chen, H.-Y.; Hou, J. H.; Zhang, S. Q.; Liang, Y. Y.; Yang, G. W.; Yang, Y.; Yu, L. P.; Wu, Y.; Li, G. *Nature Photonics,* 2009, *3,* 649.

[15] Chen, L.-M.; Xu, Z.; Hong, Z.; Yang, Y. *Journal of Materials Chemistry.* 2010, *20,* 2575.

[16] Scharber, M. C.; Mühlbacher, D.; Koppe, M.; Denk, P.; Waldauf, C.; Heeger, A. J.; Brabec, C. J. *Adv. Mater.* 2006, *18,* 789.

[17] Dennler, G.; Scharber, M. C.; Brabec, C. J. *Adv. Mater.* 2009, *21,* 1323.

[18] Pettersson, L. A. A.; Roman, L. S.; Inganas, O. J. *Appl. Phys.* 1999, *86(1),* 487.

[19] Peumans, P.; Yakimov, A.; Forrest, S. R.; *J. Appl. Phys.* 2003, *93,* 3693.

[20] Hänsel, H.; Zettl, H.; Krausch, G.; Kisselev, R. *Adv. Mater.* 2003, *15,* 2056.

[21] Huang, J.; Yu, J. S.; Lin, H.; Jiang, Y. D. *J. Appl. Phys.* 2009, *105,* 073105.

[22] Park, S. H.; Roy, A.; Beaupré, S.; Cho, S.; Coates, N.; Moon, J. S.; Moses, D.; Leclerc, M.; Lee, K.; Heeger, A. J. *Nature Photonics* 2009, *3,* 297.

[23] Chamberlain, G. A. *Solar Cells* 1983, *8(1),* 47.

[24] Marcus, R. A.; Sutin, N. *Biochim. Biophys. Acta.,* 1985, *811,* 265.

[25] Rand, B. P.; Burk, D. P. *Phys. Rev. B,* 2007, *75,* 115327.

[26] Morteani, A. C.; Friend, R. H.; Silva, C. *Chem Phys Lett,* 2004, *391,* 81-84.

[27] Kodama, Y.; Ohno, K. *Appl. Phys. Lett.* 2010, *96,* 034101.

[28] Cheyns, D.; Poortmans, J.; Heremans, P. *Phys. Rev. B,* 2008, *77,* 165332.

[29] Yoo, S.; Domercq, B.; Kippelen, B. *Appl. Phys. Lett.,* 2004, *85,* 5427-5429.

[30] Sokel, R.; Hughes, R. C. *J. Appl. Phys.* 1982, *53,* 7414.
[31] Mihailetchi, V. D.; Koster, L. J. A.; Hummelen, J. C.; Bolm, P. W. M. *Phys. Rev. Lett.* 2004, *93,* 216601.
[32] Yoo, S.; Domercq, B.; Kippelen, B. *J. Appl. Phys.,* 2005, *97,* 103706.
[33] Morteani, A. C.; Sreearunothai, P.; Herz, L. M. *Phys. Rev. Lett.,* 2004, *92,* 247402.
[34] Morteani, A. C.; Friend, R. H.; Silva, C. *J. Chem. Phys.,* 2005, *122,* 244906.
[35] Mazhari, B. *Sol. Energy Mater. Sol. Cells,* 2006, *90,* 1021-1033.
[36] Huang, J.; Yu, J. S.; Lin, H.; Jiang, Y. D. *Chinese. Sci. Bull.* 2010, *55,* 1-8.
[37] Fukuzumi, S.; Kojima, T. *J. Mater. Chem.* 2008, *18,* 1427.
[38] Umeyama, T.; Takamatsu, T.; Tezuka, N.; Matano, Y.; Araki, Y.; Wada, T.; Yoshikawa, O.; Sagawa, T.; Yoshikawa, S.; Imahori, H. *J. Phys. Chem. C* 2009, *113,* 10798.
[39] Kerp, H. R.; Donker, H.; Koehorst, R. B. M.; Schaafsma, T. J.; Faassen, E. E.; *Chem. Phys. Lett.* 1998, *298,* 302.
[40] Harima, Y.; Furusho, S.; Okazaki, K.; Kunugi, Y.; Yamashita, K. *Thin Solid Films* 1997, *300,* 213.
[41] Stübinger, T.; Brütting, W. *J. Appl. Phys.* 2001, *90,* 3632.
[42] Vivo, P.; Ojala, M.; Chukharev, V.; Efimov, A.; Lemmetyinen, H. *J. Photochem. Photobiol., A* 2009, *203*, 125.
[43] Peumans, P.; Forrest, S. R. *Appl. Phys. Lett.* 2001, *79,* 126.
[44] Xue, J.; Rand, B. P.; Uchida, S.; Forrest, S. R. *Adv. Mater.* 2005, *17,* 66.
[45] Xue, J.; Uchida, S.; Rand, B. P.; Forrest, S. R. *Appl. Phys. Lett.* 2004, *85,* 5757.
[46] Kahn, A.; Koch, N.; Gao, W.Y. *Journal of Polymer Science Part A* 2003, *41,* 2529–2548.
[47] Steudel, S; Myny, K.; Arkhipov, V.; Deibel, C.; De Vusser, S; Genoe, J.; Heremans, P. *Nat. Mater* 2005, *4,* 597–600
[48] Rand, B. P.; Xue, J.; Uchida, S.; Forrest, S. R. *J. Appl. Phys.* 2005, *98,* 124902.
[49] Terao, Y.; Sasabe, H.; Adachi, C. *Appl. Phys. Lett.* 2007, *90,* 103515.
[50] Wang, H.; Zhu, F.; Yang, J.; Geng, Y.; Yan, D. *Adv. Mater.* 2007, *19,* 2168.
[51] Li, N.; Lassiter, B. E.; Lunt, R. R.; Wei, G.; Forrest, S. R. *Appl. Phys. Lett.* 2009, *94,* 023307.
[52] Hideyuki, T.; Takeshi, Y.; Katsuhiko, F.; Tetsuo, T. *Appl. Phys. Lett.* 2006,*88,* 253506.
[53] Hirokazu, T.; Hiroshi, T.; Masaki, T.; Kazumi, M. *Appl. Phys. Lett.* 2000, *76,* 873.
[54] Mutolo, K. L.; Mayo, E. I.; Rand, B. P.; Forrest, S. R.; Thompson, M. E. *J. Am. Chem. Soc.* 2006, *128,* 8108.
[55] Ma, B.; Woo, C. H.; Miyamoto, Y.; Fréchet, J. M. *J. Chem. Mater.* 2009, *21,* 1413.
[56] Kim, D. Y.; So, F.; Gao, Y. *Sol. Energy Mater. Sol. Cells* 2009, *93,* 1688.
[57] Pasquier, A. D.; Unalan, H. E.; Kanwal, A.; Miller, S.; Chhowalla, M. *Appl. Phys. Lett.* 2005, *87,* 203511.
[58] Verreet, B.; Schols, S.; Cheyns, D.; Rand, B. P.; Gommans, H.; Aernouts, T.; Heremans, P.; Genoe, J. *J. Mater. Chem.* 2009, *19,* 5295.
[59] Hong, Z. R.; Lessmann, R.; Maennig, B.; Huang, Q.; Harada, K.; Riede, M.; Leo, K. *J. Appl. Phys.* 2009, *106*, 064511.
[60] Florent, M.; Ajay, K. P.; Simon, J.-J.; Philippe, T.; Ludovic, .; Nunzi, J.-M. *J. Appl. Phys.* 2007, *102,* 034512.

[61] Martín, G.; Rojo, G.; Agulló-López, F.; Ferro, V. R.; Vega, J. M. G.; Martínez-Díaz, M. V.; Torres, T.; Ledoux, I.; Zyss, J. *J. Phys. Chem. B* 2002, *106,* 13139.
[62] Kallmann, H.; Pope, M. *J. Chem. Phys.* 1959, *30,* 585.
[63] Pochettino, A. Acad. Lincei Rend. 1906, *15*, 355.
[64] Geacintov, N.; Pope, M.; Kallmann, H. *J. Chem. Phys.* 1966, *45,* 2639.
[65] Perez, M. D.; Borek, C.; Forrest, S. R.; Thompson, M. E. *J. Am. Chem. Soc.* 2009, *131,* 9281.
[66] Chu, C.-W.; Shao, Y.; Shrotriya, V.; Yang, Y. *Appl. Phys. Lett.* 2005, *86,* 243506.
[67] Pandey, A. K.; Shaw, P. E.; Samuel, I. D. W.; Nunzi, J.-M. *Appl. Phys. Lett.* 2009, *94,* 103303.
[68] Kinoshita, Y.; Hasobe, T.; Murata, H. *Appl. Phys. Lett.* 2007, *91,* 083518.
[69] Lloyd, M. T.; Mayer, A. C.; Tayi, A. S.; Bowen, A. M.; Kasen, T. G.; Herman, D. J.; Mourey, D. A.; Anthony, J. E.; Malliaras, G. G. *Org. Electron.* 2006, *7,* 243.
[70] Mayer, A. C.; Lloyd, M. T.; Herman, D. J.; Kasen, T. G.; Malliaras, G. G. *Appl. Phys. Lett.* 2004, *85,* 6272.
[71] Sarangerel, K.; Ganzorig, C.; Fujihira, M.; Sakomura, M.; Ueda, K. *Chem. Lett.* 2008, *37,* 778.
[72] Pandey, A. K.; Nunzi, J.-M. *Adv. Mater.* 2007, *19,* 3613.
[73] Gregg, B. A.; Cormier, R. A. *J. Am. Chem. Soc.* 2001, *123,* 7959.
[74] Chan, M. Y.; Lai, S. L.; Fung, M. K.; Lee, C. S.; Lee, S. T. *Appl. Phys. Lett.* 2007, *90,* 023504.
[75] Wang, S.; Mayo, E. I.; Perez, M. D.; Griffe, L.; Wei, G.; Djurovich, P. I.; Forrest, S. R.; Thompson, M. E. *Appl. Phys. Lett.* 2009, *94,* 233304.
[76] Silvestri, F.; Irwin, M. D.; Beverina, L.; Facchetti, A.; Pagani, G. A.; Marks, T. J. *J. Am. Chem. Soc.* 2008, *130,* 17640.
[77] Merritt, V. Y.; Hovel, H. J. *Appl. Phys. Lett.* 1976, *29,* 414.
[78] Piechowski, A. P.; Bird, G. R.; Morel, D. L.; Stogryn, E. L. *J. Phys. Chem.* 1984, *88,* 934.
[79] Yu, G.; Gao, J.; Hummelen, J. C.; Wudl, F.; Heeger, A. J. *Science* 1995, *270,* 1789.
[80] Melzer, C.; Koop, E. J.; Mihailetchi, V. D.; Blom, P. W. M. *Adv. Funct. Mater.* 2004, *14,* 865.
[81] Mihailetchi, V. D.; Koster, L. J. A.; Blom, P. W. M.; Merzer, C.; De Boer, B.; Duren, J. K. J.; Janssen, R. A. J.; *Adv. Funct. Mater.* 2005, *15,* 795.
[82] Brabec, C. J.; Shaheen, S. E.; Winder, C.; Sariciftci, N. S. *Appl. Phys. Lett.* 2002, *80,* 1288.
[83] Kobashi, M.; Takeuchi, H. *Macromolecules* 1998, *31,* 7273.
[84] Villers; B. T.; Tassone, C. J.; Tolbert, S. H.; Schwartz, B. J. *J. Phys. Chem. C,* 2009, *113,* 18979
[85] Sensfuss, S.; Al-Ibrahim, M.; Konkin, A.; Nazmutdinova, G.; Zhokhavets, U.; Gobschb, G.; Egbe, D. A. M.; Klemm, E.; Roth, H.-K. *Organic Photovoltaics IV, edited by Zakya H. Kafafi, Paul A. Lane, Proceedings of SPIE Vol. 5215 (SPIE, Bellingham, WA, 2004) · 0277-786X/04/$15 · doi:10.1117/12.505628.*
[86] Mozer, A. J.; Denk, P.; Scharber, M. C.; Neugebauer, H.; Serdar, S. N. *J. Phys. Chem. B,* 2004, *108,* 5235–5242
[87] Hou, J. H.; Tan, Z. A.; Yan, Y.; He, Y. J.;. Yang, C. H ;. Li , Y. F. *J. Am. Chem. Soc.* 2006, *128,* 4911.

[88] Zhang, F. L.; Mammo, W.; Andersson, L. M.; Admassie, S.; Andersson, M. R.; Inganäs, O. *Adv. Mater.* 2006, *18,* 2169.
[89] Mühlbacher, D.; Scharber, M.; Morana, M.; Zhu, Z G.; Waller, D.; Gaudiana, R.; Brabec, C. *Adv. Mater.* 2006, *18,* 2884.
[90] Liang , Y.; Feng , D.; Wu , Y.; Tsai , S.; Li , G.; Ray , C.; Yu , L. *J. Am. Chem. Soc.* 2009 , *131* , 7792.
[91] Price, S. C.; Stuart, A. C.; You, W. *Macromolecules,* 2010, *43,* 4609.
[92] Ma, W.; Yang, C.; Gong, X.; Lee, K.; Heeger, A. J. *Adv. Funct. Mater.* 2005, *15,* 579.
[93] Irwin, M. D.; Buchholz, B.; Hains, A. W.; Chang, R. P. H.; Marks, T. J. *Proc. Natl. Acad. Sci. USA* 2008, *105,* 2783.
[94] Koster, L. J. A.; Mihailetchi, V. D.; Blom, P. W. M. *Appl. Phys. Lett.* 2006, *88,* 093511.
[95] Roncali, J. *Chem. Rev.* 1997, *97,* 173.
[96] Mullekom, H. A. M.; Vekemans, J. A. J. M.; Havinga, E. E.; Meijer, E. W. *Mater. Sci. Eng.* 2001, *32,* 1.
[97] Svensson , M.; Zhang , F.; Veenstra , S. C.; Verhees , W. J. H.; Hummelen , J. C.; Kroon , J. M.; Inganäs , O.; Andersson, M. R. *Adv. Mater.* 2003, *15* , 988.
[98] Wang, X.; Perzon, E.; Delgado, J. L.; Cruz, P.; Zhang, F.; Langa, F.; Andersson, M. R.; Inganäs, O. *Appl. Phys. Lett.* 2004, *85,* 5081.
[99] Gadisa , A.; Wang , X.; Admassie , S.; Perzon , E.; Oswald , F.; Langa , F.; Andersson , M. R.; Inganäs , O. *Org. Electronics* 2006, *7* , 195.
[100] Zhang , F.; Jespersen , K. G.; Björström , C.; Svensson , M.; Andersson , M. R.; Sundström , V.; Magnusson , K.; Moons , E.; Yartsev , A.; Inganäs , O. *Adv. Funct. Mater. 2006, 16* , 667.
[101] Zhu , Z.; Waller , D.; Gaudiana , R.; Morana , M.; Mühlbacher , D.; Scharber , M.; Brabec , C. *Macromolecules* 2007, *40* , 1981.
[102] Coppo , P.; Turner . M. L.; *J. Mat. Chem.* 2005, *15*, 1123.
[103] Tsami , A.; Bunnagel , T. W.; Farrell , T.; Scharber , M.; Choulis , S.; Brabec , C. J.; Scherf , U. *J. Mat. Chem*. 2007, *17* , 1353.
[104] Inganäs, O.; Zhang, F. L.; Tvingstedt, K.; Andersson, L. M.; Hellström, S.; Andersson, M. R. *Adv. Mater*. 2010, *22*, E100.
[105] Andersson, V.; Tvingstedt, K.; Inganäs, O. *J. Appl. Phys*. 2008, *103,* 094520.
[106] Pettersson, L. A. A.; Roman, L. S.; Inganäs, O. *J. Appl. Phys.* 1999, *86*, 1.
[107] Han, S.; Lim, S.; Kim, H.; Cho, H.; Yoo, S. *DOI: 10.1109/JSTQE.2010.2041637.*
[108] Niggemanna, M.; Zieglera, T.; Glatthaarb, M.; Riedea, M.; Zimmermannb, B.; Gomberta, A. *Proc. of SPIE Vol. 6197 61970D-1*
[109] Slooff, L. H.; Veenstra, S. C.; Kroon, J. M.; Moet, D. J. D.; Sweelssen, J.; Koetse, M. M. *Appl. Phys. Lett.* 2007, *90*, 143506.
[110] Wang , E.; Wang, L.; Lan , L.; Luo , C.; Zhuang , W.; Peng , J.; Cao , Y. *Appl. Phys. Lett.* 2008, *92* , 033307.
[111] Chen, M.; Hou, J.; Hong, Z.; Yang, G.; Sista, S.; Chen, L.; Yang, Y. *Adv. Mat.* 2009, *21*, 1.
[112] Blouin, N.; Michaud, A.; Leclerc, M. *Adv. Mat*. 2007, *19*, 2295.
[113] Blouin, N.; Michaud, A.; Gendron, D.; Wakim, S.; Blair, E.; Plesu, R. N.; Bellette, M.; Durocher, G.; Tao, Y.; Leclerc, M. *J. Am. Chem. Soc*. 2008, *130*, 732.
[114] Katz, H. E.; Johnson, J.; Lovinger, A. J.; Li, W. *J. Am. Chem. Soc.* 2000, *122*, 7787.

[115] Struijk, C. W.; Sieval, A. B.; Dakhorst, J. E. J.; Dijk, M.; Kimkes, P.; Koehorst, R. B. M.; Donker, H.; Schaafsma, T. J.; Picken, S. J.; Craats, A. M.; Warman, J. M.; Zuilhof, H.; Sudholter, E. J. R. *J. Am. Chem. Soc.* 2000, *122*, 11057.

[116] Thompson, B. C.; Fréchet, J. M. J. *Angew. Chem. Int. Ed.* 2008, 47, 58.

[117] Liang , Y.; Wu , Y.; Feng , D.; Tsai , S.-T.; Son , H.-J.; Li , G.; Yu , L. *J. Am. Chem. Soc.* 2009, *131* , 56.

[118] Mitsumoto, R. et al., *J. Phys. Chem. A* 1998, *102*, 552.

[119] Hill, I. G.; Kahn, A. *J. Appl. Phys.*1999, *86*, 4515.

[120] Benning, P. J.; Poirier, D. M.; Ohno, T. R.; Chen, Y.; Jost, M. B.; Stepniak, F.; Kroll, G. H.; H. J.; Fure, W.J.; Smalley, R. E. *Phys. Rev. B* 1992, *45*, 6899.

[121] Pfuetzner, S.; Meiss, J.; Petrich, A.; Riede, M.; Leo, K. *Appl. Phys. Lett.* 2009, *94*, 223307.

[122] Ma, W.; Yang, C. Y.; Gong, X.; Lee, K.; Heeger, A. J. *Adv. Funct. Mater.* 2005, *15,* 1617.

[123] Mihailetchi, V. D.; Duren, J. K. J.; Blom, P.W.M.; Hummelen, J. C.; Janssen, R. A. J.; Kroon, J. M.; Rispens, M. T.; Verhees, W. J. H.; Wienk, M. M. *Adv. Funct. Mater.* 2003, *13*, 43.

[124] Anthopoulos, T. D.; Leeuw, D. M.; Cantatore, E.; Hof, P.; Alma, J.; Hummelen, J. C. *J. Appl. Phys.* 2005, *98*, 054503.

[125] Lenes, M.; Wetzelaer, G.-J. A. H.; Kooistra, F. B.; Veenstra, S. C.; Hummelen, J. C.; Blom, P. W. M. *Adv. Mater.* 2008, *20*, 2116.

[126] He, Y. J.; Chen , H.-Y.; Hou , J. H.; Li , Y. F. *J. Am. Chem. Soc.* 2010, 132, 1377 .

[127] He, Y. J.; Zhao, G.J.; Peng , B.; Li, Y. F. *Adv. Funct. Mater. 2010, XX, 1–7, DOI: 10.1002/adfm.201001122*

[128] Ross, R. B.; Cardona, C. M.; Guldi, D. M.; Sankaranarayanan, S. G.; Reese, M. O.; Kopidakis, N.; Peet, J.; Walker, B.; Bazan, G. C.;Keuren, E.; Holloway, B. C.; Drees, M. *Nat. Mater.* 2009, *8* , 208 .

[129] Brabec, C. J.; Zerza, G.; Cerullo, G.; Silvestri, S.; Luzzati, S.; Hummelen, J. C.; Sariciftci, N. S. *Chem. Phys. Lett.* 2001, *340*, 232.

[130] Hoppe, H.; Sariciftci, S. *Adv. Polym. Sci.* 2007, *12*, 121.

[131] Hummelen, J. C.; Knight, B. W.; LePeq, F.; Wudl, F.; Yao, J.; Wilkins, C. L. *J. Org. Chem.* 1995, *60*, 532.

[132] Kooistra, F. B.; Knol, J.; Kastenberg, F.; Popescu, L. M.; Verhees, W. J. H.; Kroon, J. M.; Hummelen, J. C. *Org. Lett.* 2007, *9,* 551.

[133] Brabec, C. J.; Cravino, A.; Meissner, D.; Sariciftci, N. S.; Fromherz, T.; Rispens, M. T.; Sanchez, L.; Hummelen, J. C. *Adv. Funct. Mate*r. 2001, *11*, 374.

[134] Zheng, L.; Zhou, Q.; Deng, X.; Yuan, M.; Yu, G.; Cao, Y. *J. Phys. Chem. B* 2004, *108*, 11921.

[135] Backer, S.; Sivula, K.; Kavulak, D. F.; FrSchet, J. M. J. *Chem. Mater*. 2007, *19,* 2927.

[136] Popescu, L. M.; Rt Hof, P.; Sieval, A. B.; Jonkman, H. T.; Hummelen, J. C. *Appl. Phys. Lett*. 2006, *89*, 213507.

[137] Zhao, G.; He, Y.; Li, Y. *Adv. Mater*. Inpress. DOI: 10.1002/adma.201001339

[138] Lenes, M.; Shelton, S. W.; Sieval, A. B.; Kronholm, D. F.; Hummelen , J. C.; Blom , P. W. M. *Adv. Funct. Mater*. 2009, *19*, 3002.

[139] Lee, T.-W.; Chung, Y. *Adv. Funct. Mater*. 2008, *18*, 2246.

[140] Brumbach, M.; Veneman, P. A.; Marrikar, F. S.; Schulmeyer, T.; Simmonds, A.; Xia, W.; Lee, P.; Armstrong, N. R. *Langmuir* 2007, *23*, 11089.
[141] Hains, A. W.; Marks, T. J. *Appl. Phys. Lett.* 2008, *92*, 023504.
[142] Takahashi, K.; Suzaka, S.; Sigeyama, Y.; Yamaguchi, T.; Nakamura, J.; Murata, K. *Chem. Lett.* 2007, *36,* 762.
[143] Yan, H.; Lee, P.; Armstrong, N. R.; Graham, A.; Evmenenko, G. A.; Dutta, P.; Marks, T. J. *J. Am. Chem. Soc.* 2005, *127*, 3172..
[144] Ionescu-Zanetti, C.; Mechler, A.; Carter, S. A.; Lal, R. *Adv. Mater*. 2004, *16,* 385.
[145] Kemerink, M.; Timpanaro, S.; Kok, M. M.; Meulenkamp, E. A.; Touwslager, F. J. *J. Phys. Chem. B* 2004, *108*, 18820.
[146] Ni, J.; Yan, H.; Wang, A.; Yang, Y.; Stern, C. L.; Metz, A. W.; Jin, S.; Wang, L.; Marks, T. J.; Ireland, J. R.; Kannewurf, C. R. *J. Am. Chem. Soc*. 2005, *127*, 5613.
[147] Han, S.; Shin, W. S.; Seo, M.; Gupta, D.; Moon, S.-J.; Yoo, S. *Org. Electron.* 2009, *10,* 791.
[148] Shrotriya, V.; Li, G.; Yao, Y.; Chu, C.-W.; Yang, Y. *Appl. Phys. Lett.* 2006, *88*, 073508.
[149] Peumans, P.; Bulović, V.; Forrest, S. R. *Appl. Phys. Lett.* 2000, *76*, 2650.
[150] Chan, M. Y.; Lee, C. S.; Lai, S. L.; Fung, M. K.; Wong, F. L.; Sun, H. Y.; Lau, K. M.; Lee, S. T. *J. Appl. Phys*. 2006, *100*, 094506.
[151] Rand, B. P.; Li, J.; Xue, J.; Holmes, R. J.; Thompson, M. E.; Forrest, S. R. *Adv. Mater.* 2005, *17*, 2714.
[152] Takanezawa, K.; Tajima, K.; Hashimoto, K. *Appl. Phys. Lett.* 2008, *93*, 063308
[153] Gilot, J.; Barbu, I.; Wienk, M. M.; Janssen, R. A. J. *Appl. Phys. Lett.* 2007, *91*, 113520.
[154] Zhao, D. W.; Sun, X. W.; Jiang, C. Y.; Kyaw, A. K. K.; Lo, G. Q.; Kwong, D. L. *Appl. Phys. Lett*. 2008, *93*, 083305.
[155] Hadipour, A.; Boer, B.; Wildeman, J.; Kooistra, F. B.; Hummelen, J. C.; Turbiez, M. G. R.; Wienk, M. M.; Janssen, R. A. J.; Blom, P. W. M. *Adv. Funct. Mater*. 2006, *16*, 1897.
[156] Gilot, J.; Wienk, M. M.; Janssen, R. A. J. *Appl. Phys. Lett.* 2007, *90*, 143512.
[157] Bailey-Salaman, R. F.; Rand, B. P.; Forrest, S. R. *Appl. Phys. Lett.* 2007, *91*, 013508.
[158] Brumbach, M.; Placencia, D.; Armstrong, N. R. *J. Phy. Chen. C.* 2008, *112,* 3142-3151.
[159] Mattheus, C. C.; Michaelis, W.; Kelting, C.; Durfee, W. S.; Wöhrle, D.; Schlettwein, D. *Synth. Met.* 2004, *146*, 335.
[160] Placencia, D.; Wang, W.; Shallcross, R. C.; Nebesny, K. W.; Brumbach, M.; Armstrong, N. R. *Adv. Funct. Mater*. 2009, *19*, 1913.
[161] Mizuguchi, J.; Rihs, G.; Karfunkel, H. R. *J. Phys. Chem*. 1995, *99*, 16217.
[162] Popovic, Z. D.; Khan, M. I.; Atherton, S. J.; Hor, A. M.; Goodman, J. L. *J. Phys. Chem. B* 1998, *102,* 657.
[163] Yamaguchi, S.; Sasaki, Y.; *J. Phys. Chem. B* 1999, *103*, 6835.
[164] Tsushima, M.; Ikeda, N.; Yonehara, H.; Etori, H.; Pac, C.; Ohno, T.; *Coord. Chem. Rev.* 2002, *229*, 3.
[165] Norton, J. E.; Brédas, J.-L. *J. Chem. Phys*. 2008, *128,* 034701.
[166] Conboy, J. C.; Olson, E. J. C.; Adams, D. M.; Kerimo, J.; Zaban, A.; Gregg, B. A.; Barbara, P. F. *J. Phys. Chem. B*. 1998, *102*, 4516.
[167] Nakai, K.; Ishii, K.; Kobayashi, N. *J. Phys. Chem. B* 2003, *107*, 9749.
[168] Yang, F.; Forrest, S. R. *ACS Nano* 2008, *2,* 1022.

[169] Yang, F.; Shtein, M.; Forrest, S. R. *Nat. Mater*. 2005, *4,* 37.
[170] Yoo, S.; Potscavage, W. J.; Domercq, B.; Han, S. H.; Li, T. D.; Jones, S. C.; Szoszklewicz, R.; Levi, D.; Riedo, E.; Marder, S. R.; Kippelen, B. *SolidState Electron. 51*, 1367.
[171] Hong, Z. R.; Maennig, B.; Lessmann, R.; Pfeiffer, M.; Leo, K.; Simon, P. *J. Appl. Phys*. 2007, *90*, 203505.
[172] Uchida, S.; Xue, J. G.; Rand, B. P.; Forrest, S. R. *Appl. Phys. Lett.* 2004, *84*, 4218.
[173] Yang, F.; Sun, K.; Forrest, S. R. *Adv. Mate*r. 2005, *19*, 4166.
[174] Yang, F.; Shtein, M.; Forrest, S. R. *J. Appl. Phys*. 2005, *98*, 014906.
[175] Sullivan, P.; Heuta, S.; Schultes, S. M.; Jones, T. S. *Appl. Phys. Lett.* 2004, *84*, 1210.
[176] Maennig, B.; Drechsel, J.; Gebeyehu, D.; Simon, P.; Kozlwski, F.; Werner, A.; Li, F.; Grundman, S.; Sonntag, S.; Koch, M.; Leo, K.; Pfeiffer, M.; Hoppe, H.; Meissner, D.; Sariciftci, N. S.; Riedel, I.; Dyakonov, V.; Parisi, J. *Appl. Phys. A*. 2004, *79*, 1.
[177] Yakimov, A.; Forrest, S. R. *Appl. Phys. Lett*. 2004, *80*, 1667.
[178] Triyana, K.; Yasuda, T.; Fujita, K.; Tsutsui, T. T*hin Solid Films* 2005, *447*, 198.
[179] Xue, J.; Rand, B. P.; Uchida, S.; Forrest, S. R. *Adv. Mater*. 2005, *17*, 66.
[180] Cheyns, D.; Rand, B. P.; Heremans, P. *Appl. Phys. Lett*. 2010, *97*, 033301.
[181] Hiramoto, M.; Suezaki, M.; Yokoyama, M. *Chem. Lett*. 1990, *19*, 327.
[182] Drechsel, J.; Männig, B.; Kozlowski, F.; Pfeiffer, M.; Leo, K.; Hoppe, H. *Appl. Phys. Lett.* 2005, *86*, 244102.
[183] Cheyns, D.; Gommans, H.; Odijk, M.; Poortmans, J.; Heremans, P. *Sol. Energy Mater. Sol. Cells* 2007, *91*, 399.
[184] Schueppel, R.; Timmreck, R.; Allinger, N.; Mueller, T.; Furno, M.; Uhrich, C.; Leo, K.; Riede, M. *J. Appl. Phys.* 2010, *107,* 044503.
[185] Gilot, J.; Wienk, M. M.; Janssen, R. A. J. *Adv. Mater*. 2010, *22*, E67.
[186] Rand, B. P.; Peumans, P.; Forrest, S. R. *J. Appl. Phys.* 2004, *96*, 7519.
[187] Meiss, J.; Allinger, N.; Riede, M. K.; Leo, K. *Appl. Phys. Lett.* 2008, *93*, 103311.
[188] Tong, X.; Bailey-Salzman, R. F.; Wei, G.; Forrest, S. R. *Appl. Phys. Lett*. 2008, *93,* 173304.
[189] Falkenberg, C.; Uhich, C.; Olthof, S.; Maennig, B.; Riede, M. K.; Leo, K. *J. Appl. Phys.* 2008, *104*, 034506.
[190] Suemori, K.; Miyata, T.; Yokoyama, M.; Hiramoto, M. *Appl. Phys. Lett.* 2005, *86*, 063509.
[191] Oishi, Y.; Hiramoto, M.; Yokoyama, M. *Extended Abstracts of the 48th Spring Meeting 2001, Japan Society of Applied Physics and Related Societies,* 29a-ZG-7.
[192] Matsunobu, G.; Oishi, Y.; Yokoyama, M.; Hiramoto, M. *Appl. Phys. Lett.* 2002, *81*, 1321.
[193] Hiramoto, M.; Suemori, K.; Yokoyama, M. *Jpn. J. Appl. Phys.* 2002, *41*, 2763.
[194] Peumans, P.; Uchida, S.; Forrest, S. R. *Nature* 2003, *425,* 158.
[195] Chu, C.W.; Shrotriya, V.; Li, G.; Yang, Y. *Appl. Phys. Lett*. 2006, *88*, 153504.
[196] Huang, J.; Yu, J. S.; Guan, Z. Q.; Jiang, Y. D. "Improvement in open circuit voltage of organic solar cells by inserting thin phosphorescent iridium complex layer", *Appl. Phys. Lett. (in press)*
[197] Lai, S. L.; Lo, M. F.; Chan, M. Y.; Lee, C. S.; Lee, S. T. Appl. Phys. Lett. 2009, 95, 153303.

[198] Chan, M. Y.; Lai, S. L.; Fung, M. K.; Lee, C. S.; Lee, S. T. Appl. Phys. Lett. 2007, 90, 023504.

[199] Kinoshita, Y.; Hasobe, T.; Murata, H. *Appl. Phys. Lett.* 2007, *91*, 083518.

[200] Yang, F.; Lunt, R. R.; Forrest, S. R. *Appl. Phys. Lett.* 2008, *92*, 053310.

[201] Kinoshita, Y.; Hasobe, T.; Murata, H. *Appl. Phys. Lett.* 2007, *91*, 083518.

[202] Wang, N. N.; Yu, J. S.; Zang, Y.; Huang, J.; Jiang, Y. D. *Sol. Energy Mater. Sol. Cells* 2010, *94*, 263.

[203] Wang, N. N.; Yu, J. S.; Lin, H.; Jiang, Y. D. *Chin. J. Chem. Phys.* 2010, *23*, 84.

[204] Wang, N. N.; Yu, J. S.; Zang, Y.; Jiang, Y. D. *Chin. Phys. B* 2010, *19*, 038602.

[205] Kawano, K.; Adachi, C. Appl. Phys. Lett. 2010, 96, 053307.

[206] Brabec, C. J.; Dyakonov, V.; Parisi, J.; Sariciftci, N. S. *in: Organic Photovoltaics: Concepts and Realization, Springer, Heidelberg 2003.*

[207] Kim, S-S.; Na, S-I.; Jo, J.; Tae, G.; Kim, D-Y. *Adv. Mater.* 2007, *19*, 4410.

[208] Chen, L-M.; Hong, Z.; Li, G.; Yang, Y. *Adv. Mater.* 2009, *21*, 1434.

[209] Krebs, F. C. *Sol. Eng. Mater. Sol. Cells,* 2009, *93*, 394

[210] Hoth, C. N.; Choulis, S. A.; Schilinsky, P.; Brabec, C. J. *Adv. Mater.* 2007, *19,* 3973.

[211] Aernouts, T.; Aleksandrov, T.; Girotto, C.; Genoe, J.; Poortmans, J. *Appl. Phys. Lett.* 2008, *92,* 033306.

[212] Brabec, C. J.; Cravino, A.; Meissner, D.; Sariciftci, N. S.; Minze, T. F.; Rispens, T.; Sanchez, L.; Hummelen, J. C. *Adv. Funct. Mater.* 2001, *11*, 374.

[213] Pudas, M.; Hagberg, J.; Leppavuori, S. *Progr. Org. Coatings*, 2004, *49*, 324–335.

[214] Blom, P. W. M.; Mihailetchi, V. D.; Koster, L. J. A.; Markov, D. E. *Adv. Mater.* 2007, *19*, 1551.

[215] Erb, T.; Zhokhavets, U.; Gobsch, G.; Raleva, S.; Stuhn, B.; Schilinsky, P.; Waldauf, C.; Brabec, C. J. *Adv. Funct. Mater.* 2005, *15*, 1193.

[216] Li, G.; Shrotriya, V.; Yao, Y.; Huang, J.; Yang, Y. *J. Appl. Phys.* 2005, *98*, 043704.

[217] Mihailetchi, V. D.; Xie, H. X.; Boer, B.; Koster, L. J. A.; Blom, P. W. M. *Adv. Funct. Mater.* 2006, *16*, 699

[218] Yang, X.; Loos, J.; Veenstra, S. C.; Verhees, W. J. H.; Wienk, M. M.; Kroon, J. M.; Michels, A. J.; Janssen, R. A. J. *Nano Lett.*, 2005, *5*, 579.

[219] H. Hoppe, T. Glatzel, M. Niggemann, W. Schwinger, F. Schaeffler, A. Hinsch, M. Ch, Lux-Steiner, N. S. Sariciftci, *Thin Solid Films* 2006, *511*, 587.

[220] Padinger, F.; Rittberger, R. S.; Sariciftci, N. S. *Adv. Funct. Mater.* 2003, *13*, 85.

[221] Ko, C.-J.; Lin, Y.-K.; Chen, F.-C. *Adv. Mater.* 2007, *19*, 3520.

[222] Li, G.; Shrotriya, V.; Huang, J.; Yao, Y.; Moriarty, T.; Emery, K.; Yang, Y. *Nat. Mater.* 2005, *4*, 864.

[223] Li, G.; Shrotriya, V.; Yao, Y.; Huang, J.; Yang, Y. *J. Mater. Chem.* 2007, *17*, 3126.

[224] Li, G.; Yao, Y.; Yang, H.; Shrotriya, V.; Yang, G.; Yang, Y. *Adv. Funct. Mater.* 2007, *17*, 1636.

[225] Shrotriya, V.; Yao, Y.; Li, G.; Yang, Y. *Appl. Phys. Lett.* 2006, *89*, 063505.

[226] Chen, H.-Y.; Lo, M. K. F.; Yang, G. W.; Monbouquette, H. G.; Yang, Y. *Nature Nanotechnology,* 2008, *3,* 543

[227] Chu, C.-W.; Yang, H.; Hou, W.-J.; Huang, J.; Li, G.; Yang, Y. *Appl. Phys. Lett.* 2008, *92*, 103306.

[228] Peet, J.; Soci, C.; Coffin, R. C.; Nguyen, T. Q.; Mikhailovsky, A.; Moses, D.; Bazan, G. C. *Appl. Phys. Lett.* 2006, *89*, 252105.

[229] Peet, J.; Kim, J. Y.; Coates, N. E.; Ma, W. L.; Moses, D.; Heeger, A. J.; Bazan, G. C. *Nat. Mater.* 2007, *6*, 497.

[230] Yao, Y.; Hou, J. H.; Xu, Z.; Li, G.; Yang, Y. *Adv. Funct. Mater.* 2008, *18*, 1783–1789.

[231] Chen, H.-Y.; Yang, H.; Yang, G. W.; Sista, S.; Zadoyan, R.; Li, G.; Yang, Y. *J. Phys. Chem. C,* 2009, *113,* 7946.

[232] Chen, F.-C.; Tseng, H.-C.; Ko, C.-J. *Appl. Phys. Lett.* 2008, *92*, 103316.

[233] Li, L. G.; Lu, G. H.; Yang, X. N. *J. Mater. Chem.* 2008, *18,* 1984.

[234] Rughooputh, S. D. D. V.; Hotta, S.; Heeger, A. J.; Wudl, F.; *J. Polym. Sci. Part B* 1987, *25*, 1071.

[235] Moulé, A. J.; Meerholz, K. *Adv. Mater.* 2008, *20*, 240.

[236] Duren, J. K. J.; Yang, X. N.; Loos, J.; Bulle-Lieuwma, C. W. T.; Sieval, A. B.; Hummelen, J. C.; Janssen, R. A. J. *Adv. Funct. Mater.* 2004, *14,* 425.

[237] Hou, J. H.; Chen, H.-Y.; Zhang, S.; Li, G.; Yang, Y. *J. Am. Chem. Soc.* 2008, *130,* 16144.

[238] Chen, M.-H.; Hou, J. H.; Hong, Z.; Yang, G. W.; Sista, S.; Chen, L.-M.; Yang, Y. *Adv. Mater.* 2009, *21,* 4238.

[239] Ross, R. B.; Cardona, C. M.; Guldi, D. M.; Sankaranarayanan, S. G.; Reese, M. O.; Kopidakis, N.; Peet, J.; Walker, B.; Bazan, G. C. *Nat. Mater.* 2009, *8*, 208.

[240] Ma, C.-Q.; Fonrodona, M.; Schikora, M. C.; Wienk, M. M.; Janssen, R. A. J.; Bauerle, P. *Adv. Funct. Mater.* 2008, *18*, 3323.

[241] Sahin, Y.; Alem, S.; Bettingnies, R.; Nunzi, J. M. *Thin Solid Films* 2005, *476*, 340.

[242] Song, M. Y.; Kim, K. J.; Kim, D. Y. *Sol. Energy Mater. Sol. Cells* 2005, *85*, 31.

[243] Jong, M. P.; Ijzendoom, L. J.;Voigt, M. J. A. *Appl. Phys. Lett.* 2000, *77*, 2255.

[244] Greczynski, G.; Kugler, T.; Keil, M.; Osikowicz, W.; Fahlman, M.; Salaneck, W. R. *J. Electron SpectrOPV. Relat. Phenom.* 2001, *121,* 1.

[245] Campoy-Quiles, M.; Ferenczi, T.; Agostinelli, T.; Etchegoin, P. G.; Kim, Y.; Anthopoulos, T. D.; Stavrinou, P. N.; Nelson, J. *Nat. Mater.* 2008, *7*, 158.

[246] White, M. S.; Olson, D. C.; Shaheen, S. E.; Kopidakis, N.; Ginley, D. S. *Appl. Phys. Lett.* 2006, *89*, 143517.

[247] Hau, S. K.; Yip, H.-L.; Baek, N. S.; Zou, J.; O'Malley, K.; Jen, A. K.-Y. *Appl. Phys. Lett.* 2008, *92*, 253301.

[248] Zou, J. Y.; Yip, H.-L.; Hau, S. K.; Jen, A. K.-Y. *Appl. Phys. Lett.* 2010, *96*, 203301.

[249] Waldauf, C.; Morana, M.; Denk, P.; Schilinsky, P.; Coakley, K.; Choulis, S. A.; Brabec, C. J. *Appl. Phys. Lett.* 2006, *89*, 233517.

[250] Steim, R.; Choulis, S. A.; Schilinsky, P.; Brabec, C. J. *Appl. Phys. Lett.* 2008, *92*, 093303.

[251] Mor, G. K.; Shankar, K.; Paulose, M.; Varghese, O. K.; Grimes, C. A. *Appl. Phys. Lett.* 2007, *91*, 152111.

[252] Baek, W.-H.; Choi, M.; Yoon, T.-S.; Lee, H. H.; Kim, Y.-S. *Appl. Phys. Lett.* 2010, *96*, 133506.

[253] Li, G.; Chu, C.-W.; Shrotriya, V.; Huang, J.; Yang, Y. *Appl. Phys. Lett.* 2006, *88*, 253503.

[254] Liao, H.-H.; Chen, L.-M.; Xu, Z.; Li, G.; Yang, Y. *Appl. Phys. Lett.* 2008, *92*, 173303.

[255] Xu, Z.; Chen, L.-M.; Yang, G. W.; Huang, C.-H.; Hou, J. H.; Wu, Y.; Li, G.; Hsu, C.-S.; Yang, Y. *Adv. Funct. Mater.* 2009, *19,* 1227.

[256] Lim, Y.-F.; Lee, S.; Herman, D. J.; Lloyd, M. T.; Anthony, J. E.; Malliaras, G. G. *Appl. Phys. Lett.* 2008, *93*, 193301.
[257] Zhang, H. M.; Ouyang, J. Y. *Appl. Phys. Lett.* 2010, *97*, 063509.
[258] Zhao, D. W.; Liu, P.; Sun, X. W.; Tan, S. T.; Ke, L.; Kyaw, A. K. K. *Appl. Phys. Lett.* 2009, *95*, 153304.
[259] Huang, J.; Li, G.; Yang, Y. *Adv. Mater.* 2008, *20*, 415.
[260] Chen, Y. C.; Yu, C. Y.; Chen, C. P.; Chan, S. H.; Ting, C. *J. Sol. Energy Eng.* 2010, *132*, 021103.
[261] Würfel, P. *in Physics of Solar Cells, Wiley-VCH, Berlin, Germany 2004.*
[262] Green, M. A.; Emery, K.; Hisikawa, Y.; Warta, W. *Prog. Photovolt. Res. Appl.* 2007, *15*, 425.
[263] Dennler, G.; Scharber, M. C.; Ameri, T.; Denk, P.; Forberich, K.; Waldauf, C.; Brabec, C. J. *Adv. Mater.* 2008, *20*, 579.
[264] Dennler, G.; Prall, H.-J.; Koeppe, R.; Egginger, M.; Autengruber, R.; Sariciftci, N. S. *Appl. Phys. Lett.* 2006, *89*, 73502.
[265] Kawano, K.; Ito, N.; Nishimori, T.; Sakai, J. *Appl. Phys. Lett.* 2006, *88*, 73514.
[266] Sista, S.; Park, M. H.; Hong, Z. R.; Wu, Y.; Hou, J. H.; Kwan, W. L.; Li, G.; Yang, Y. *Adv. Mater.* 2010, *22*, 380.
[267] Yao, Y.; Liang, Y. Y.; Shrotriya, V.; Xiao, S. Q.; Yu, L. P.; Yang, Y. *Adv. Mater.* 2007, *19*, 3979.
[268] Hadipour, A.; Boer, B.; Blom, P.W. M. *J. Appl. Phys.* 2007, *102*, 074506.
[269] Dennler, G.; Forberich, K.; Ameri, T.; Waldauf, C.; Denk, P.; Brabec, C. J. *J. Appl. Phys.* 2007, *102,* 123109.
[270] Yu, B.; Zhu, F.; Wang, H.; Li, G.; Yan, D. H. *J. Appl. Phys.* 2008, *104,* 114503.
[271] Namkoong, G.; Boland, P.; Lee, K.; Dean, J. J. Appl. Phys. 2010, 107, 124515.

In: Organic Semiconductors
Editors: Maria A. Velasquez
ISBN: 978-1-61209-391-8

Chapter 2

CERTAIN PROBLEMS IN THE PRACTICAL CHARACTERIZATION OF MOLECULAR FILMS AND ASSOCIATED INTERFACES BY DIRECT PHOTOEMISSION

Ján Ivančo
Institute of Physics, Slovak Academy of Sciences, Slovak Republic

ABSTRACT

The prerequisite of the further progress in organic electronics is the understanding of fundamental processes governing the overall performance of organic devices, so to speak, the controlling the properties of both organic films and involved interfaces, the entities which constitute the organic devices.

Among other techniques, photoemission characterization directly assesses the electronic structure of both the molecular films and the associated interfaces, and it thereby belongs to the most popular surface and interface analytical techniques.

This work focuses on certain problems encountered while applying the direct photoemission technique for the basic characterization of both the vacuum-sublimed molecular films and the interfaces formed between the films and the metal, organic, and inorganic substrates. Fundamental electronic parameters, such as the work function, the ionization energy, and the interfacial energy level alignment, and their interplay affected by the film structure and morphology will be addressed in particular, with the reference to the molecular orientation in the films. It will be demonstrated that the work function of molecular films is an essential parameter governing the energy level alignment between the film and the substrate. A simple model for the prediction of the injection interfacial barrier based on the work functions of materials brought into the contact is proposed. Also, it will be demonstrated that neglecting to consider the work function may obstruct an unambiguous interpretation of electronic properties of the involved interface, and it may even implicate a spurious detection of the band bending in the molecular film.

Keywords: Photoemission, Molecular films, Organic semiconductors, Organic-metal interface, Organic-organic interface, Organic-inorganic interface, Surface and interface analytical studies, Molecular orientation, Ionization energy, Work function, Energy level alignment, Band bending

1. Introduction

Electronically and optically active organics have been sneaking into today's electronics. Organic optical devices were already becoming commercially successful; on the other hand, the physics behind the electronic properties of interfaces is still not firmly established and understood to exploit their potential functional capabilities of organic electronics. In many aspects, its development benefits from extensive—in comparison to inorganic semiconductors—and continuously increasing number of new molecules have been examined in the role of active layers in the molecular devices.

While properties of both organic films and associated interfaces determine the overall performance of organic electronic devices, their influence may be difficult to extract and determine from the characterization of transport properties performed on completed devices. Therefore, understanding and controlling fundamental relations on atomic level is necessary. Organic electronics has also introduced a novel class of interfaces, namely these formed between molecular films and metals or inorganic semiconductors. The "novel" is understood in terms of electronic properties of such interfaces that have to be controlled.

Surface-science approach in investigations on organic films benefits from well-described and defined systems under study; it focuses on molecules with low molecular weight represented by several model molecules sublimed on atomically clean substrates and characterized *in situ* in ultra-high vacuum. Although not intended to be technologically competitive for the production of organic devices, the surface-science approach permits the understanding of fundamental chemical and electronic issues in the device design and performance [1].

Among the most traditional techniques, photoemission may probe fundamental electronic, chemical, and morphological properties; the issues have been addressed in great details by many photoemission studies, yet opposed opinions on important aspects still persist. Photoemission characterizations are commonly performed during the growth of molecular films. The approach allows a detailed examination of the electronic and chemical properties at the onset of the interface formation and thus the determination of factors governing the interface properties between organic films and contacts.

In this review, we will point out certain problems encountered in the standard characterization of organic films and related interfaces by photoemission, and we accentuate principles to be considered in the analyses of photoemission data necessary for their unambiguous interpretation. Namely, the basic electronic parameters such as the band alignment (BA), the work function (WF), and the ionization energy (IE), and particularly their relation, will be discussed in terms of their dependence on the molecular orientation. At variance with the current approach, the importance of the WF examination for the characterization of molecular films will be stressed. We propose that the mechanism for the alignment of energy levels at interfaces is governed by the WF*s* of materials brought into the contact. Further, we demonstrate that the band bending in molecular films frequently deduced from photoemission measurements is in fact an artifact occurring owing to the WF varying with the film thickness.

1.1. List of Molecules

Table 1. The molecules and/or molecular films tackled in the review arranged alphabetically according to their acronyms.

Acronym	Name	Chemical formula	IE (eV)
α-NPD	*N,N*′-diphenyl-*N,N*′-bis(1-naphthyl)-1,1′-biphenyl-4,4′-diamine	$C_{44}H_{32}N_2$	5.45(a)
6T	sexithiophene	$C_{24}H_{14}S_6$	5.0(a), 5.8(m)
6P	sexiphenyl	$C_{36}H_{22}$	5.7-6.6(b)
Alq_3	tris(8-hydroxyquinoline)aluminum	$C_{27}H_{18}AlN_3O_3$	5.95(a)
BCP	Bathocuproine	$C_{26}H_{20}N_2$	6.4(a), 6.35(l)
BP2T	2,5-bis(4-biphenylyl) bithiophene		5.3(k)
C_8	octane-1-thiolate	C_8H1_8S	
CBP	4,4′-*N,N*′-dicarbazolyl-biphenyl	$C_{36}H_{24}N_2$	6.35(a)
CuPc	copper phthalocyanine	$C_{32}H_{16}CuN_8$	5.05(a), 5.15(l)
DH4T	α,ω -dihexyl-quaterthiophene	$C_{20}H_{26}S_4$	4.9(c)
DIP	diindenoperylene	$C_{32}H_{16}$	5.0, 5.4(n)
F_4CuPc	tetrafluoro copper phthalocyanine	$C_{32}H_{12}F_4CuN_8$	5.55(d), 5.7(e)
F_4TCNQ	2,3,5,6-tetrafluoro-7,7,8,8-tetracyanoquinodimethane	$C_{12}F_4N_4$	8.4(a)
$F_{16}CuPc$	hexadecafluoro copper phthalocyanine	$C_{32}F_{16}CuN$	6.35(a), 6.09, 6.44(k)
DiMe-PTCDI	*N,N*′-dimethyl-3,4,9,10-perylenetetracarboxylic diimide	$C_{26}H_{14}O_4N_2$	6.58(d)
H_2Pc	metal-free phthalocyanine	$C_{32}H_{18}N_8$	4.96(d)
HBC	hexa-*peri*-hexabenzocoronene	$C_{42}H_{18}$	
NADPO	amphiphilic substituted 2,5-diphenyl-1,3,4-oxadiazole	$C_{20}H_{32}O_4N_4$	7.9(h)
NPB	*N,N*′-bis(1-naphthyl)-*N,N*′-diphenyl-1,1′-biphenyl-4,4′-diamine	$C_{44}H_{32}N_2$	5.2(i)
NTCDA	1,4,5,8-naphthalene tetracarboxylic dianhydride	$C_{14}H_4O_6$	8.05(a)
PEN	Pentacene	$C_{22}H_{14}$	5.0(a)
PFP	perfluoro-pentacene	$C_{22}F_{14}$	5.6-6.3(f)
PTCBI	3,4,9,10-perylenetetracarboxylic bisbenzimidazole	$C_{36}H_{16}O_2N_4$	6.25(a)
PTCDA	3,4,9,10-perylenetetracarboxylicdianhydride	$C_{24}H_8O_6$	6.85(a)
PTCDI	3,4,9,10-perylenetetracarboxylic diimide	$C_{24}H_{10}N_2O_4$	6.42(d)
TCNQ	Tetracyanoquinodimethane	$C_{12}H_4N_4$	7.85(a)
NiPc	nickel phthalocyanine	$C_{32}H_{16}NiN_8$	4.9(g)
ZnPc	zinc phthalocyanine	$C_{32}H_{16}ZnN_8$	5.29(j)

(a) Ref. [2], (b) Ref. [3], (c) Ref. [4], (d) Ref. [5], (e) Ref. [6], (f) Ref. [7], (g) Ref. [8], (h) Ref. [9], (i) Ref. [10], (j) Ref. [11], (k) Ref. [12], (l) Ref. [13], (m) Ref. [14], (n) Ref. [15]

1.2. Energy Diagram of an Organic Device

Figure 1 shows the energy diagram of an organic film sandwiched between metal electrodes under applied voltage, U. Overall transport properties of the device are determined both by the organic film and by the interface barriers, φ_e and φ_h, which determine injection/extraction efficiencies of charge carriers. The barriers can be deduced either from the characterization of overall transport properties performed on the final device or measured directly, e.g., by photoemission during the interface formation, which can probe relative positions of molecular energy levels with respect to the Fermi level of metal electrode at the interface, so called the band alignment (BA); while the φ_h is ascertainable by the direct photoemission (PE), the φ_e can be probed by the inverse photoemission. This review focuses on the former method. Admittedly, the details of the energy diagram may depend on the employed model, as it will be shown later.

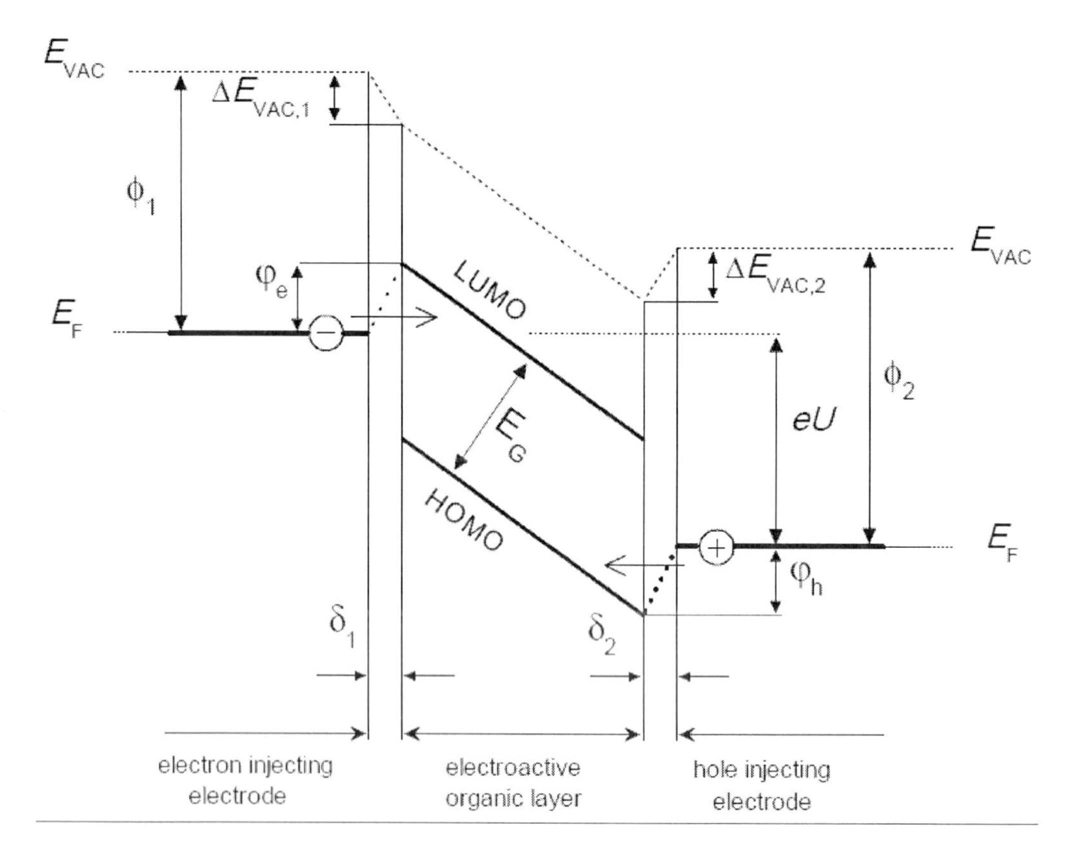

Figure 1. A schematic energy diagram of an electroactive organic layer sandwiched between electrodes under bias, U. The ϕ_i are the respective work functions of the metal electrodes *1* and *2*, and the φ_i and φ_h are the injection barriers for major charge carriers. The E_F, E_{VAC}, and E_G are the Fermi level, the vacuum level, and the band gap. The HOMO and LUMO state for the highest occupied and the lowest unoccupied molecular orbitals. The $\Delta E_{VAC,i}$ (i = 1,2) are the vacuum level changes over the respective interface region thickness, δ_i.

2. Electronic Properties of Molecular Films and Associated Interfaces

2.1. Valence Band, DOS

Figure 2 illustrates the VB electronic structure of a molecular film represented here by CuPc. The VB of molecules consists of shallow energy levels reflecting the both intra- and intermolecular interactions, the former being represented by the electronic structure of atoms constituting the molecule and their interaction. The intermolecular interaction has a weak influence on the electronic structure of a molecular film, as relative positions of particular molecular orbitals forming the VB are weakly sensitive to distinct neighborhoods of a molecule [16]; the molecules in the gas phase can be regarded as the isolated ones in contrast to the molecules embedded in a molecular film. As the shallowest orbitals are shaped by π-bonds of a molecule, the VB region near the Fermi level is referred to as the (upper) π-band.

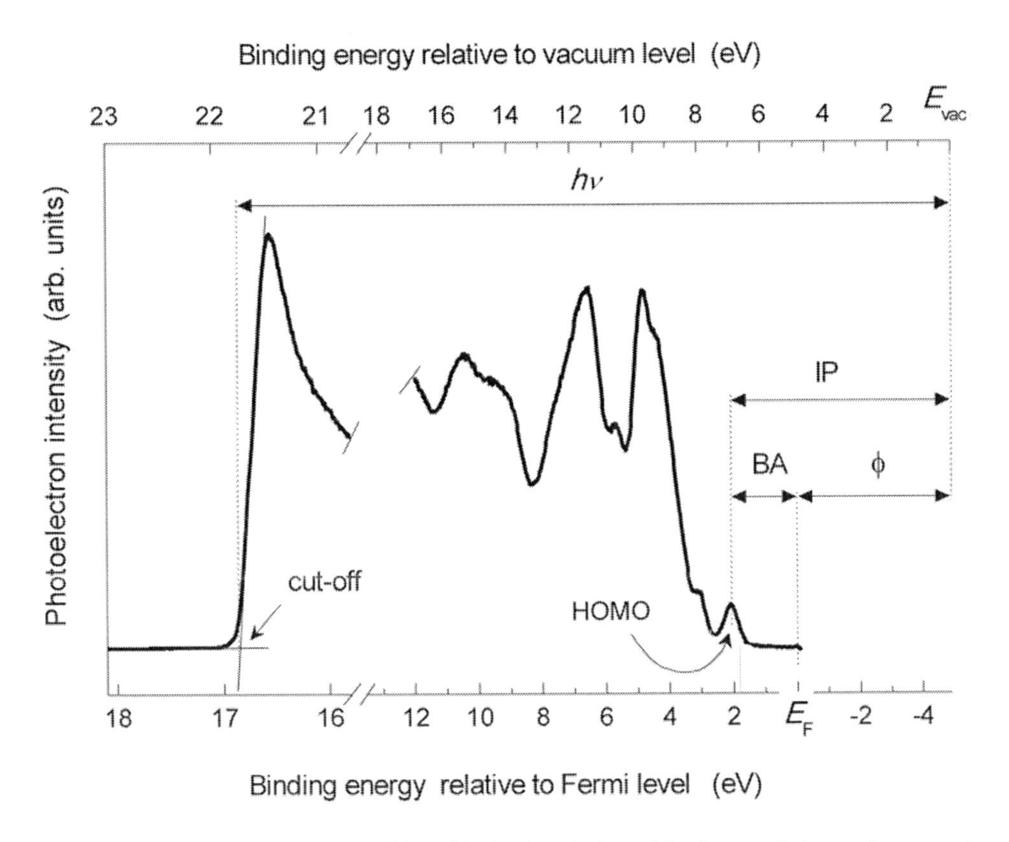

Figure 2. The valence band of CuPc film with the description of fundamental electronic parameters. The $h\nu$ corresponds to the photon energy of the incident beam, and the BA, IE, and ϕ state for the band alignment, the ionization energy, and the work function, respectively.

Although not the most prominent, the most important molecular orbital as far as the electronic properties of interfaces are concerned (Fig. 1), is the highest occupied molecular orbitals (HOMO), i.e., the molecular orbital with the lowest binding energy (BE). Consequently, the photoemission studies have prevailingly focused on examination of the

HOMO *position* with respect to the Fermi level of a substrate denoted the band alignment (BA) (Fig. 2) for various molecular films grown on a variety of substrates. The BA illustrated in Fig. 2 is the hole injection barrier, φ_h, introduced in Fig. 1. The great importance for the interface engineering would be the prediction of the BA solely from electronic parameters of constituents such as the organic film and contacting material, the latter acting either in the role of the substrate or the top contact. In spite of many experimental studies on properties of various interfaces, contradicting models have emerged. The issue will be discussed in Chaps. 2.4. and 3.4.

For the sake of the completeness, we note that the fine structure of the HOMO was recently reported to reflect the transport properties via the charge mobility-molecular vibration coupling [17-19]. Such and the extended characterizations of the upper π-band with the angle- and photon energy-resolved ultraviolet photoemission spectroscopy yielding the energy band dispersion (e.g., Refs. [20-22] are beyond the scope of this review. Of the recent reviews on organic films, the reader is also referred to Ref. [23].

Unlike the isolated and thus well-determined core levels, the identification of the molecular orbitals may be obscured by their mutual overlapping due to their proximity and due to both vaguely determined shallow energy levels of particular elements and their chemical shifts. Therefore, the VB is rarely attempted to reveal the chemical structure of a molecule. The bunch of rather undetermined MOs is referred to as the density of states (DOS). Apart from the HOMO, the rest of the DOS attracts less attention in experimental works and little is known about the electronic structure of the valence band. Considering that the numerical methods play an increasing role in the molecular design, an experimental feedback to *ab initio* theoretical calculations of the molecular electronic structure is necessary. The photoemission characterization by means of tunable photon energy allows to some extent the discrimination of the molecular orbitals in the VB in terms of their association with constituent atoms; this applies for molecules such as, e.g., phthalocyanines consisting of metal atoms next to carbon and nitrogen, which have distinct dependences of photoemission cross-section on the incidence photon energy; Fig. 3 illustrates the decomposition of the CuPc valence band into particular molecular orbitals and indicates their association with the constituent atoms [16].

The high-binding energy (BE) region in Fig. 2 reveals the secondary electrons cut-off indicating the onset of photoelectron emission. As indicated in the figure, the cut-off position is determined by its linear extrapolation. The cut-off corresponds to the minimal exciting energy required for an electron located at the Fermi level to escape the solid surface with its final kinetic energy being zero, i.e., the electron work function (WF). Naturally, such electrons have zero pass energy while entering into the analyzer, resulting in low detecting signal, as the transfer function of an analyzer decreases with decreased pass energy. In practical measurements, photoemission intensity of the cut-off can be increased by applying a known negative bias U_s to the sample. Electron energy is increased by eU_s and so is its pass energy, which helps to gain higher PE intensity. Obviously, the applied negative bias shifts the entire PE spectra towards high BE by $\Delta E = eU_s$. With bias, the cut-off is located at binding energy $BE_{cut} = h\nu - \phi - |U_s|$, where $h\nu$ is the photon energy; the formula allows to simply calculate the work function. The cut-off in Fig. 2 has been already corrected for the applied bias, i.e., it is shown in the real binding energy scale. The sample-detector lay-out is a key parameter at the photoelectric WF examination, the detector has to be oriented along the sample`s surface normal [24]. The work function of molecular films will be discussed in Chap. 3.

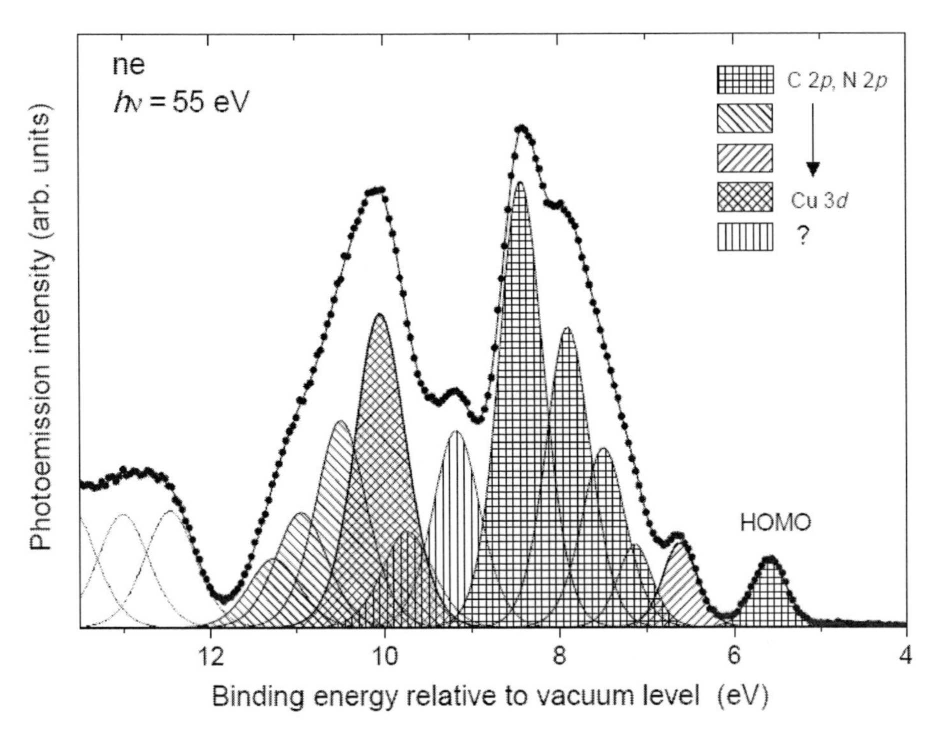

Figure 3. The deconvoluted valence band of CuPc. The legend indicates the character of particular molecular orbitals in terms of the contribution from C 2*p* and Cu 3*d* energy levels [16].

The sum of the BA and the work function, ϕ, gives the ionization energy (IE). The importance of the IE in the characterization of organic films will be discussed in Chap. 2.3.

One has to be aware that the BE scale referred to the Fermi level is in principle the relative one, for it refers to the spectrometer ground. The scale is shifted by the surface work function, ϕ, from the absolute energy scale referred to the vacuum level (note both energy scales in Fig. 2). While binding energies of isolated atoms/molecules (in the gas phase) have to be referred to the vacuum level (VL), the Fermi level reference commonly employed in characterization of solid surfaces is a matter of the convenient choice only since the Fermi level should coincide with the spectrometer`s ground. Benefits of adhering to the absolute energy scale will emerge in further discussion.

2.2. Work Function, Interfacial Dipole

Figure 4 shows a typical change of the substrate work function upon the organic film growth, here of the Al(111) surface upon the 6P growth. The description of the WF characterization was described in Chap. 2. The WF lowering, $\Delta\phi$, is

$$\Delta\phi = \phi_{subs} - \phi_{film} = \Delta E_{VAC}, \qquad (1)$$

here by about 0.45 eV, corresponds to the vacuum level shift, ΔE_{VAC}. The concept of the vacuum level (VL) misalignment due to the ID formed at the organic/substrate interface was introduced by Seki and coworkers to explain the apparent invalidity of the vacuum level alignment (VLA) [25] and since it has been widely employed for the description of various organic/substrate interfaces studied by the photoemission [14, 26-34]. Both signs of the vacuum level shift were observed [28]; the example given in Fig. 3 presents the WF decrease corresponding to the lowering of the VL. The ID manifests itself due to the *abrupt* change of the vacuum level. The dipole is inherent to the contact, and it can dramatically affect the alignment of molecular energy levels at the interfaces. The issue will be analyzed in detail in Chapters 2.4. and 3.4. In this chapter, the relation of the ID to electronic properties of the substrate and the molecular film will be discussed.

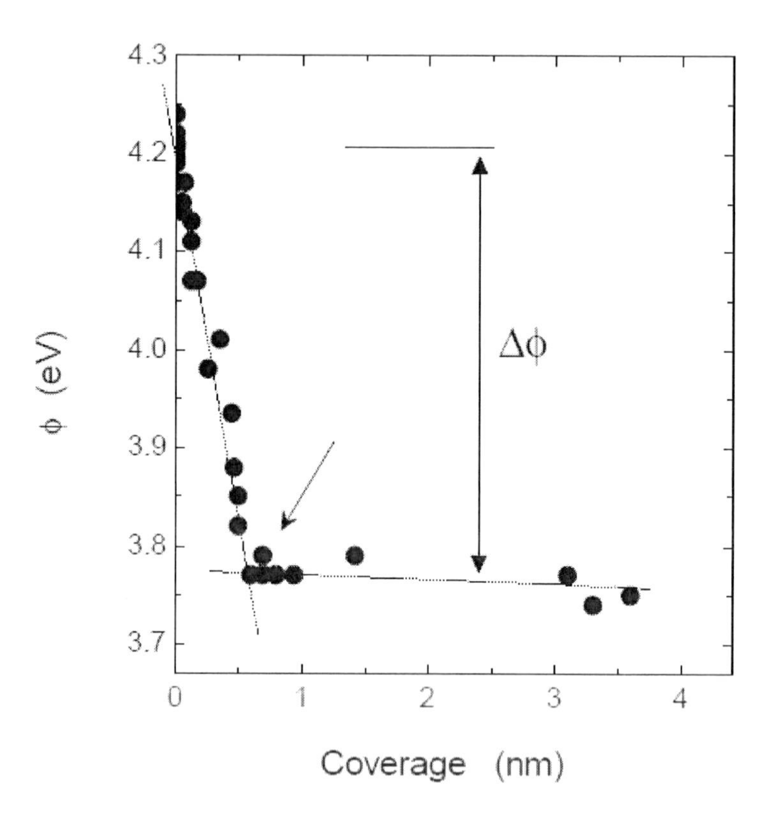

Figure 4. The work function evolution with the 6P film thickness. The arrow points the flex point indicating the onset of the interfacial dipole saturation and so the closing the substrate surface by the first monolayer [3].

According to the ID model, the ID is completed already upon the first monolayer (ML) completion, while the ML thickness is determined by the dimension of molecules, the packing density, and their predominant orientation relative to the substrate surface. In Fig. 4, the work function decreases linearly with the nominal coverage (up to about one monolayer) as it reflects the linear relation between the nominal coverage and the surface area occupation by adsorbed molecules. This is true for submonolayer coverage provided that either Frank-Van der Merwe or Stranski-Krastanov growth modes apply (the laminar growth and the island growth on the wetting layer, respectively) and the adsorption geometry is not coverage dependent. Varying packing density within first ML upon the nominal coverage may affect the linearity of the dependence. Constancy of the WF beyond the first monolayer expresses none or relatively weak charge redistribution between the next molecular layers, which reflects the weak bonding. The bonding strength between the first ML and the substrate on the one side, and between the particular molecular layers in the molecular film can be discriminated, e.g., by the thermal desorption spectroscopy [35]. Conversely, the work function measurement in coverage series may be convenient for the growth rate calibration, as the flex point (Fig. 4) in the work function *versus* coverage dependence may indicate the closed first monolayer.

Various origins of the ID detectable via the vacuum level change were suggested, such as (*i*) The permanent dipole moment of the adsorbed molecule [36], (*ii*) the charge redistribution over the junction due to the interaction of π bonds with the substrate surface [25, 28, 37], (*iii*) newly formed chemical bonds at the interface [28, 37], (*iv*) push-back effect [28] or pillow effect [32], where the tail of the electronic cloud is pushed back by the adsorbate and modifies thus the substrate WF, and (*v*) the interfacial dipole is of quantum origin due to the exchange-like effect (Pauli repulsion mechanism) [38].

Significant experimental efforts have been done to assign essential electronic parameters of molecules and substrates, which would determine the electronic structure of interfaces between them. Assignation of autonomous parameters of isolated constituents would allow the prediction of junction properties. Basically, there have been two approaches in the experimental search for pivotal parameters of constituents, namely attempts to find correlations between the ID and a single electronic parameter of (a) the molecular film and (b) the substrate. The former approach can be experimentally determined by comparing the IDs after the deposition of various molecules on the chosen specific substrate, the latter approach—*vice versa*—*via* the deposition of a specific molecule on various substrates.

Peisert et al. [39] grew various molecular films on gold foil and found no correlation between the ID and the IP, yet an apparent correlation was observed between the ID and the product IP-½ E_t, where the transport gap of semiconductors, E_t, was adopted from Ref. [40]. In fact, the observed correlation suggests that the ID is determined—as far as the adsorbed molecule contribution is concerned—by the LUMO position of the molecular film referenced to the vacuum level. This is explicitly claimed in Ref. [5], where the linear correlation between the ID and the electron affinity of four various phthalocyanines, namely H_2Pc, CuPc, F_4CuPc, and F_{16}CuPc, has been reported. Yan et al. [41] reported the linear relations between the ID and the product [ϕ_M - (IP-E_g/2)]. Such correlation would indicate that the ID is correlated with the *difference* between the ϕ_M and the LUMO of the organic film, i.e., with the barrier to electron injection from the metal to the organic.

Surprisingly, whereas the ID increases upon the fluorination of Pc according to Ref. [5], the opposite trends have been reported for the same molecules in Ref. [39]. The discrepancy is apparently occurring due to different substrates, namely H-passivated Si(111) and gold foil, respectively. Blumenfeld et al. [42] calculated the ID and thus the interfacial electronic structure on the basis of molecular properties such as the dipole moment, the polarizibility and the size.

The search for the electronic parameter of the substrate determining the ID has been more frequent and many studies reported the linear dependences of the ID on the substrate work function [13, 25, 28, 43-46]. However, the claimed linear correlations for specific molecule were based on typically three measured values only, with various slope parameters, $S = \Delta E_{vac}/\Delta\phi_M$, ranging from 0 to 1 depending on the molecule. Furthermore. some molecules were reported to reveal two distinct linear dependences; for example, PEN has shown the slope parameter either zero or +1, if grown on substrates with their WF smaller or higher than the IE of PEN, respectively [47], while quite reverse behavior has been observed for BCP, i.e., the slope parameters of +1 and zero for small and high WF of the substrates, respectively [13].

Based on better statistics, the linear correlation with the slope parameter of -0.7 has been reported between the ID and the electron affinity of the semiconducting substrate; PTCDA on various surface reconstructions of GaAs with the electron affinity ranging over the about 1.4-eV interval were employed in the role of the substrate [5, 48]. On the other hand, no correlation between the ID and the substrate work function has been detected (Fig. 5) when growing 6T on eight various substrates with work function ranging from about 2 to 5 eV [14].

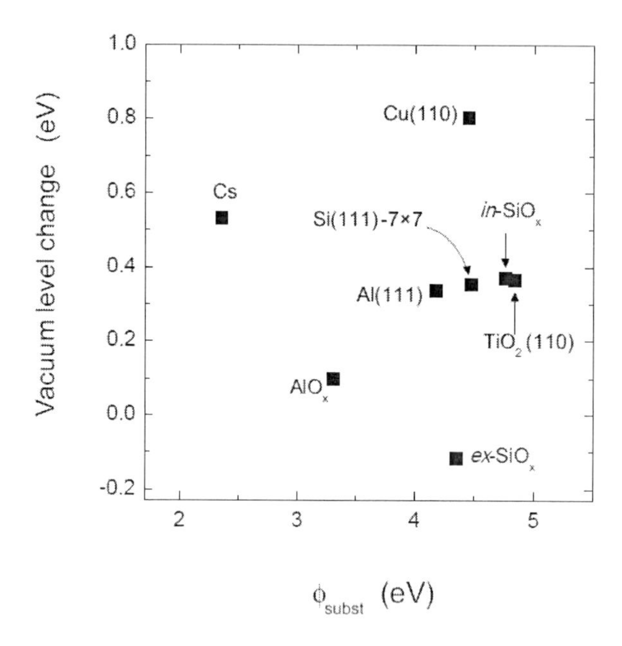

Figure 5. The correlation plot of the substrate work function, ϕ_{subs}, and the VL change, $\Delta E_{vac} = \phi_{subs}-\phi_{film}$ [14].

2.3. Ionization Energy

Equivalent to the first ionization potential of elements, which is considered to be the material constant, the IE of a molecular film is the minimal energy necessary for the removal of an electron from the molecular film and moving it to infinity. Consequently, the IP corresponds to the binding energy of the HOMO related to the vacuum level, and it is evaluated by summing the BA and the WF of the molecular film (Fig. 2).

Note that the HOMO position is identified with its peak energy in this study, even though the low BE onset of the HOMO approximated by its leading edge is frequently employed. The use of the onset has been substantiated in relation with transport properties, that is to say both the HOMO and the LUMO onsets were argued to better represent transport levels of organic semiconductors (see, e.g., Refs. [5] and [49]). We advocate the HOMO position to be identified with its peak since the peak energy is—in comparison to the leading edge—less influenced by experimental factors such as instrumental resolution or sample temperature. Virtually, unlike, e.g., onsets of the secondary electrons and the Fermi edge, the HOMO may show no leading edge, i.e., no linear portion. Thereby, the usage of the "leading edge" of the HOMO particularly at low intensities may result in an uncertain determination of its binding energy.

The IE of a molecule in the condensed phase, i.e., forming the molecular film, is lower typically by about 1 eV in comparison to the IE of the isolated molecule (in gas phase) due to the extra-molecular screening reflecting the neighborhood of the molecule [50]. Also, a molecule embedded in the bulk of the molecular film and that resided on its surface experience distinct surroundings, which is reflected in the PE spectra [51].

Table 1 presents also the IE*s* of the listed molecules measured on the corresponding molecular films. The IE*s*, which were commonly determined according to the low-BE onset of the HOMO, range from 5.0 to 8.4 eV. Constituent elements of molecules other than carbon and hydrogen may affect the IE dramatically; in the first approximation, the presence of N, O, or F in the molecule tends to increase its IE. Apparently, the location of these atoms within the molecule matters, the mentioned atoms augment the IE provided they are located at the periphery of the molecule which are mostly planar; phthalocyanines (Pc*s*) have generally the low IE similar to, e.g., purely carbon-based pentacene, in spite of 8 constituent nitrogen atoms located, however, between the central metal atom and peripheral carbon and hydrogen atoms. In contrast, TCNQ with its four terminal nitrogen atoms has the higher IE. Substitution of hydrogen in the C-H bonds by fluorine leads to substantial increase of the IE, as seen for fluorine-substituted TCNQ [2], Pc*s* [5], PEN [52], and alkanethiols with different end groups [53].

We have recently prepared 6P films with their IE depending on the growth conditions [3]. This is illustrated in Fig. 6, where the upper π band of the 6P films grown at RT and 395 K are contrasted. For the spectra are referenced to the vacuum level, the HOMO offset indicates the difference in the ionization potential. The distinct IE suggests that the IE of the molecular film is not a material constant. Distinct IE*s* have been associated with differences in conjugation length induced by distinct molecular packing and conformation due to the torsional freedom in the films, which may be interrelated; the twist angle between neighboring benzene rings may vary depending on the intermolecular interaction. The torsional angle is maximal for an isolated molecule (in the gas phase) due to the spherical hindrance of hydrogen atoms, while molecule embedded in the condensed film may be nearly planar. The varying inter-ring angle would directly affect molecular π conjugation and thus the width of the π band [54]. The alternating twist of sexiphenyl π-rings on Ag(111) was evidenced by STM combined with DFT calculations [55].

Figure 7 shows the IE evolution of 6T film with the film thickness for the films grown at LNT and RT. The film grown at LNT displays the IE roughly independent of the thickness suggesting the invariant electronic structure during the growth. It is reasonable to suppose that the film grown at LNT is poorly ordered as the molecular arrangement at low temperature is obstructed by the low surface diffusion of molecules. In contrast, the growth at RT displays significant thickness dependence of the IE by about 0.7 eV (Ref. [34]) suggesting major changes in the electronic structure with the film thickness. Notable, the IE of the film at the onset of the growth, i.e., up to several monolayers thick, is independent of the growth temperature. This also suggests that the conjugation in the films grown at RT improves with the film thickness due to improved molecular order. Similar trends were reported for the NADPO [9].

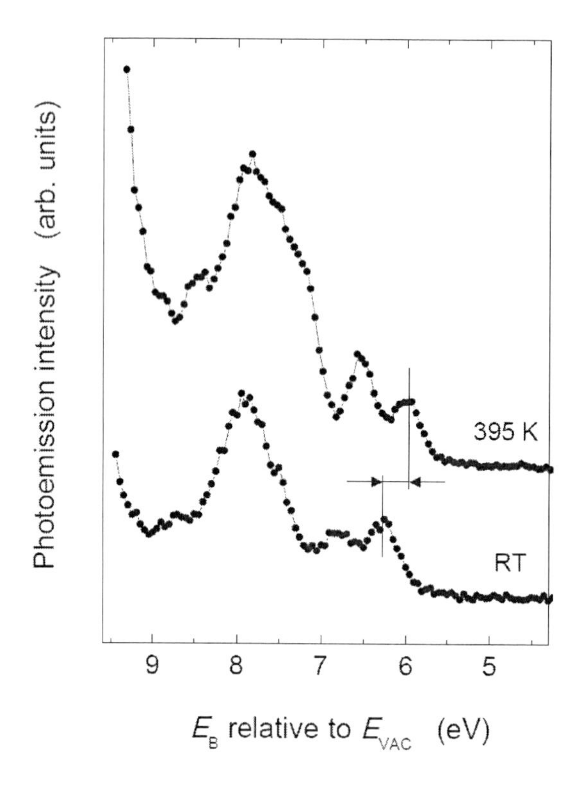

Figure 6. The upper π band of a 6P film grown at RT and elevated temperature. The HOMO positions are indicated by vertical ticks. The schematic formula of the 6P is in the upper part [3].

The further studies reported varying IE for other molecular films such as 6T [34, 56], DH6T [56], and even for rigid molecules with no torsional freedom, such as PFP [52], CuPc [57-59], F_{16}CuPc [58] and PEN [60]. The most recent works related the variations of the IE with the MO, either lying down or standing upright [56, 58].

Consequently, the importance of the determining both the BA and the WF for the characterization of molecular films and their interfaces is obvious. If the work function is not determined, the electronic distinctness of the molecular film may be unnoticed. The importance of the IE argues for the necessity of the vacuum level (VL) reference for comparison purposes of the spectra. Compared to the habitually employed Fermi level, the VL reference not only enables the comparison solid state spectra with the gas phase spectra, the latter being necessarily referenced to VL, it also eliminates rigid energy shift arising from the interfacial dipole and thus enables the comparison of films grown on different substrates.

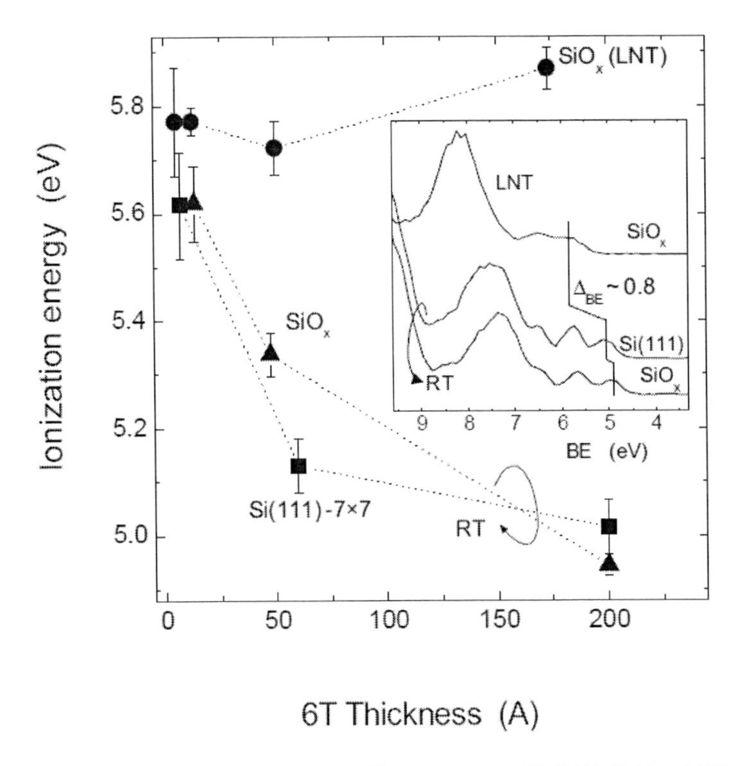

Figure 7. Thickness dependence of the IE for 6T films grown on Si (111)-7×7 and SiO_x substrates at room temperature (RT), and on SiO_x grown near the liquid nitrogen temperature (LNT). The inset shows UPS spectra of the π bands of the corresponding systems referred to the vacuum level for the maximal 6T thicknesses. The HOMO positions are indicated by vertical lines [14].

2.4. Energy Level Alignment

If two dissimilar materials are in contact, their distinct energy level positions result in the formation of an interfacial electronic barrier, which is in fact an electronic discontinuity seen by charge carriers. Such interfacial barriers established between the molecular film and metals, φ_e and φ_h, are shown in Fig. 1. The latter barrier can be directly probed by the direct photoemission; this is illustrated in Fig. 2, where the energy span between the HOMO of the molecular film and the Fermi level of the substrate, habitually denoted the band alignment (BA) or band offset, corresponds to the hole injection barrier, φ_h.

The simplest theoretical approach for the alignment of energy levels at the interface is shown in Fig. 8(b); two isolated materials characterized by their WF and IE [Fig. 8(a)] align according to the common vacuum level. The approach is characterized by

$$BA = IE - \phi_{subs}. \quad (2)$$

Validity of Eq. 2 would imply that the BA is the linear function of the ϕ_{subs} with the slope of -1 for each molecular film characterized by its IE. Notable, the vacuum level alignment (VLA) model is equivalent to Schottky-Mott rule proposed for the determination of the barrier height at metal-semiconductor interfaces [61-62]. Therefore, the dependence according to Eq. (2) and thus having the slope parameter $S_{BA} = \Delta BA/\Delta\phi_{subs} = -1$ is referred to Schottky-Mott limit, while the BA insensitive to ϕ_{subs}, *i.e.*, S = 0, is referred to Bardeen limit. Although the Schottky-Mott rule has been shown unworkable, it has been widely used for comparison purposes.

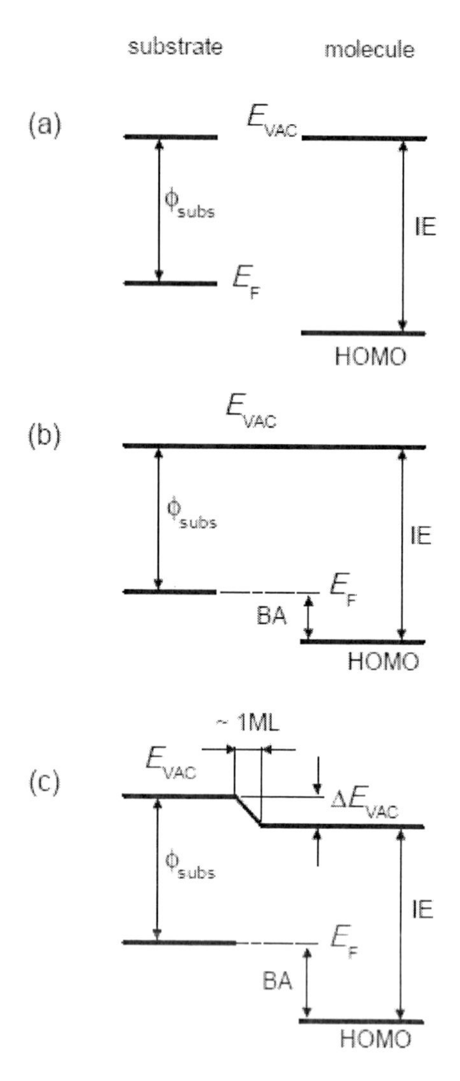

Figure 8. Energy band diagrams of (a) an isolated molecule and a substrate, and after the establishment of intimate contact without (b) and with (c) the interface dipole. See the text for details.

Similar to the search for the ID (Chap. 2.2), two basic approaches have been followed to unravel a relationship between the BA and parameters of the molecule and substrate brought into the contact; (a) growth of various molecular films on a specific substrate with the known WF, and (b) the growth of the particular molecular film on various substrates, while the latter approach has greatly prevailed. Frequently, the BA linearly dependent on the substrate WF, however with the slope parameters ranging between 0 and -1 for various molecules, has been reported [2, 13, 26, 28, 39, 43-47, 63, 64]. It should be mentioned that the reported dependences have been typically based on three experimental points only; thereby the evidence of the linear relationship is less reliable. Fukagawa et al. [47] have grown PEN on various substrates and observed that the BA is either inversely proportional or insensitive to ϕ_{subs}, i.e., the Schottky (VLA) or Bardeen limit, provided that $\phi_{subs} < IE_{PEN}$ or $\phi_{subs} > IE_{PEN}$, respectively. In contrast, BCP revealed qualitatively reversed behavior, namely Bardeen and Schottky-Mott limit on substrates with low and high WF, respectively [47]. The specific dependences have been claimed to be due to the gap states. On the other hand, no linear dependences were claimed, e.g., for bithiophene [65] and 6T (Fig. 9) grown on variety of substrates.

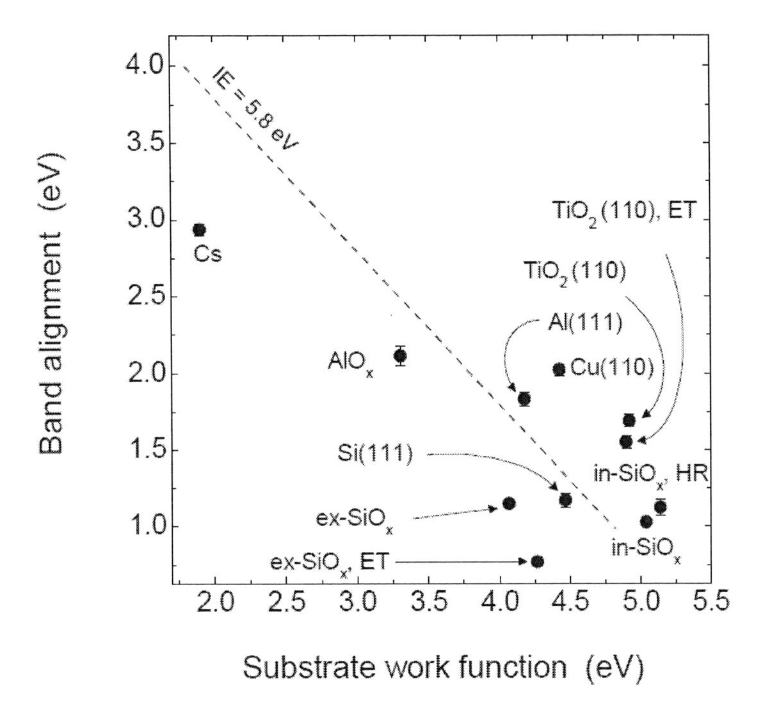

Figure 9. The band alignment (BA) of the sexithiophene films grown on a variety of substrates with the work function ϕ_{subs}. The straight dashed line represents the Schottky-Mott relation [Eq. (2)] for 6T with the IE of 5.8 eV [14].

Disobedience of the VLA model has been suggested to be due to the formation of the interfacial dipole [25, 66], which may significantly affect the BA. The energy diagram involving the ID is shown on Fig. 8(c), and it obeys the following formula

$$BA = IE + ID - \phi_{subs}, \quad (3)$$

where ID $\equiv \Delta E_{VAC} = \phi_{film} - \phi_{subs}$. The ϕ_{film} is the work function measured after the film growth (Fig. 4). The validity of Eq. 3 can be experimentally tested using the formula

$$BA = IE - \phi_{film}. \quad (4)$$

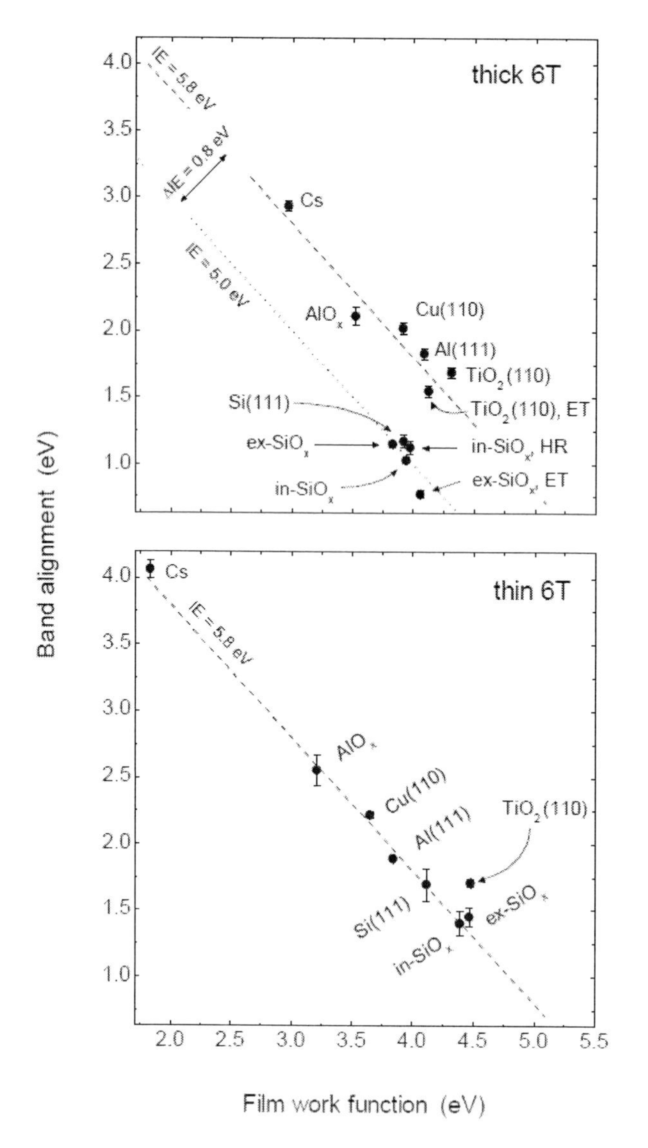

Figure 10. Plot of the band alignment versus the work function of (a) thick and (b) thin sexithiophene films grown on various substrates [14]. The dashed lines represent the Schottky-Mott limit (Eq. 2) with the IE indicated next to the lines.

This is shown in Fig. 10 (upper panel) where the data from Fig. 9 are replotted. At first look, no correlation is observed. However, the closer inspection reveals two sets of the data each grouping along a linear dependence according to Eq. 4. The linear dependences are spaced by 0.8 eV corresponding to the IE change, in other words, two distinct values of the IE have to be taken into the account in Eq. 4. This example points out the importance of realizing that the IE of molecular films can depend both on the growth conditions and the film thickness and that the determination of interfacial properties can be obscured by extrinsic effects such as the properties of molecular films. Thereby, the films with the same IE have to be prepared to allow the meaningful verification of Eq. 4 for any molecule-on-substrate system. The same IE for a particular molecule however grown on various substrates may be attained for ultra-thin films grown at low temperatures, as thicker films may show various IE depending on growth conditions such as the substrate surface, the film thickness, growth temperature, and the growth rate (Fig. 7).

In Fig. 10, bottom panel, Eq. 4 is verified using the data obtained on the same systems as in the upper panel, however, with the ultra-thin 6T films in the range up to several monolayers grown at liquid nitrogen temperature (LNT). The resulting correlation is very good, indicating the validity of Eq. 3.

3. Molecular-Orientation Resolved Electronic Properties

3.1. Molecular Orientation

Molecules may form organic crystals and crystalline films. Even though the molecules within the unit cell may show different orientations owing to the herringbone structure of molecular films, there is an average molecular orientation pertinent to a film; in the present study, the term "molecular orientation" expresses the average orientation in terms of the molecular plane and/or its backbone with respect to the surface normal. According to this notification, the molecules with their molecular planes parallel or near parallel to the substrate surface are lying down, while the molecules with their molecular planes and/or backbones parallel or near parallel with the surface normal are oriented upright. Asymmetry of molecules induces the highly anisotropic electronic properties of molecular crystals being highly correlated with the molecular orientation. For example, transport properties of organic devices are dramatically affected by the molecular orientation relative to the current flow [67]. To control the MO, the mechanism of the organic film growth has been widely studied via the molecule deposition on a variety of substrates, and the profound influence of ordered substrates on the film structure and packing has been frequently demonstrated. However, this has mostly concerned thin films, but further growth towards thick films resulted in the different film structure and morphology weakly or not dependent on the substrate. Three basic growth modes have been described for inorganic films, namely the Frank-Van der Merwe, Stranski-Krastanov, and Volmer-Weber growth modes employed to characterize the laminar growth, the island growth on a wetting layer, and the island growth, respectively. The growth of molecular films is complicated by the asymmetry of molecules, which may dramatically affect electronic properties. Few most common scenarios are sketched in Fig. 11. Except Fig. 11(b), all illustrated growth modes correspond to the Frank-Van der Merwe (laminar) growth mode, however, it will be shown below that both the evolution and the final electronic properties of the films are distinct. Figure 11(b) represent the Stranski-Krastanov growth mode with the molecules lying-down.

In the framework of common models considering the growth mode in terms of the molecule-substrate interaction strength via molecular π-bonds, the strong (weak) interaction represented by the molecule-on-metal (molecule-on-oxidized surface) case favors the lying-down (upright) molecular orientation, Figs. 11(a) and 11(b), respectively. Several reported inconsistencies of the interaction strength model may be exemplified, namely the lying CuPc growth on Au(111) but upright CuPc on polycrystalline Au [68], the lying 6T on $TiO_2(110)$-1×1 but upright on TiO_2 [69] and Au, Ag, and Cu films [70], and lying PTCDA on CuPc [71]. According to the interaction strength model, the lying CuPc should have been attained on both Au substrates (the distinct molecular orientation was argued to be due to different roughness), while upright 6T and PTCDA molecular orientations should be attained on the TiO_2 and CuPc substrates owing to the weak interaction with oxidized and organic substrates, respectively [68, 69].

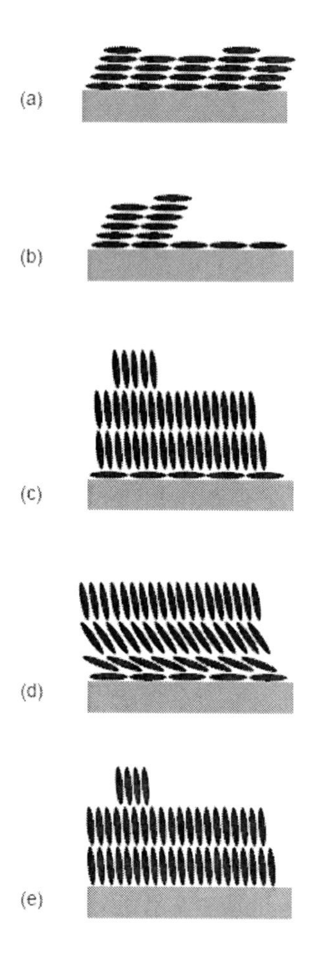

Figure 11. Sketches of most common growth modes of molecular films with consideration of the molecular orientation (see text).

In addition, the interaction strength model is unable to rationalize the lying-down orientation of molecules beyond the first monolayer of the molecular film sketched in Fig. 11(a), as the weak molecule-molecule interaction should result to universal upright orientation by the second layer upon the closing the first layer [Fig. 11(c)]. However, this is not the case and moderately thick films with the lying-down geometry can be accomplished on the proper substrate. On the other hand, the lying-down geometry, if achieved, usually applies for thin films only, and the molecular orientation changes towards the upright if the film thickness is further increased [Fig. 11(d)]. Thereby such molecular films become inhomogeneous in terms of the molecular orientation. The orientational transition has been frequently observed, provided that the molecular orientation was examined, and it can occur either abruptly already by the second layer [Fig. 11(c)] (Refs. [72-74]), or gradually within the nominal film thickness ranging from several nanometers up to several tenths of a nanometer [Fig. 11(d)] (Ref. [34, 75-79]). The higher growth temperature favors the upright molecular orientation [73, 80]. The gradual change of the molecular orientation associated with thicker films was reported also for polymers [81], and it may be a general growth phenomenon, although it has been often unnoticed. Importantly, the orientational transition affects the film electronic properties; this will be discussed in the next chapters.

The organic film growth was proposed to view via its kinetics, and the orientation-specific re-evaporation was claimed to govern the molecular orientation [73]. Recently, the growth of different crystallite orientations has been argued to be controlled by the kinetics mediated by the (dis)order of the substrate surface rather than exclusively by chemical interaction between the molecule and the substrate. Unless a chemical reaction between the surface and the adsorbed molecules occurs, molecules freely diffuse over the surface until they are incorporated in the organic crystal. The necessary condition for getting lying-down molecular orientation in the film is the presence of a director forcing molecules to diffuse over the surface along preferential directions, as opposed to randomly oriented diffusion [69, 82, 83]. The director, which may be a preferential azimuthal direction at the substrate surface manifesting via lower potential energy for the molecular diffusion along the direction, or unioriented steps on vicinal surfaces [75] can exist owing to the surface reconstruction. The coarse surface reconstruction or that intentionally mutilated, e.g., by slight sputtering, has a dramatic effect on the film growth; instead of lying-down adsorption geometry, the molecules orient upright in contrast to the interaction strength model [69]. The weak interaction of a molecule to the substrate and its preferentially oriented diffusion results in the growth governed by attractive intermolecular interaction [84]; the diffusion of molecules in a particular direction ensures the maximal overlap of the π bonds of encountered molecules and results in the embedding of molecules in the film with the lying-down orientation. Molecules diffusing over the surface randomly in terms of azimuthal orientations form energetically more favorable crystals made of upright molecules. In other words, to form ordered molecular crystals, the weak interaction of molecules to the substrate is necessary in contradiction to the interaction strength model [84].

Obviously, molecules lay flat down while diffusing on any kind of planar surface, the ordered one or the disordered one and the lying-down and upright molecular orientations discussed above are meant in terms of the final orientation of embedded molecules. Thus the upright orientation occurs either (*i*) at the elevated growth temperature, when the azimuthally oriented surface barriers become too small for warm molecules, or (*ii*) at high growth rate, when high flow of impinging molecules buries aforedeposited molecules and thus impedes their diffusion, or (*iii*) on unordered or poorly ordered surfaces. The last listed scenario takes place also with the increasing thickness of the ordered molecular film, whereas the surface order of molecular films deteriorates with the film thickness due to growth imperfectness, and hence the directors gradually vanish. On the other hand, distinct molecular orientation need not affect the crystal structure except its orientation with respect to the substrate, as suggested by the same vibronic progressions [85].

By the identification of the substrate surface morphology as primary factor for the molecular orientation in the film, the diffusion model rationalizes distinct molecular orientation obtained on chemically identical surfaces; the lack of the preferential diffusion, as it occurs, e.g., on amorphous or polycrystalline surfaces, results in upright-oriented molecules in the film, while the molecules in the film lie down on the reconstructed/ordered surface. The chemical origin of the substrate, either it is a metal or an oxide or an organic, plays a minor role unless a chemical reaction between the adsorbed molecule and the substrate occurs.

3.2. Ionization Energy

Table 1 presents IE*s* of selected oligomer molecules. We have shown in Chap. 2.3. that IE of molecular films is not the material constant, and distinct IE by up to 0.8 eV were observed. The most recent works related the IE differences with the molecular orientation, either lying-down or standing upright [15, 56, 58].

3.3. Work Function

The growth of a *thick* 6T film at LNT reveals a surprising behavior [Fig. 12(a)]; the WF suggests its saturation after the interfacial dipole formation up to several nanometers similar to the 6P (Fig. 4); yet the WF again diminishes with the further growth by about 0.4 eV (Ref. [34]). The analogous WF evolution with the film thickness have been observed, e.g., for the NiPc [8], CuPc [59, 75, 86-88], 6T and thiophene [89], and D4HT [4]. The WF drop occurring beyond the ID-related WF drop was associated with varying molecular orientation from the lying-down towards the upright [4, 34]. Yamane et al. suggested that the gradual WF change is due to the summation of incremental dipoles that have been arising due to the gradually changed molecular tilt angle [87].

That the work functions of thin and thick (otherwise chemically identical) films differ need not be surprising as the film *surfaces* are formed either by lying-down or by upright molecules, respectively, which implies distinct terminations of the film surface; the carbon is the most prevailing element at the surface in the former case, while hydrogen dominates in the latter one. Thereby, the WF*s* of about 4.4 eV and 4.0 eV correspond to the films with lying-down and upright 6T molecules, respectively. We note in this context that even single-crystal elemental surfaces can demonstrate the work function variation of several tenths of an eV depending on the probed crystal facet. For example, the work functions of W (111) and W (110) being of 4.47 eV and of 5.25 eV, respectively, differ by 0.78 eV [90].

Returning back to Fig. 12, note that all films, i.e., irrespective of the growth temperature, the substrate properties, and the growth rate, show at their maximum thickness the same final WF of about 4 eV. However, the thickness dependences suggest that the final upright molecular orientation is achieved at different film thicknesses. For example, the gradual work function drop for the growths at RT and 120°C—in contrast to the *abrupt* drop of the LNT-grown film—suggests that the molecules tend to arrange upright from the onset of the growth; the growth evolutions correspond to sketches in Figs. 11(e) and 11(d), respectively.

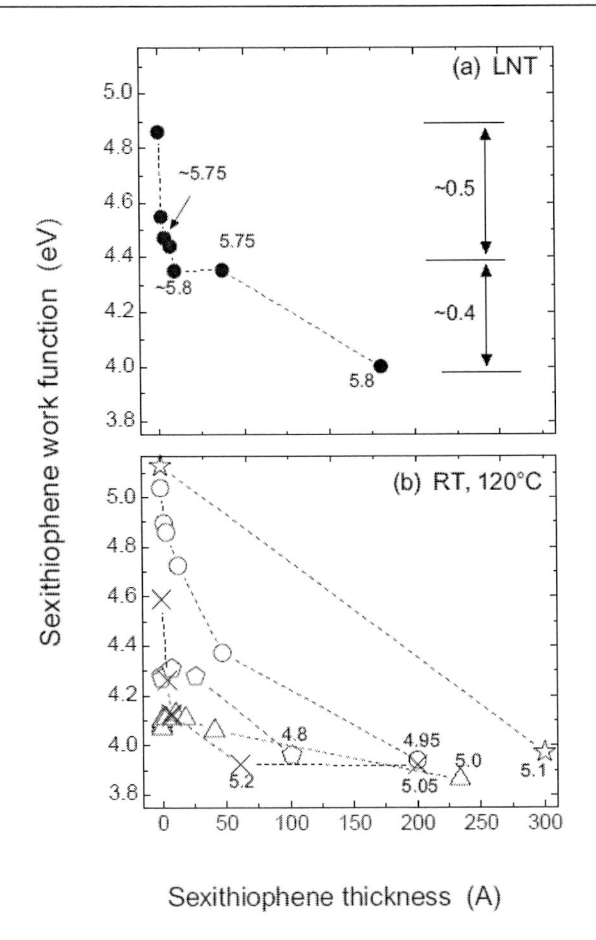

Figure 12. Work function diagrams of the 6T-growth on silicon-related surfaces at LNT (a) and RT and 120°C (b). In the panel (b), the circles, triangles and crosses indicate the growth on the UHV, native oxide, and the Si(111)-7×7, respectively. The pentagons and stars represent the growth on the native oxide at 120°C and the growth on the UHV oxide at high deposition rate, respectively. Numbers near to the points give ionization potential values of the corresponding films. Straight dotted lines serve as guidelines only [34].

Consequently, any WF change $\Delta\phi$ related with the growth of molecular films has to be partitioned into two contributions: (1) due to the formation of the 1st monolayer, $\Delta\phi_{1stML}$, accounted for the ID formation, and (2) due to the varying molecular orientation, $\Delta\phi_{MO}$

$$\Delta\phi = \Delta\phi_{1stML} + \Delta\phi_{MO}. \qquad (5)$$

If the lying-down molecular orientation is preserved for at least several monolayers beyond the first monolayer (ML), the work function evolution may display the plateau after the initial drop. With further increasing film thickness, the molecular orientation may change towards upright and the varying molecular orientation induces the further change of the work function, $\Delta\phi_{MO}$. For a given molecules, the orientational transition depends on the growth conditions such as the substrate surface morphology, growth rate, and growth temperature. The WF plateau need not be always noticeable; typically, this may occur either if the molecules are oriented upright in the entire film, i.e., including the first ML [Fig. 11(e)], or if the first ML consists of lying molecules, but the next layers are already formed by upright molecules [Fig. 11(c)] [74, 91, 92]. In the latter case, the true magnitude of the interfacial dipole may be difficult to estimate unless the precise calibration of the coverage is achieved. Moreover, the absence of the clear distinction between the first ML-related WF change and the molecular orientation-induced WF change gives a rise to discussion on the mixing the ID and the band bending [91, 92], the latter being deduced from $\Delta\phi_{MO}$. The reported band bending will be shown to be an artifact of the varying molecular orientation (Chapter 3.5). The coherent view on factors affecting the WF evolution is given in Chaps 3.4. and 3.5.

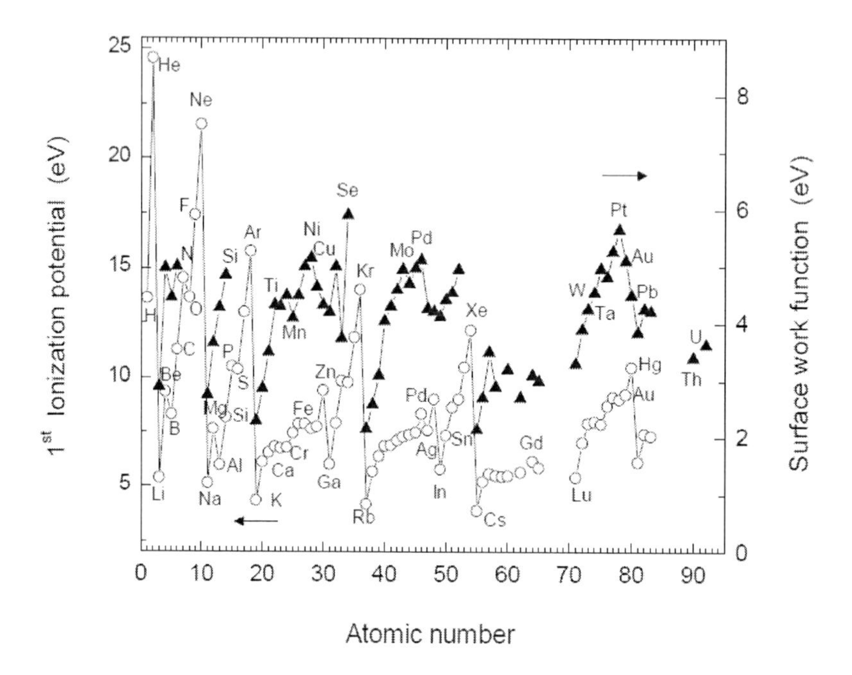

Figure 13. The first ionization potential (empty circles) and the work function (full triangles) plots for elements; the correlation plot reported in Ref. [96] is replotted here using the latter values adopted from Refs. [90] and [97]. If more values of the work function of a specific element have been available, the value without an indication of the surface reconstruction has been adopted. The element labeling is incomplete and some of labels are only indicated to aid the orientation in the plot. The solid line connecting the data points are to guide the eye.

Interestingly, the $\Delta\phi_{MO}$, if detected, frequently shows the value of about 0.4 eV for unsubstituted (hydrogen terminated) molecules, such as 6T [34,, 56], NiPc [8], CuPc [59, 86, 87, 93, 94], D4HT [4], HBC [79]. Chen et al. measured systematically both WF*s* and MO*s* of CuPc films and observed the WF of 3.95 eV and 4.35 eV, i.e., differing by 0.4 eV, for the films built from lying-down and upright molecules [94]. We presume that the value of 0.4 eV is related with the difference of the WF owing to distinct surface terminations of the polycrystalline molecular films, namely either via the π bonds of lying molecules or through hydrogen terminating the carbon bonds. The surface termination presumably determinates the surface dipole affecting the WF. In Fig. 12, $\Delta\phi_{1stML}$ = -0.5 eV and $\Delta\phi_{MO}$ = -0.4 eV have the same sign. Generally, the sign of both contributions may differ; the increasing WF with the thickness of fluorine-substituted molecular films such as F_{16}CuPc [39] and PFP [52] suggests the positive $\Delta\phi_{MO}$ due to varying molecular orientation towards upright. The trend can be explained qualitatively by the surface element-controlled WF: The surface of the molecular film made of upright-oriented perfluorinated molecules is dominated by terminating fluorine. Although the WF of fluorine could not be obviously determined due to its gaseous state, based on the correlation existing between the first ionization potential free atoms and the WF of the corresponding surfaces shown in Fig. 13 (Ref. [95, 96]), the IP of fluorine being the third highest among the elements after the He and Ne suggests the high WF of the fluorine-terminated surface.

The effect of the varying molecular orientation on the PE spectra has been often neglected. The point to be stressed is that electronic properties of molecular films develop to a certain extent with their thickness, the changes are responsive to morphological and structural changes in molecular films. Thereby, the molecular film thickness is a needful parameter for any examination of electronic properties. Besides, the nominal film thickness, as suggested by microbalance measurements, may significantly differ from the real film thickness on the sample due to departure from the laminar growth mode commonly observed for molecular films [Fig. 11(b)].

3.3.1. Intrinsic Work Function of a Molecular Film

According to the general wisdom, the WF of inorganic materials is the material constant. Yet, the surface morphology may affect the WF of otherwise chemically identical surface, as already mentioned earlier. The WF of elemental surfaces is usually obtained by the characterization of self-supported samples, where the thickness is high enough not to affect the WF value. In practice, when merely WF is issue, the semi-infinite substrates are commonly substituted by the corresponding *films* mechanically supported by a substrate. Importantly, the film should be thick enough to manifest the surface properties of semi-infinite bulk material. For example, a platinum film evaporated on the silicon substrate with the WF of 4.4 eV shows the WF of 5.6 eV corresponding to the WF of platinum for thickness beyond about 12 ML [98]. Clearly, the WF of the Pt film in the intermediate thickness region, i.e., below 12 ML, indicates that either the Pt film is not electronically developed and/or the influence of the substrate has not been eliminated yet. The transition region may extend to larger overlayer thicknesses, and it has been attributed to a violation of the local charge neutrality in films [99]. Therefore, it is invalid to account the WF of the platinum film without considering the substrate in the intermediate thickness region and instead of speaking about the evolution of the *film* WF, the more appropriate description is such that the varying WF expresses the WF of the silicon *substrate upon* the Pt film growth. For Pt thickness beyond 12 ML, the WF shows the WF of Pt; neither the substrate nor the film thickness play any role as far as the electronic properties of the film are concerned.

Regarding the inorganic elemental materials, it has been well understood that a minimal film thickness is necessary to attain the corresponding WF established for bulk material. In contrast, the WF of molecular films are often examined for small thicknesses where the film electronic structure has not been yet accomplished, and consequently the corresponding WF is intermediate and understood as the parameter depending on the substrate.

It is reasonable to presume that—similar to surfaces of inorganic electronic materials—a surface of each semi-infinite organic material with charge carriers has to be describable by its intrinsic WF, ϕ_{int}. In contrast to the WF of a thin film, which cannot be considered without the substrate, the intrinsic WF does not depend on the substrate and is determined by the organic material itself. The existence of the intrinsic WF also means that the Fermi level position for the given organic film is specific, which is the inferable condition. Although apparently abundant, the term "intrinsic" is employed in this study, for it would distinguish the work function being the material constant from the still developing work function of moderately thin films.

At least two intrinsic WF can exist for each molecular film formed by asymmetric molecules, $\phi_{int,\ lying}$ and $\phi_{int,\ upright}$, each representing two extreme situations as far as the molecular orientation near and at the surface is concerned; the different molecular orientation implicates distinct termination of the molecular film, thus distinct chemistry of the surface. In Fig. 12, it was demonstrated that the WF functions of all *thick* 6T films have saturated at the value of about 4.0 eV irrespective the substrate work function. Given the structural characterization of the films, we assume that this value corresponds to the intrinsic WF of the 6T film formed or at least terminated by upright-oriented molecules, $\phi_{int,upright}$. On the other hand, the value 4.4 eV would correspond to the WF of the film formed by lying-down molecules, $\phi_{int,\ lying}$. Provided that the molecular orientation had not been examined, we suppose that—given the reported behaviors of various systems—the organic film thickness beyond about 10 nm may be sufficient to attain the intrinsic WF. The thickness of ca 10 nm is a consequence of the orientational transition towards the upright orientation frequently spontaneously accomplished at the thickness of ca 5 to 10 nm. Yet the minimal thickness required for attainment of the intrinsic WF may be substantially lower provided that the particular MO is preserved through the entire film. Moreover, morphological issue of the film has to be carefully considered; for example the Stranski-Krastanov and the Volmer-Weber growth modes result in high surface roughness, and the real film thickness may dramatically differ from the nominal one. Consequently, the measured WF is distorted/obscured and its magnitude corresponds to the weighted average of the patchy surface.

In spite of plethora of photoemission studies on organic film growth, the usable data in the literature for the examination of the intrinsic WF of various molecular films are relatively scarce. This is mainly due to the fact that investigations have mostly focused to the BA and ID examination, which were observed to saturate already for ultrathin films, and the WF of organics has been habitually considered as a subsidiary parameter and thus its examination was consequently often omitted.

In the following compilation, several plausible examples collected through reported studies sustain the notion of the intrinsic WF of molecular films. Concerning the data selection, certain limitations must be recognized. The molecular films with the sufficient thickness of ~6 to 10 nm have been qualified. Moderately thick films were chosen, too, provided that the WF could be crosschecked with the same films grown, however on different substrates. On the other hand, nominally thick molecular films with the substrate features still visible in the spectra were excluded, since such surfaces were electronically inhomogeneous and the measured film WF was apparently heavily affected by the insufficiently buried substrate [7, 52, 100, 101]. Obviously, the misrepresentative work function implicates the misrepresentative ionization energy. Such approach in the data selection limits dramatically the collection of eligible data, and it eventuates to the most popularly investigated and thereby the most reported molecules. Consequently, the complied values discussed in next paragraphs and listed in Table 2 must be simply accepted as being the best available to our knowledge. The selected data are admittedly non-exhaustive, and the notion of the intrinsic WF requires further verification.

The 10 nm-thick NiPc films displayed the same work function of 4.0 eV, in spite of the 0.8-eV difference between the work function of Au and Ag polycrystalline substrates being of 5.2 eV and 4.4 eV, respectively [8]. Although the molecular orientation was not examined in the study, the thickness dependences of the WF resembling that in Fig. 12(a) suggest that the molecular orientation changed from lying-down to upright. The 6T films grown at 130 K on three substrates with their WF ranging from 5.4 eV to 3.9 eV displayed the WF between 4.3 and 4.5 eV [89]. Lau et al. [102] examined CuPc/F_{16}CuPc heterostructures prepared using both deposition sequences. The final WF*s* of about 20–nm-thick films for both heterostructures correspond to the assumed intrinsic WFs reported by other authors (Table 2). The WF of various phthalocyanine*s* grown on distinct substrates, namely on polycrystalline gold with WF = 5.3 eV [39] and hydrogenated Si(111) with the affinity of 5.2 eV and the WF = 4.22 eV [5], have also striking similarities. In spite of the dramatic difference in both the WF of the substrates and the IDs, the WFs of thick molecular films are similar. Admittedly, there is rather large difference of 0.47 eV for the F_{16}CuPc, which is presumably owing to the distinct molecular orientation induced by different thicknesses of the examined films, namely the declared 5 to 10 nm and 15 nm. The MO influence is supported by recent results by Wee and collaborators [58]; the authors systematically examined WF of various molecules depending on their orientation and found the WF = 4.7 eV and 5.3 eV for the lying and upright F_{16}CuPc molecules, respectively. The 12 nm-thick PTCDA grown on GaAs (100) with various surface modifications displayed the final WF of about 4.7 eV, irrespective the substrate WF ranging from 4.52 to 5.39 eV [48]. The WF of 12.8-nm-thick PTCBI films displayed the same magnitude of about 4.5 eV irrespective the employed substrate (Au, Ag, or Mg) [37].

Noticeably, if the $\phi_{int,upright}$ ~4.0 eV of the 6T were due to the corresponding film termination, as proposed earlier, any thick film (i.e., the thick to ensure the upright orientation) made of molecules with the hydrogen termination should display the similar WF. Indeed, such a trend may be observed and ϕ ~ 4.0 eV for thick films were reported for molecules such as 6P [103], H_2Pc [5], CuPc [5, 75, 104], NiPc [8], Alq [64], α-NPD [11], BCP [11].

Analogous to the correlation between the IP and WF for elements (Fig. 13), the correlations exists for molecular films, too. This is shown in Fig. 14, where the molecular film-related data were adopted from Table 2 whereas the element-related data replot those presented in Fig. 13. The dashed lines indicate the corresponding linear fits, while As, C, and Hg—the only labeled elements in the graph—were excluded from the fitting due to their large deviation. The linear fits follow the equations

$$\phi_{int} = 1.88 + 0.45 \times IE \quad (6)$$

$$\phi = 0.08 + 0.56 \times IP \quad (7)$$

for the molecular films and elements, respectively. The correlation plot for molecular films does not discriminate the molecular orientation as there is not enough reliable data explicitly relating the WF and the molecular orientation for specific organics. Noticeably, the scattering of WF-IE dependence for organic films is comparable to that for the elements. PTCDA has the IE of ca 6.6 eV, the highest values among the molecules listed here. Yet this value is exceeded by the IE of TCNQ, NTCDA, and F_4TCNQ [2], hence, the commensurably high WF*s* are expected. This will be further discussed in the next chapter.

Table 2. The compiled values of the measured WFs (bold characters) for several molecular films. If the WF values were not reported explicitly in the literature, they were calculated using the formula WF = IE-BA. Below each WF value, the corresponding IE is provided (standard characters). The WF*s* relevant to the molecular geometry—lying (L) down or upright (U)—are discriminated provided that the molecular orientation was examined. The assumed ϕ_{int} are given on the bottom. The indices refer to the film thickness and the substrate or a film acting in the role of the substrate, and their WF. The compiled IE*s* were determined according to the onset of the HOMO except the values labeled by asterisk, which were determined according to the HOMO peak position.

Ref.		NiPc	H_2Pc	CuPc	F_4CuPc	$F_{16}CuPc$	PTCDA	6T, DH4T	BP2T	PTCBI	PEN	PFP	α-NPD	BCP	CBP	Alq_3	PTCDI	DiMe-PTCDI
[2]				5.05		6.35	6.85	5.0		6.25	5.0		5.5	6.4	6.35	5.9		
[8]		**4.0**[l] 4.9																
[104]				**3.9**[a] 5.0	**4.55**[a] 5.7	**4.95**[a] 6.1												
[6]				**4.1**[o] 5.0 **4.1**[y] 5.0	**4.7**[o] 5.7													
[5]			**4.04**[b] 4.96	**3.87**[b] 4.82	**4.70**[b] 5.55	**5.42**[b] 6.32	**4.94**[gg] 6.94										**4.67**[hh] 6.42	**4.55**[ii] 6.58
[58]	L			**4.35**[f] 5.2		**4.7**[c] 5.9	**4.5**[ff] 6.4, 6.45 **4.8**[d] 6.55 **4.95**[h] 6.40											
	U			**3.95**[g] 4.8		**5.3**[e] 6.6												
[48]							**4.67**[n] 6.61											
[27]							**5.0**[m] 6.8						**4.0**[bb] 5.7			**3.95**[cc] 5.95		
[102]				**4.24**[j] 4.82		**5.3**[j] 6.66												
[37]										**4.6**[q] 6.2 **4.5**[q] 6.2 **4.5**[q] 6.0								
[105]						**5.1**[k] 6.3												

Table 2. (continued)

Ref.		NiPc	H_2Pc	CuPc	F_4CuPc	$F_{16}CuPc$	PTCDA	6T, DH4T	BP2T	PTCBI	PEN	PFP	α-NPD	BCP	CBP	Alq_3	PTCDI	DiMe-PTCDI
[11]													**4.18**[t] 5.4	**4.18**[u] 6.4 **4.14**[v] 6.4	**4.72**[x] 6.16 **4.45**[w] 6.21			
[106]											**4.33**[s] 4.91							
[75]				**4.0**[dd] 4.85 *														
[107]											**4.5**[z] 4.85							
[7]	L											**4.95**[aa] 5.6						
	U											**4.95**[aa] 6.3						
[34]	L							**4.35**[ee] 5.75*										
	U							**3.96**[r] 4.95* **3.94**[r] 5.05*										
[89]	L							**4.28** 5.8 **4.5**[p] 5.9										
	U																	
[4]	U							**4.05**[jj] 4.9										
[12]						**6.44**[nn] **6.09**[oo] 5.13, 4.8			**5.27**[ll] **5.3**[mm] 4.35 4.47									
[10]																**4.0**[kk] 5.7		
assumed ϕ_{int}	L			*4.4*		*4.7*	*4.6*	*4.35*										
	U	*4.0*	*4.04*	*3.9*	*4.7*	*5.3*		*3.95*	*5.3*	*4.6*	*4.5*	*4.95*	*4.1*	*4.15*	*4.45*	*4.0*	*4.67*	*4.55*

(a) 5-10 nm Pc/Au, $\phi_{Au} = 5.3$ eV (Ref. [104])
(b) 15 nm Pc/*p*-Si(111), $\chi_{Si} = 5.2$ eV, $\phi_{Si} = 4.22$ (Ref. [5])
(c) 4 nm $F_{16}CuPc$/HOPG, ϕ_{HOPG} = ca 4.3 eV, (Ref. [58])
(d) 8 nm PTCDA/4 nm $F_{16}CuPc$/HOPG, $\phi_{F16CuPc} = 4.7$ eV, (Ref. [58])
(e) 4 nm $F_{16}CuPc/C_8$-SAM, $\phi_{C8\text{-}SAM} = 4.3$ eV, (Ref. [58])
(f) 5 nm CuPc/HOPG, ϕ_{HOPG} = *n.a.*, (Ref. [58])
(g) 4 nm CuPc/SiO_2, ϕ_{SiO2} = *n.a.*, (Ref. [58])
(h) 8 nm PTCDA/4 nm $F_{16}CuPc$/ C_8-SAM, $\phi_{F16CuPc} = 4.7$ eV, (Ref. [58])
(i) 20 nm $F_{16}CuPc$/CuPc, $\phi_{CuPc} = 4.24$ eV, (Ref. [102])

(j) 20 nm CuPc/F_{16}CuPc, $\phi_{F16CuPc}$ = 5.3 eV, (Ref. [102])
(k) 10 nm F_{16}CuPc/Au, ϕ_{Au} = 5.0 eV, (Ref. [105])
(l) 10 nm NiPc/Au, 10 nm NiPc/polycrystalline Ag, ϕ_{Au} = 5.2 eV, ϕ_{Ag} = 4.4 eV, (Ref. [8])
(m) 10 nm PTCDA/Au, ϕ_{Au}= 5.2 eV, (Ref. [27])
(n) 12 nm PTCDA/GaAs, ϕ_{GaAs}= 4.52-5.39 eV (5 different surfaces), (Ref. [48])
(o) 9 nm CuPc/Au(100), 9 nm F_4CuPc/Au(100), ϕ_{Au} = 5.3 eV, (Ref. [6])
(p) 5 nm 6T/Au, 5 nm 6T/PFDT/Au, 5 nm 6T/ODT/Au at 130 K, ϕ_{Au} = 5.1, ϕ_{PFDT} = 5.4, ϕ_{ODT} = 3.9 eV, (Ref. [89])
(q) 12.8 nm PTCBI/Au, 6.3 nm PTCBI/Ag, 6.3 nm PTCBI/Mg, ϕ_{Au}=5.0, ϕ_{Ag}=4.3, ϕ_{Mg}=3.8 eV, (Ref. [37])
(r) 20 nm 6T/Si(111)-7×7, 20 nm 6T/SiO_2, ϕ_{Si} = 4.6 eV, ϕ_{SiO2} = 5.05 eV, (Ref. [34])
(s) 6.4 nm PEN/SiO_2, ϕ_{SiO2} = 4.33 eV, (Ref. [106])
(t) 10 nm α-NPD/Au, ϕ_{Au} = n.a., (Ref. [11])
(u) n.a. nm BCP/α-NPD/Au, $\phi_{\alpha\text{-NPD}}$=4.18 eV, (Ref. [11])
(v) n.a. nm BCP/doped α-NPD/Au, $\phi_{\alpha\text{-NPD}}$=4.74 eV, (Ref. [11])
(w) n.a. nm CBP/ZnPc/Au, ϕ_{ZnPc} = 4.45 eV, (Ref. [11])
(x) n.a. nm CBP/doped ZnPc/Au, $\phi_{d\text{-}ZnPc}$ = 5.05 eV, (Ref. [11])
(y) 11 nm CuPc/GeS(100), ϕ_{GeS} = 4.6 eV, (Ref. [86])
(z) 12.8 nm PEN/PEDOT:PSS, ϕ_{PEDOT} = 5.3 eV, (Ref. [107])
(aa) 7 nm PFP/Au, ϕ_{Au}=5.3 eV, (Ref. [7])
(bb) 10 nm α-NPD/Au, ϕ_{Au}=5.2 eV, (Ref. [27])
(cc) 10 nm Alq_3/Au, ϕ_{Au}=5.2 eV, (Ref. [27])
(dd) 3.6 nm CuPc/H-Si(111), ϕ_{Si} = 4.28 eV, (Ref. [75])
(ee) 5 nm 6T/SiO_2 at 90K, ϕ_{SiO2} = 4.83 eV, (Ref. [34])
(ff) 10 nm PTCDA/5 nm CuPc/HOPG, ϕ_{CuPc} = 4.35 eV, 10 nm PTCDA/ 5 nm CuPc/SiO_2, ϕ_{CuPc} = 3.95 eV, (Ref. [58])
(gg) 15 nm PTCDA/S-GaAs(100), ϕ_{GaAs} = 5.4 (Ref. [5])
(hh) 15 nm PTCDI/S-GaAs(100), ϕ_{GaAs} = 5.28, (Ref. [5])
(ii) 15 nm DiMe-PTCDI/S-GaAs(100), ϕ_{GaAs} = 5.17, (Ref. [5])
(jj) 10.4 nm DH4T/Au, ϕ_{Au}=5.25 eV, (Ref. [4])
(kk) 15 nm Alq_3/Au, ϕ_{Au}= n.a., (Ref. [10])
(ll) 25.6 nm BP2T/F_{16}CuPc, $\phi_{F16CuPc}$ = 4.8 eV (Ref. [12])
(mm) 20 nm BP2T/Au, ϕ_{Au} = 5.3 eV (Ref. [12])
(nn) 25.6 nm F_{16}CuPc/BP2T, ϕ_{BP2T} = 4.35 eV (Ref. [12])
(oo) 20 nm F_{16}CuPc/Au, ϕ_{Au} = 5.3 eV (Ref. [12])

In contrast to the unsubstituted (hydrogen-terminated) molecules, where the upright molecules show typically lower WF than the lying-down molecules (see, e.g., 6T [33], CuPc, PTCDA [58, 94]), the fluorine-substituted Pc*s* show the opposite trend, i.e., the WF increase upon the orientational change towards upright [58, 94]. This is due to relatively low WF of the hydrogen-terminated surface compared to the fluorine-terminated surface; the presence of fluorine in a molecule increases its IP as a rule and thus does the ϕ_{int}, the effect being the most pronounced with fluorine forming the surface, thus for the upright molecules.

It is conceivable that IE change due to the varying molecular orientation and/or the film structure may be simply implicated by the WF of the film following the correlation dependence according to Eq. (6). This seemingly holds for 6T, where $\Delta\phi$ and ΔIE are of 0.4 and 0.8 eV, respectively [34], however, it does not apply for CuPc and F_{16}CuPc, where the WF change and the IE change on the account of the molecular orientation are about the same [58]. Apparently, the issue requires further investigation.

3.4. The Role of the Intrinsic WF in Energy Level Alignment

The above-exemplified trends in the vacuum level misalignment determined on *thick* films suggest that the total change of the vacuum level corresponds to the difference between the substrate WF and the intrinsic WF of the molecular film.

$$\Delta E_{vac} = \phi_{subs} - \phi_{int} \,. \quad (8)$$

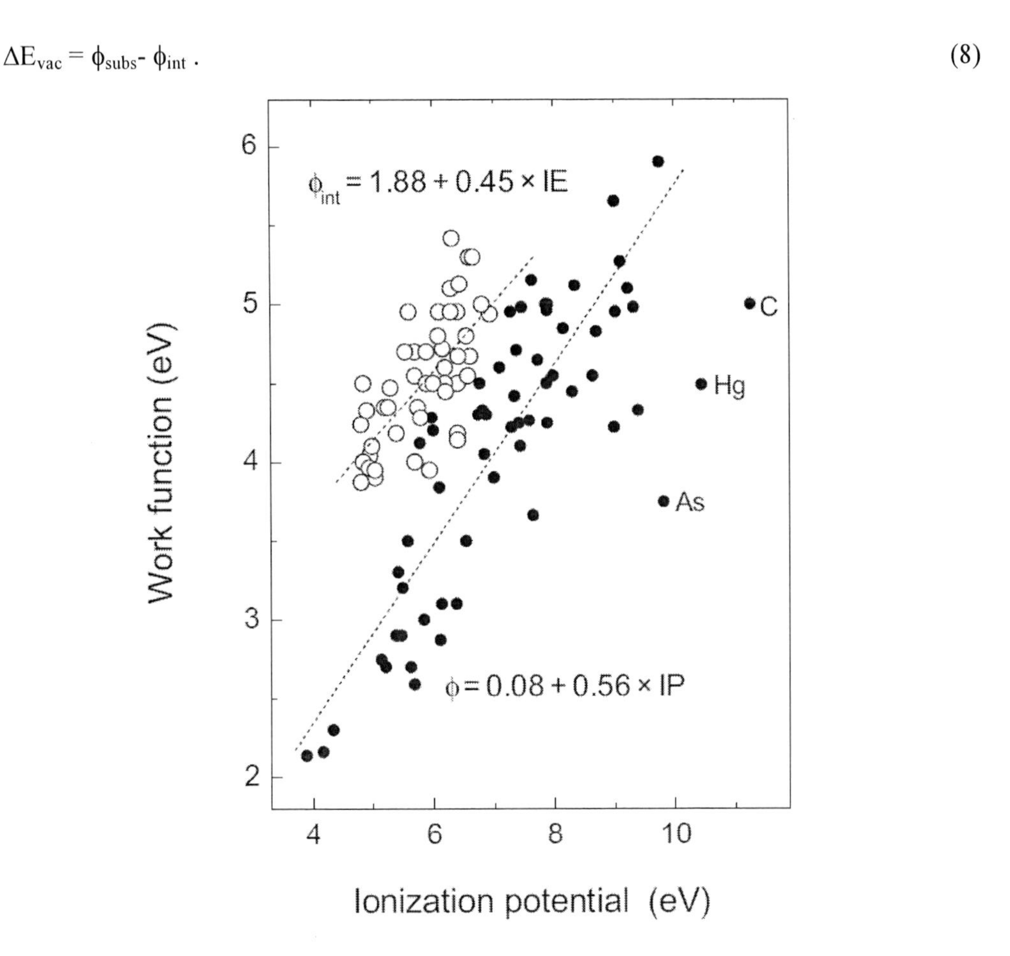

Figure 14. The WF-IP and WF-IE correlation plots for the elements (full symbols) and for the molecular films (empty symbols). The lines express the least square fits.

It is reasonable to presume that the same mechanism for the energy level alignment according to the WF difference is valid universally for all film thicknesses:

$$\Delta E_{vac} = \phi_{subs} - \phi_{film}(\phi_{subs}, t, \phi_{int}), \quad (9)$$

where *t* states for the film thickness and $\phi_{film} \rightarrow \phi_{int}$ for $t \rightarrow \infty$.

We want to stress that even though Eq. (8) is formally identical to Eq. (1) which has been employed for the determination of the VL change, Eq.(8) expresses the driving mechanism for the VL change. Hitherto we propose a novel rule governing the alignment of energy levels, the “WF rule”: The vacuum level of the substrate changes upon the film growth by the difference between the ϕ_{subs} and ϕ_{film}, the latter converging to ϕ_{int} with the increasing film thickness. In other words, the overall vacuum level change for, e.g., organic film with moderate and high thicknesses is predictable provided the ϕ_{int} of the examined organic film is known. However, unless the thickness dependence of ϕ_{int} is unraveled, the prediction is limited to moderately thick and thick films only.

Knowing the rule for the ΔE_{VAC} determination, the BA can be deduced employing the equation

$$BA = IE - \phi_{int} \quad (10).$$

The graphical sketch of the WF-rule model [Eq. (10)] for a particular molecular film characterized by IE grown on various substrates is depicted on Fig. 15. The straight solid line represents the measured BA values. The dotted line with the slope of -1 indicates the Schottky-Mott relation omitting the interfacial dipole, i.e., the vacuum level alignment. The arrows exemplify the total vacuum level changes occurring for the thick films grown on substrates with $\phi_{subs,1}$ and $\phi_{subs,2}$. The point of intersection between the measured line and the Schottky-Mott relation occurs at ϕ_{int} and no ΔE_{vac} is observed provided that $\phi_{int} \approx \phi_{subs}$.

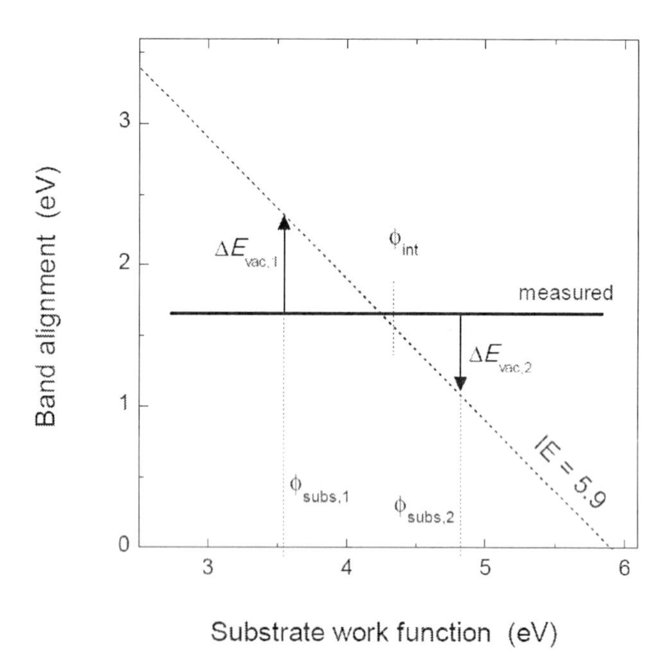

Figure 15. The graphical representation of the WF rule. The dashed line illustrates the Schottky-Mott rule for a molecular film with the IE of 5.9 eV. The solid horizontal line expresses the insensitivity of the BA to the substrate work function; the arrows refer to particular vacuum level changes, ΔE_{vac}, upon the growth at the substrates with different work functions, ϕ_{subs}. The intersection between the measured line and the S-M rule dependence corresponds to the intrinsic WF of the molecular film, ϕ_{int}.

The BA independent on the substrate WF (the slope parameter $S = dBA/d\phi_{subs} = 0$) were reported for several molecules, admittedly, the molecular films with $-0.8 < S < 0$ were observed, too [2, 28]. In the latter case, we presume—forasmuch the reported experimental data lack often details such as the film morphology—that the departure from the situation characterized by S = 0 as far as the BA is concerned is due to the influence of the substrate not entirely covered by the thick organic overlayer. If the film thickness is not sufficient (either nominally or due to the heavy islanding [Fig. 11(b)]) the substrate work function still affects the measured WF and hence the apparent BA is higher/lower for the molecular film grown on the substrate with the low/high work function. In other words, the absolute value of the S parameter increases if the intrinsic WF has not been accomplished.

The WF characterization of molecular films has been frequently understood merely for the ID determination and investigations focused on the evaluation of WF *change*. In the present model, the absolute value of the WF matters, as the total vacuum level change is derived from the difference between the intrinsic WF*s* of the substrate and the film, both being absolute magnitudes, too.

According to Eq. (6), both signs of the vacuum level change can arise. The downward change of the vacuum level occurs if molecular films with low ϕ_{int}, such as 6P, 6T, Pc, PEN, α-NPD are grown on substrates with higher WF, such as Ag, Au, Ni, etc. In contrast, the upward shift of the vacuum level can be observed provided that molecular films with high ϕ_{int}, such as $F_{16}CuPc$ or PTCDA are grown on substrates with low and moderate WF (e.g., Ag, Al, Mg). Regarding the former case, the downward shift of E_{vac} can change upward upon the doping of the low ϕ_{int} film with the high ϕ_{int} molecular film; for example, the doping of α-NPD on Au by F_4TCNQ [2]. The more common downward change of the vacuum level can be explained through the higher popularity of high-WF substrates combined with "standard" molecular films with moderate and low WF, while the rather uncommon upward WF change has been due to relatively less investigated, e.g., perfluorinated (fluorine substituted), molecular films or NTCDA, both having the high intrinsic WF.

Noticeably, the F_4TCNQ has the highest IE of 8.4 eV [2] among the listed molecules, which gives WF of 5.6 eV according to the correlation equation Eq. (6) for molecular films, the value ranking to the highest one among the both elements and organics. This also implicates that the F_4TCNQ grown on any substrate, except Pt and Se (Fig. 13), will increase the substrate WF. Recently, the increase of the gold WF by 0.29 eV upon even the ultrathin F_4TCNQ-film growth has been reported [108].

The proposed model, which quantifies the vacuum level change as the difference between the substrate and the film work functions, can explain the infrequently observed VL change at organic heterointerfaces. Since the charges over molecules are rather confined, no vacuum level change is expected at heterointerfaces. However, the VL change of 0.4-0.5 eV arises for interfaces between organics with the large difference in their ϕ_{int}, such as between, e.g., PTCDA and CuPc or Alq_3 (Ref. [63]), for PTCBI/BCP (Ref. [63]), C_{60}/CuPc (Ref. [109]), NTCDA/Alq_3, CuPc, BCP (Ref. [110]), CuPc/$F_{16}CuPc$ (Ref. [58, 102]). Heterointerfaces formed by molecular films with the same or similar ϕ_{int}, such as ZnPc/CBP, ZnPc/BCP, α-NPD/BCP, α-NPD/CBP (Ref. [110]), and PEN/CuPc (Ref. [111]) show no or negligible VL change. Obviously, no VL change arises also if intrinsic WF of organic films equals to the WF of the inorganic substrate; for example, the PEN on the highly oriented pyrolitic graphite (HOPG) shows no VL change considering that $\phi_{PEN} \sim \phi_{HOPG} \sim 4.4$ eV [111].

The WF-rule model can also explains non-monotonous evolution of the ΔE_{VAC} with the film thickness [52]; in this work, PFP was grown on Ag. The vacuum level abruptly dropped by 0.42 eV after the first monolayer and remained constant with further growth. The E_{vac}, however, increased by 0.12 eV at the thickness of 5 nm. The similar non-monotonous evolution of the WF was observed for the $F_{16}CuPc/Au$ structure [39]. Such a curious behavior of the WF can occur if the orientational transition takes place in the perfluorinated molecular film and the following inequality holds:

$$\phi_{int,lying} < \phi_{subs} < \phi_{int,\ upright}$$

The mirrored WF evolution, i.e., the WF increase followed by the WF decrease upon the film growth may occur for an unsubstituted molecular film provided that the following equation holds

$$\phi_{int,lying} > \phi_{subs} > \phi_{int,\ upright}.$$

Indeed, such behavior of the WF was reported for the growth of titanyl phthalocyanine film on HOPG 112].

The charge neutrality level (CNL) model [31] introduces the so-called CNL being an intrinsic property of the molecule. The ELA is determined by the alignment of the CNL with the Fermi level of the metal substrate. The alignment is determined both by CNL and by the substrate through the ϕ_M and the slope parameter S. Vazquez et al. [33] calculated the CNL for several molecules; namely for PTCDA (4.8 eV), PTCBI (4.4 eV), CBP (4.2 eV), CuPc (4.0 eV), α-NPD (4.2 eV), BCP (3.8 eV), and Alq_3 (3.8 eV). The CNL values given in parenthesis are referred to the vacuum level and they are similar to intrinsic WF*s* given in Table 2. The differences may be due to film thickness; in fact, the calculations were based on the one molecule positioned near the substrate surface suggesting that the intrinsic WF has not been yet attained.

The integrity of the WF rule can be illustrated via its validity for the classic Schottky contact—the metal/semiconductor junction obtained by evaporation of metal on semiconductor, where the WF of both materials, ϕ_M and ϕ_S, are known. The barrier at the *real* metal/*n*-semiconductor junction is given by [66]

$$\phi_B = \phi_M - \chi + \Delta, \qquad (11)$$

where χ and Δ are the semiconductor affinity and the interfacial dipole, respectively. Notably, Eq. (11) is the Schottky-Mott relation extended by the ID, as the original relation assuming no dipole has been found unworkable.

The manipulation of Eq. (11) gives

$$\Delta = (\phi_B + \chi) - \phi_M = \phi_S - \phi_M. \qquad (12)$$

According to Eq. (12), the ID is determined by the difference between the WF of the semiconductor (the substrate) and the metal (the evaporated film). For the WF of a metal is generally accepted material constant; it is also intuitively understood as the magnitude to which the final WF of the metal/semiconductor contact has to converge for the thick metal film. In other words, the WF rule expresses the boundary conditions for the metal/semiconductor system in terms of its characterization. Presumably, the driving mechanism for energy level alignment at any interface between electronic materials is universal, thereby at organic/substrate junctions, too.

In the previous models on organic/metal interfaces [28, 37], the charge exchange between the metal and the molecule determines the vacuum level misalignment. In this regard, ϕ_{subs} and IE of the molecular film seem to be incommensurate parameters for the description of the interface considering that the former is the surface property whereas the latter parameter characterizes the bulk. In the present model, the ΔE_{vac} is the consequence of difference between the substrate WF and the film WF, i.e., the driving mechanism for the ELA is determined by the surface properties of both materials forming the contact. The view is intuitively consistent, if one considers that the contact is formed by two *surfaces* and thus the interfacial properties should be determined by means of parameters describing the surfaces brought into the contact.

In this context, a comment on the picture of the interfacial dipole is necessary. According to the common concept [25, 66], the ID is formed at the interface between the substrate and the adsorbed layer and the ID is saturated after the 1-3 ML formation. The ID is detected via the vacuum level change, and that does not change with the film thickness (unless due to the varying molecular orientation discussed earlier in this review). Let`s discuss the validity of the ID model for a typical metal-on-semiconductor contacts, such as the Pt on silicon introduced earlier. The ID is formed at the onset of the interface formation and with the further growth of the metal film the magnitude of vacuum level saturates at the VL corresponding to the thick metal film translated as its WF. The picture is apparently inconsistent, as the VL of the metal layer, i.e., the surface property, continues to reflect the interface properties between the metal layer and the substrate. Thereby, a justifiable question arises, that is, how the spectrometer can detect the ID buried under the thick metal layer or, in other words, why the interface property manifested by the vacuum level change is not screened out by the thick overlayer. To reconcile the raised inconsistency, we propose that the VL change upon the molecular film growth is not due to the interfacial dipole, but it is rather induced by the surface dipole of the overlayer, which can be understood as the dipole at the layer-vacuum interface [113]. Obviously, the surface dipole arises already with the first monolayer, however, there is a minimal thickness—exceeding the monolayer thickness—specific for each particular material necessary to evolve the SD inherent to that of a bulk material. This also implies that the interfacial properties between the non-interacting substrate and the film do not govern the vacuum level changes, just the opposite: the electronic properties of the formed interface such as the BA are derived from the WF differences of two *surfaces* forming the contact.

3.5. Band Bending

The terminology and interpretation of observed phenomena in organic semiconductors have mimicked concepts developed for the inorganic semiconductors in many aspects, while the underlying mechanism in conjugated molecular films is often different. The operation principle of the Metal-oxide-semiconductor (MOS) structure is based on a field effect employed for the control of free charge carrier concentration in the surface region of the semiconductor. This is achieved via surface potential modulation by an external field. The varying surface potential is also understood in terms of the bending of energy bands at the semiconductor surface, i.e., controlling the Fermi level position with respect to the energy bands of the semiconductor. Depending on the sign of the band bending (BB), depletion or accumulation of free charge carriers occurs.

Since an organic field-effect transistor (OFET) was successfully demonstrated about two decades ago [114, 115], many photoemission studies on organic film growth have reported BB in molecular films observed by photoemission (see, e.g., Refs. [8, 86, 102, 116-118]). The BB was deduced from the high-binding energy shift of photoemission features with the increased film thickness. The shifted HOMO for four thicknesses suggesting the upward BB is indicated in Fig. 16(a). Here, the $\Delta\phi$ indicates the vacuum level change due to the first monolayer with the thickness δ, and BE_i (i = 1..4) are the binding energies of the HOMO observed at the film thickness of t_i. The interpretations have presumed that the situation is equivalent to that occurring at the surface or interface of inorganic semiconductor shown in Fig. 16(b). However, there is an essential difference between the both situations in terms of their characterizations sketched at the bottom of the figures; whereas the BB in the inorganic semiconductor is evaluated by a single measurement, the position of photoemission features for the molecular film is measured by means of set of measurements during the film growth, whereas the constancy of the WF is not fulfilled; beyond the vacuum level change $\Delta\phi_{1stML}$ upon the first monolayer growth, the vacuum level is not constant (indicated by $E_{VAC,1stML}$), but it varies according to $E_{VAC,\ film}$ with the film thickness [Fig. 16(a)] owing to the varying molecular orientation.

The binding energy E_B of a core level is given by the equation [120]

$$E_B = h\nu - U_{an} - \phi_{an}, \tag{13}$$

where $h\nu$ is the energy of exciting radiation, E_B is the binding energy of a core level referred to the Fermi level, U_{an} is the voltage applied to the analyzer to detect an emitted electron, and ϕ_{an} is the work function of the energy analyzer. Even though Eq. (13) does not explicitly involve the WF of the substrate, the positions of photoemission features of an adsorbate depend on the substrate WF. This is shown in Fig. 17, where the principal energy diagrams for the photoemission characterization for an adsorbate represented by its core level E_B on two electronically distinct surfaces *1,2* with the work function $\phi_{S,1}$ and $\phi_{S,2}$ are contrasted. Since the ionization energy of the adsorbate IE = $E_B + \phi_s$ is constant on both surfaces, the analyzer detects the difference in the E_B equal to the work function difference. Thus, the high BE shift ΔE_B is equivalent to the work function difference $\Delta\phi_S$.

During the molecular film growth, the adsorbing molecules successively build particular layers. Note that (n-1)th represents a substrate for the nth layer. If the molecular orientation varies with the film thickness, the concomitant work function change is implied. Accordingly, the photoemission spectra of nth molecular layer will be rigidly shifted relative to the spectra of the (n-1)th layer by their work function difference. Eventually, the summed shift corresponds to the total work function change due to varying molecular orientation. Thus, the shift of the photoemission spectra is induced by the varying work function instead of the band bending [119].

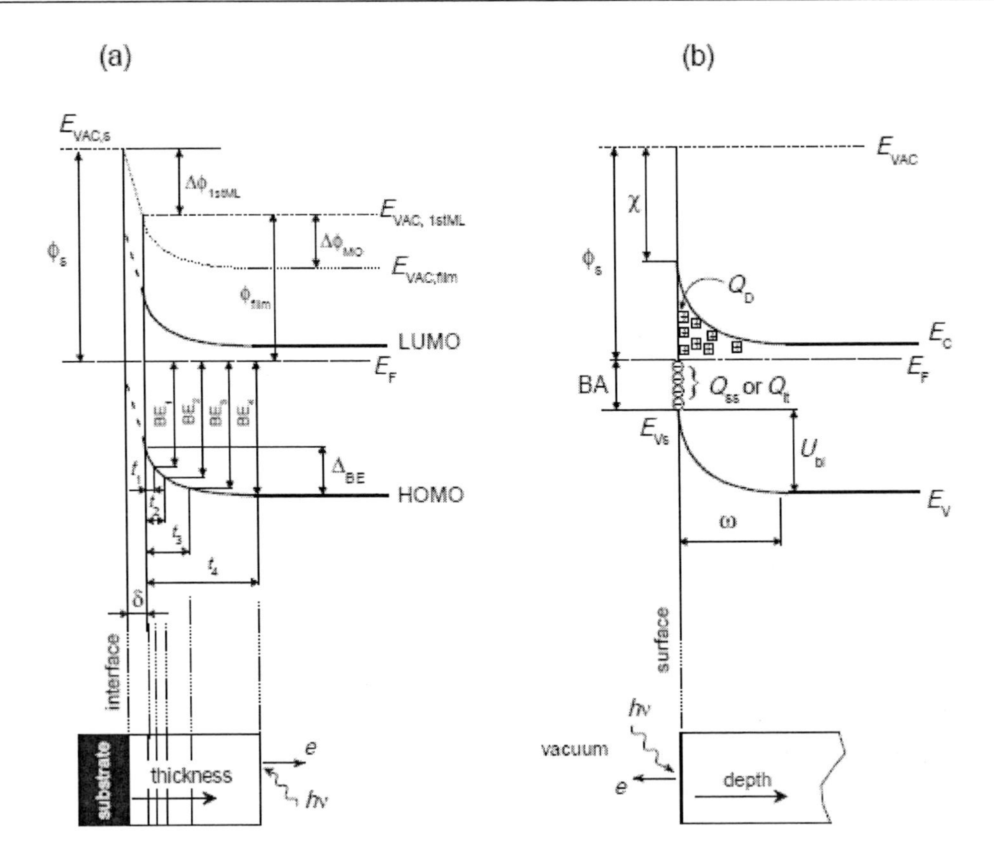

Figure 16. (a) The approach habitually employed for the deduction on the band bending in organic semiconductors. The ϕ_{subs} and ϕ_{film} are the substrate and the film work function, Δ and δ are the interface dipole and the thickness of the dipole layer, ΔBE is the total shift of energy levels interpreted as the band bending, BE_i (i=1...4) are the HOMO positions observed for the film thickness, t_i. (Ref. [119]). (b) Energy diagram an n-type semiconductor with upward band bending near the surface; the ϕ_s is the semiconductor work function, χ is the electron affinity, Q_{ss} and Q_{it} are the charge in surface and/or interface states, respectively, Q_D is the depletion zone charge, w is the depletion zone width, U_{bi} is the build-in potential (i.e., the band bending magnitude), and the BA is the band alignment, i.e., the binding energy of the valence band edge on the surface, E_{Vs}. The bottom pictograms sketch the experimental set ups.

The plausible examples confirming the effect of the substrate work function on the photoemission spectra can be found in Ref. [25], where ZnTPP on various metal (Mg, Ag, Au, Al) substrates were examined; upon the oxygen exposure, the ZnTPP spectra shifted by amount equal to the expected work function changes due to the oxidation of the corresponding substrate. Koller et al. [121] examined bithiophene on nanoscopically patterned substrates; the photoemission spectra of bithiophene were aligned according to local work function of the substrate.

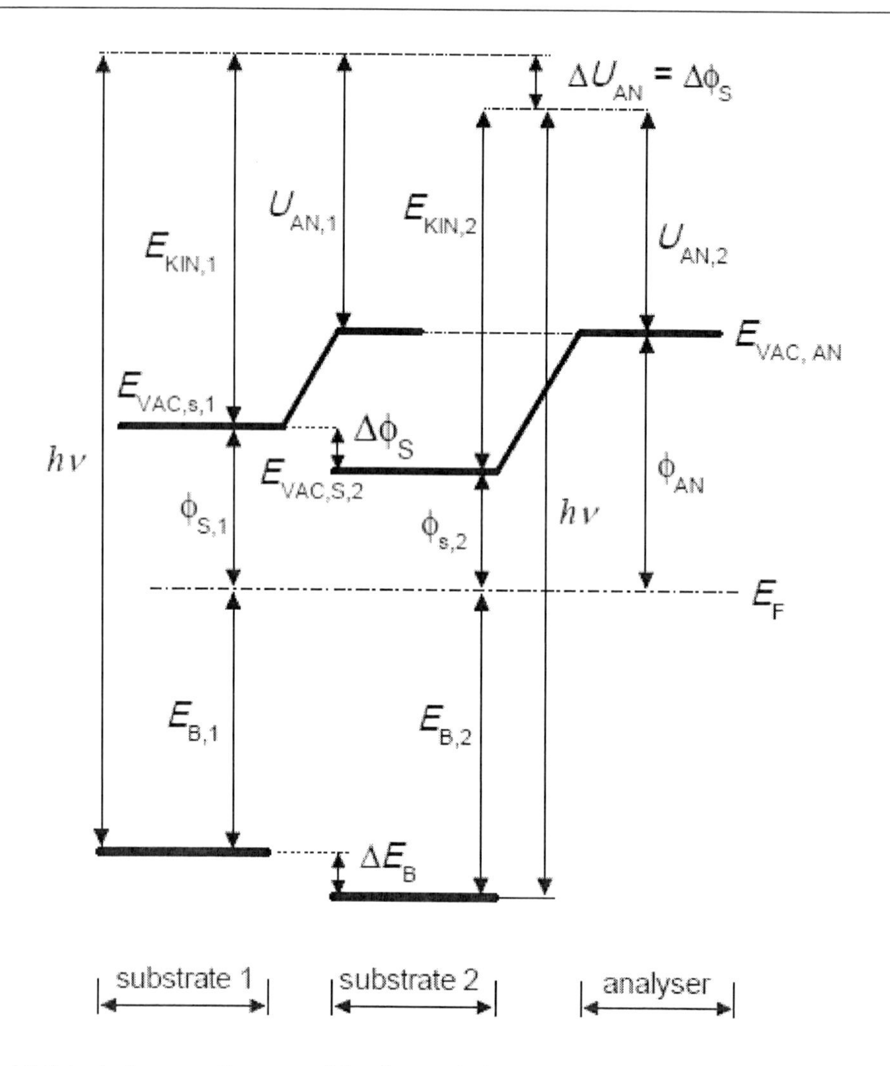

Figure 17. Principal energy diagrams of the photoemission measurement contrasting an adsorbate on surfaces with work function $\phi_{s,1}$ and $\phi_{s,2}$ (see text).

The spurious BB can be observed at organic-organic interfaces as well. Lau et al. [102] grew CuPc/F_{16}CuPc heterostructures prepared by both deposition sequences. They observed the low-BE shift of F_{16}CuPc-related features with the F_{16}CuPc growth over CuPc and the high-BE shift of CuPc–related features with the CuPc growth over F_{16}CuPc. The low- and high-BE shifts were interpreted as the opposite band bending occurring within the thickness range of 15 nm. This has allowed calculating the charge carrier density, which was estimated to be of about 10^{18} cm^{-3}. Given the earlier discussion, the both the up- and down-band bending are presumably artifacts due to varying WF; irrespective the substrate (bottom organic) WF, the WF of the top film has converged towards its intrinsic WF with the increased thickness. Since the intrinsic WF of F_{16}CuPc is larger than the intrinsic WF of the CuPc (Table 2), the CuPc/F_{16}CuPc structure displays the gradual WF decrease with the CuPc film thickness, while the WF of the F_{16}CuPc/CuPc structure will gradually increase with the F_{16}CuPc thickness. The WF decrease and increase implicate the high- and low-BE shifts, which were mistakenly interpreted as the band bending.

CONCLUSION

Electronic properties of molecular films and associated interfaces examined by the photoemission have been reviewed. The fundamental electronic parameters of molecular films such as the ionization energy and the work function have been distinguished in terms of the molecular orientation. Similar to elemental surfaces, which can be recognized by their work function, molecular films have been demonstrated to display the intrinsic work function, too. The importance of the work function evaluation for the characterization of molecular films has been accentuated; on the one hand, the work function evaluation allows the determination of the ionization energy of the molecular film, on the other hand, the difference between the work functions of the substrate and the overlayer was demonstrated to govern the energy level alignment at the interface and thereby the charge injection barrier.

ACKNOWLEDGMENT

The financial support from MNT-ERA-Net 2007-009-SK project, Grant Agency VEGA Bratislava (project No. 2/0041/11), and from ASFEU Bratislava (project No. 26240220011 OPVaV-2008/4.2/01-SORO) is acknowledged.

REFERENCES

[1] Stadler, P.; Track, A.M.; Ullah, M.; Sitter, H.; Matt, G. J.; Koller, G.; Singh, T.B.; Neugebauer, H.; Serdar Sariciftci, N.; Ramsey, M.G. *Org. Electron*. 2010, 11, 207-211.

[2] Kahn, A.; Koch, N.; Gao, W. *J. Polym. Sci. B: Polym. Phys*. 2003, 41, 2529-2548.

[3] Ivanco, J.; Winter, B.; Netzer, F.P.; Ramsey, M.G. *Adv. Mater*. 2003, 15, 1812-1815.

[4] Liu, X.; Knupfer, M.; Huisman, B.H. *Surf. Sci.* 2005, 595, 165-171.

[5] Zahn, D.R.T.; Gavrila, G.N.; Gorgoi, M. *Chem. Phys*. 2006, 325, 99-112.

[6] Peisert, H.; Knupfer, M.; Fink, *J. Surf. Sci.* 2002, 515, 491-498.

[7] Koch, N.; Vollmer, A.; Duhm, S.; Sakamoto, Y.; Suzuki, T. *Adv. Mater*. 2007, 19, 112-116.

[8] Petraki, F.; Papaefthimiou, V.; Kennou, S. *Org. Electron*. 2007, 8, 522-528.

[9] Casu, M.B.; Imperia, P.; Wong, J.E.; Schrader, S. *Synth. Met*. 2003, 138, 131-134.

[10] Forsythe, E.W.; Choong, V.-E.; Le, T.Q.; Gao, Y. *J. Vac. Sci. & Technol. A* 1999, 17, 3429-3432.

[11] Gao, W.; Kahn, A. *Appl. Phys. Lett* 2003, 82, 4815-4817.

[12] Gao, Y.; Ding, H.; Wang, H.; Yan, D. *Appl. Phys. Lett* 2007, 91, 142112-3.

[13] Sakurai, T.; Toyoshima, S.; Kitazume, H.; Masuda, S.; Kato, H.; Akimoto, K. *J. Appl. Phys*. 2010, 107, 043707-6.

[14] Ivanco, J.; Netzer, F.P.; Ramsey, M.G. *J. Appl. Phys*. 2007, 101, 103712-7.

[15] Huang, Y.L.; Chen, W.; Huang, H.; Qi, D. C.; Chen, S.; Gao, X.Y.; Pflaum, J.; Wee, A.T.S. *J. Phys. Chem.* C 2009, 113, 9251-9255.

[16] Toader, T.; Gavrila, G.; Braun, W.; Ivanco, J.; Zahn, D.R.T. *Phys. Stat. Sol. (b)* 2009, 246, 1510-1518.

[17] Ueno, N.; Kera, S. *Progr. Surf. Sci.* 2008, 83, 490-557.
[18] Kera, S.; Yamane, H.; Ueno, N. *Progr. Surf. Sci.* 2009, 84, 135-154.
[19] Yamane, H.; Nagamatsu, S.; Fukagawa, H.; Kera, S.; Friedlein, R.; Okudaira, K.K.; Ueno, N. *Phys. Rev. B* 2005, 72, 153412-4.
[20] Berkebile, S.; Puschnig, P.; Koller, G.; Oehzelt, M.; Netzer, F.P.; Ambrosch-Draxl, C.; Ramsey, M.G. *Phys. Rev. B* 2008, 77, 115312-5.
[21] Koller, G.; Berkebile, S.; Oehzelt, M.; Puschnig, P.; Ambrosch-Draxl, C.; Netzer, F.P.; Ramsey, M.G. *Science* 2007, 317, 351-355.
[22] Gavrila, G.N.; Mendez, H.; Kampen, T.U.; Zahn, D.R.T.; Vyalikh, D.V.; Braun, W. *Appl. Phys. Lett.* 2004, 85, 4657-4659.
[23] Gao, Y. *Mater. Sci. Engn.* R 2010, 68, 39-87.
[24] Helander, M.G.; Greiner, M.T.; Wang, Z.B.; Lu, Z.H. *Appl. Surf. Sci.* 2010, 255, 2602-2605.
[25] Narioka, S.; Ishii, H.; Yoshimura, D.; Sei, M.; Ouchi, Y.; Seki, K.; Hasegawa, S.; Miyazaki, T.; Harima, Y.; Yamashita, K. *Appl. Phys. Lett* 1995, 67, 1899-1901.
[26] Hill, I.G.; Rajagopal, A.; Kahn, A.; Hu, Y. *Appl. Phys. Lett* 1998, 73, 662-664.
[27] Rajagopal, A.; Wu, C.I.; Kahn, A. *J. Appl. Phys.* 1998, 83, 2649-2655.
[28] Ishii, H.; Sugiyama, K.; Ito, E.; Seki, K. *Adv. Mater.* 1999, 11, 26.
[29] Kera, S.; Yabuuchi, Y.; Yamane, H.; Setoyama, H.; Okudaira, K.K.; Kahn, A.; Ueno, N. *Phys. Rev. B* 2004, 70, 085304-6.
[30] Yamane, H.; Honda, H.; Fukagawa, H.; Ohyama, M.; Hinuma, Y.; Kera, S.; Okudaira, K.K.; Ueno, N. *J. Electron Spectr. Rel. Phenom.* 2004, 137-140, 223-227.
[31] Vázquez, H.; Flores, F.; Oszwaldowski, R.; Ortega, J.; Pérez, R.; Kahn, A. *Appl. Surf. Sci.* 2004, 234, 107-112.
[32] Vazquez, H.; Dappe, Y.J.; Ortega, J.; Flores, F. *J. Chem. Phys.* 2007, 126, 144703-8.
[33] Vazquez, H.; Gao, W.; Flores, F.; Kahn, A. *Phys. Rev. B* 2005, 71, 041306.
[34] Ivanco, J.; Krenn, J.R.; Ramsey, M.G.; Netzer, F. P.; Haber, T.; Resel, R.; Haase, A.; Stadlober, B.; Jakopic, G. *J. Appl. Phys.* 2004, 96, 2716-2724.
[35] Müllegger, S.; Winkler, A. *Surf. Sci.* 2006, 600, 1290-1299.
[36] De Renzi, V.; Rousseau, R.; Marchetto, D.; Biagi, R.; Scandolo, S.; del Pennino, U. *Phys. Rev. Lett.* 2005, 95, 046804.
[37] Hill, I. G.; Schwartz, J.; Kahn, A. *Org. Electron.* 2000, 1, 5-13.
[38] Bagus, P. S.; Staemmler, V.; Wöll, C. *Phys. Rev. Lett.* 2002, 89, 096104.
[39] Peisert, H.; Knupfer, M.; Fink, *J. Appl. Phys. Lett.* 2002, 81, 2400-2402.
[40] Hill, I. G.; Kahn, A.; Soos, Z. G.; Pascal, J. R. A. *Chem. Phys. Lett.* 2000, 327, 181-188.
[41] Yan, L.; Watkins, N. J.; Zorba, S.; Gao, Y.; Tang, C. W. *Appl. Phys. Lett* 2001, 79, 4148-4150.
[42] Blumenfeld, M. L.; Steele, M. P.; Ilyas, N.; Monti, O. L. A. *Surf. Sci.* 2010, 604, 1649-1657.
[43] Hill, I. G.; Rajagopal, A.; Kahn, A. *J. Appl. Phys.* 1998, 84, 3236-3241.
[44] Watkins, N. J.; Yan, L.; Gao, Y. *Appl. Phys. Lett* 2002, 80, 4384-4386.
[45] Yan, L.; Watkins, N. J.; Zorba, S.; Gao, Y.; Tang, C. W. *Appl. Phys. Lett* 2002, 81, 2752-2754.
[46] Koch, N. *J. Phys.: Condens. Matter* 2008, 20, 12.

[47] Fukagawa, H.; Kera, S.; Kataoka, T.; Hosoumi, S.; Watanabe, Y.; Kudo, K.; Ueno, N. *Adv. Mater.* 2007, 19, 665-668.
[48] Park, S.; Kampen, T.U.; Zahn, D.R.T.; Braun, W. *Appl. Phys. Lett.* 2001, 79, 4124-4126.
[49] Krause, S.; Casu, M.B.; Schöll, A.; Umbach, E. *New J. Phys*. 2008, 10, 16.
[50] Sato, N.; Seki, K.; Inokuchi, H. *J. Chem. Soc., Faraday Trans.* 2 1981, 77, 1621-1633.
[51] Salaneck, W.R. *Phys. Rev. Lett.* 1978, 40, 60-63.
[52] Duhm, S.; Hosoumi, S.; Salzmann, I.; Gerlach, A.; Oehzelt, M.; Wedl, B.; Lee, T.-L.; Schreiber, F.; Koch, N.; Ueno, N.; Kera, S. *Phys. Rev. B* 2010, 81, 045418.
[53] Alloway, D.M.; Graham, A.L.; Yang, X.; Mudalige, A.; Colorado, R.; Wysocki, V.H.; Pemberton, J.E.; Randall Lee, T.; Wysocki, R.J.; Armstrong, N.R. *J. Phys. Chem. C* 2009, 113, 20328-20334.
[54] Chandrasekhar, M.; Guha, S.; Graupner, W. *Adv. Mater*. 2001, 13, 613-618.
[55] Braun, K.-F.; Hla, S.-W. *Nano Lett.* 2005, 5, 73-76.
[56] Duhm, S.; Heimel, G.; Salzmann, I.; Glowatzki, H.; Johnson, R.L.; Vollmer, A.; Rabe, J.P.; Koch, N. *Nature Mater*. 2008, 7, 7.
[57] Ellis, T.S.; Park, K.T.; Hulbert, S.L.; Ulrich, M.D.; Rowe, J.E. *J. Appl. Phys*. 2004, 95, 982-988.
[58] Chen, W.; Qi, D.C.; Huang, Y.L.; Huang, H.; Wang, Y.Z.; Chen, S.; Gao, X.Y.; Wee, A.T.S. *J. Phys. Chem. C* 2009, 113, 12832-12839.
[59] Gorgoi, M.; Zahn, D.R.T. *Org. Electron*. 2005, 6, 168-174.
[60] Fukagawa, H.; Yamane, H.; Kataoka, T.; Kera, S.; Nakamura, M.; Kudo, K.; Ueno, N. *Phys. Rev. B* 2006, 73, 245310-5.
[61] Mott, N.F. *Proc. R. Soc*. London 1939, 171.
[62] Schottky, W. *Phys. Z.* 1940, 41.
[63] Hill, I.G.; Milliron, D.; Schwartz, J.; Kahn, A. *Appl. Surf. Sci.* 2000, 166, 354-362.
[64] Lee, S.T.; Hou, X.Y.; Mason, M.G.; Tang, C.W. *Appl. Phys. Lett.* 1998, 72, 1593-1595.
[65] Koller, G.; Blyth, R. I. R.; Sardar, S. A.; Netzer, F. P.; Ramsey, M. G. *Appl. Phys. Lett.* 2000, 76, 927-929.
[66] Tung, R.T. *Phys. Rev. Lett*. 2000, 84, 6078-6081.
[67] Yanagi, H.; Okamoto, S. *Appl. Phys. Lett*. 1997, 71, 2563-2565.
[68] Peisert, H.; Schwieger, T.; Auerhammer, J.M.; Knupfer, M.; Golden, M.S.; Fink, J.; Bressler, P.R.; Mast, M. *J. Appl. Phys*. 2001, 90, 466-469.
[69] Ivanco, J.; Haber, T.; Krenn, J.R.; Netzer, F.P.; Resel, R.; Ramsey, M.G. *Surf. Sci.* 2007, 601, 178-187.
[70] Okajima, T.; Narioka, S.; Tanimura, S.; Hamano, K.; Kurata, T.; Uehara, Y.; Araki, T.; Ishii, H.; Ouchi, Y.; Seki, K.; Ogama, T.; Koezuka, H. *J. Electron Spectr. Rel. Phenom*., 1996, 78, 379-382.
[71] Chen, W.; Huang, H.; Chen, S.; Chen, L.; Zhang, H.L.; Gao, X.Y.; Wee, A.T.S. *Appl. Phys. Lett*. 2007, 91, 114102-3.
[72] Winter, B.; Ivanco, J.; Netzer, F.P.; Ramsey, M.G. *Thin Solid Films* 2003, 433, 269-273.
[73] Kiguchi, M.; Yoshikawa, G.; Saiki, K. *J. Appl. Phys*. 2003, 94, 4866-4870.
[74] Duhm, S.; Glowatzki, H.; Rabe, J.P.; Koch, N.; Johnson, R.L. *Appl. Phys. Lett* 2006, 88, 203109.

[75] Toader, T.; Gavrila, G.; Ivanco, J.; Braun, W.; Zahn, D.R.T. *Appl. Surf. Sci.* 2009, 255, 3.

[76] Kowarik, S.; Gerlach, A.; Sellner, S.; Schreiber, F.; Cavalcanti, L.; Konovalov, O. *Phys. Rev. Lett.* 2006, 96, 125504-4.

[77] Peisert, H.; Biswas, I.; Zhang, L.; Knupfer, M.; Hanack, M.; Dini, D.; Cook, M.J.; Chambrier, I.; Schmidt, T.; Batchelor, D.; Chassé, T. *Chem. Phys. Lett.* 2005, 403, 1-6.

[78] Heiner, C.E.; Dreyer, J.; Hertel, I.V.; Koch, N.; Ritze, H.H.; Widdra, W.; Winter, B. *Appl. Phys. Lett.* 2005, 87, 093501-3.

[79] Glowatzki, H.; Gavrila, G. N.; Seifert, S.; Johnson, R.L.; Rader, J.; Mullen, K.; Zahn, D.R.T.; Rabe, J.P.; Koch, N. *J. Phys. Chem.* C 2008, 112, 1570-1574.

[80] Haber, T.; Muellegger, S.; Winkler, A.; Resel, R. *Phys. Rev. B* 2006, 74, 045419-9.

[81] Kline, R.J.; McGehee, M.D.; Toney, M.F. *Nature Mater.* 2006, 5, 222-228.

[82] Oehzelt, M.; Koller, G.; Ivanco, J.; Berkebile, S.; Haber, T.; Resel, R.; Netzer, F.P.; Ramsey, M.G. *Adv. Mater.* 2006, 18, 2466-2470.

[83] Koller, G.; Berkebile, S.; Ivanco, J.; Netzer, F.P.; Ramsey, M.G. *Surf. Sci.* 2007, 601, 5683-5689.

[84] Oehzelt, M.; Leonhard Grill; Berkebile, L.S.; Koller, G.; Netzer, F.P.; Ramsey, M.G. *Chem Phys Chem* 2007, 8, 1707-1712.

[85] Cerminara, M.; Tubino, R.; Meinardi, F.; Ivanco, J.; Netzer, F.P.; Ramsey, M.G. *Thin Solid Films* 2008, 516, 4247-4251.

[86] Peisert, H.; Knupfer, M.; Schwieger, T.; Auerhammer, J.M.; Golden, M.S.; Fink, J. *J. Appl. Phys.* 2002, 91, 4872-4878.

[87] Yamane, H.; Yabuuchi, Y.; Fukagawa, H.; Kera, S.; Okudaira, K.K.; Ueno, N. *J. Appl. Phys.* 2006, 99, 093705-5.

[88] Ivanco, J.; Haber, T.; Resel, R.; Netzer, F.P.; Ramsey, M.G. *Thin Solid Films* 2006, 514, 156-164.

[89] Ge, Y.; Whitten, J.E. *J. Phys. Chem.* C 2008, 112, 1174-1182.

[90] *Handbook of Chemistry and Physics*; Lide, D. R.; 75th Ed.; CRC Press: Boca Raton, 1994; pp 12-113.

[91] Paasch, G.; Peisert, H.; Knupfer, M.; Fink, J.; Scheinert, S. *J. Appl. Phys.* 2003, 93, 6084-6089.

[92] Hill, I.G.; Makinen, A.J.; Kafafi, Z.H. *Appl. Phys. Lett* 2000, 77, 1825-1827.

[93] Chen, W.; Huang, H.; Chen, S.; Gao, X.Y.; Wee, A.T.S. *J. Phys. Chem. C* 2008, 112, 5036-5042.

[94] Chen, W.; Huang, H.; Chen, S.; Huang, Y.L.; Gao, X.Y.; Wee, A.T.S. *Chem. Mater.* 2008, 20, 7017-7921.

[95] Michaelson, H.B. *J. Appl. Phys.* 1977, 48, 4729-4733.

[96] Sze, S.M.; Kwok, K.Ng. *Physics of Semiconductor Devices*; Wiley: New York; 2007.

[97] *CRC Handbook of Chemistry and Physics*; Weast, R. C.; 65th Ed.; CRC Press: Boca Raton, FL, 1984-1985.

[98] Keister, J.W.; Rowe, J.E.; Kolodziej, J.J.; Madey, T.E. *J. Vac. Sci. Technol.* B 2000, 18, 2174-2178.

[99] Hornauer, H.; Vancea, J.; Reiss, G.; Hoffmann, H.Z. *Phys. B - Conden. Mat.* 1989, 77, 399-407.

[100] Watkins, N.J.; Gao, Y. *J. Appl. Phys.* 2003, 94, 1289-1291.

[101] Koch, N.; Gerlach, A.; Duhm, S.; Glowatzki, H.; Heimel, G.; Vollmer, A.; Sakamoto, Y.; Suzuki, T.; Zegenhagen, J.; Rabe, J.P.; Schreiber, F. *J. Am. Chem. Soc.* 2008, 130, 7300-7304.

[102] Lau, K.M.; Tang, J.X.; Sun, H.Y.; Lee, C.S.; Lee, S.T.; Yan, D. *Appl. Phys. Lett.* 2006, 88, 173513-3.

[103] Seki, K.; Hayashi, N.; Oji, H.; Ito, E.; Ouchi, Y.; Ishii, H. *Thin Solid Films* 2001, 393, 298-303.

[104] Knupfer, M.; Peisert, H. *Phys. Stat. Sol.* (a) 2004, 201, 1055-1074.

[105] Shen, C.; Kahn, A. *J. Appl. Phys.* 2001, 90, 4549-4554.

[106] Watkins, N.J.; Gao, Y. *J. Appl. Phys.* 2003, 94, 5782-5786.

[107] Salzmann, I.; Duhm, S.; Opitz, R.; Johnson, R.L.; Rabe, J.P.; Koch, N. *J. Appl. Phys.* 2008, 104, 114518-11.

[108] Rangger, G.M.; Hofmann, O. T.; Romaner, L.; Heimel, G.; Broker, B.; Blum, R.-P.; Johnson, R.L.; Koch, N.; Zojer, E. *Phys. Rev. B* 2009, 79, 165306-12.

[109] Molodtsova, O.V.; Schwieger, T.; Knupfer, M. *Appl. Surf. Sci.* 2005, 252, 143-147.

[110] Kahn, A.; Zhao, W.; Gao, W.; Vázquez, H.; Flores, F. *Chem. Phys.* 2006, 325, 129-137.

[111] Blumenfeld, M.L.; Steele, M.P.; Monti, O.L.A. *J. Phys. Chem. Lett.* 2010, 1, 145-148.

[112] Fukagawa, H.; Yamane, H.; Kera, S.; Okudaira, K.K.; Ueno, N. *Phys. Rev. B* 2006, 73, 041302-4.

[113] Wandelt, K. *Appl. Surf. Sci.* 1997, 111, 1-10.

[114] Horowitz, G.; Fichou, D.; Peng, X.; Xu, Z.; Garnier, F. *Solid State Commun.* 1989, 72, 381-384.

[115] Servet, B.; Horowitz, G.; Ries, S.; Lagorsse, O.; Alnot, P.; Yassar, A.; Deloffre, F.; Srivastava, P.; Hajlaoui, R.; Lang, P.; Garnier, F. *Chem. Mater.* 1994, 6, 1809-1815.

[116] Schroeder, P.G.; France, C.B.; Parkinson, B.A.; Schlaf, R. *J. Appl. Phys.* 2002, 91, 9095-9107.

[117] Papaefthimiou, V.; Siokou, A.; Kennou, S. *J. Appl. Phys.* 2002, 91, 4213-4219.

[118] Lu, B.; Zhang, H.J.; Li, H.Y.; Bao, S.N.; He, P.; Hao, T.L. *Phys. Rev. B* 2003, 68, 125410.

[119] Ivanco, J.; Zahn, D.R.T. *J. Vac. Sci. Technol.* A 2009, 27, 1178-1182.

[120] Ertl, G.; Kuppers, J. *Low Energy Electrons and Surface Chemistry*; VCH: Weinheim, 1985.

[121] Koller, G.; Winter, B.; Oehzelt, M.; Ivanco, J.; Netzer, F.P.; Ramsey, M.G. *Org. Electron.* 2007, 8, 63-68.

In: Organic Semiconductors
Editors: Maria A. Velasquez
ISBN: 978-1-61209-391-8

Chapter 3

SOLUTION PROCESSED POLYMERS: PROPERTIES, FABRICATION AND APPLICATIONS

Harshil N. Raval[1, 2], Ravishankar S. Dudhe[1, 2], V. Seena[1, 2], Ramesh R. Navan[1, 2], Anil Kumar[2, 3], and V. Ramgopal Rao[1, 2]
[1]Department of Electrical Engineering, IIT Bombay, Powai, Mumbai, India
[2]Centre of Excellence in Nanoelectronics, IIT Bombay, Powai, Mumbai, India
[3]Department of Chemistry, IIT Bombay, Powai, Mumbai, India

1. INTRODUCTION

Solution processed polymer materials are extensively used at various stages of micro/nano fabrication processes either as structural materials or as sacrificial layers like photoresists, release layers etc. due to their special chemical, electrical, thermal and mechanical properties catering to different applications. With the discovery and development of highly conductive organic polymers ("polyacetylene" class) in 1977 by Alan G. MacDiarmid, Hideki Shirakawa, and Alan J. Heeger a new area of research came into existence - polymer electronics. For this revolutionary discovery they were jointly awarded the 2000 Nobel Prize in Chemistry. The electronic properties of these functional polymers are derived from their chemical structure, which contains the so-called "conjugated polymer chains", consisting of a strictly alternating sequence of single and double bonds resulting in delocalized electrons giving conducting/semiconducting properties to these organic materials [1 - 4]. The conducting/semiconducting properties of these materials can be controlled by adding/removing different functional groups to them. Various researchers all over the world have shown tremendous interest in using these organic semiconducting materials as active semiconductors in different organic electronic components and circuits. Results of all such activities have classified the organic semiconductor based electronic area into four broad categories:

- Organic Light Emitting Diodes
- Organic Photo Voltaic Cells

- Organic Field Effect Transistors
- Organic Semiconductor based Sensors

Research in the field of organic light emitting diodes (OLED) has been attracting various researchers since these materials have advantages such as superior performance in the aspects of contrast, thinness, lightness, power consumption, response speed, viewing angle, and cost [5 - 8]. At the same time, need for sustainable energy resources attracted researchers to the polymer conducting/semiconducting materials for organic photovoltaic applications with the advantages being the cost and low temperature processing [9, 10]. To facilitate backbone of the display and circuits for various other applications such as radio frequency identification (RF-ID) tags, signal processing circuits etc. organic field effect transistor (OFET) have been explored by various researchers worldwide [2, 11 - 15] for a variety of applications [16 - 18]. The application of organic materials for low voltage ambipolar applications has also been reported in literature [19 - 21]. Use of pentacene OFETs on flexible substrates with polymer dielectrics has also been demonstrated [22]. Moreover, the field of intelligent sensing using various solution processed organic conducting/semiconducting materials is of interest to various researchers all over the world. Sensors based on these materials are used in various applications involving biomedical applications [23, 24], chemical sensing [25], explosive vapor sensing [26 - 30], toxic gases such as NO_2 [31], humidity sensing [32] and ionizing radiation detection [33 - 35]. Another class of sensors namely "microcantilevers" have also been demonstrated using solution processed polymers such as SU8 [36, 37]. SU-8, the most commonly used polymer structural material in MEMS, is an epoxy based polymer developed by IBM. SU-8 being a high aspect ratio negative photoresist with its ability in defining layers with thickness in the range of few hundreds of nanometers to millimeters, its usage within the MEMS field has been exponentially growing during the last couple of years. SU-8 has a Young's modulus 40 times lower than that of silicon which makes it the most appropriate structural material for highly sensitive microcantilever platforms.

2. Simulation and Fabrication of Organic Electronic Circuits

Among various organic semiconductor materials currently in use, solution processed poly 3-hexylthiophene (P3HT) and vacuum evaporated pentacene have been the most extensively studied p-type organic semiconducting materials [38, 39]. Most of the work reported in the literature is based on these p-type organic semiconductors for OFETs operating at a higher range of operating voltages [15, 40]. To realize organic electronic circuits for applications such as RF-ID tags, etc. low operating voltage OFETs and circuits with inorganic high-k dielectrics as gate dielectric have also been demonstrated recently [16, 20]. Moreover, most of the results in the literature for OFETs are with common bottom gate (generally a highly doped silicon wafer) with thermally grown SiO_2 as the gate dielectric for OFETs, but for making the circuits, the gate has to be isolated and patterned for interconnections. In this section, patterned gate OFETs, inverters and bootstrapped inverters for higher output swing are discussed with high-k gate dielectrics for low voltage operation of these devices.

As transport phenomenon in organic semiconductors is still not fully understood, no reliable standard physical models exist for them, which can be used for device/circuit simulations. In this section two possible approaches for simulating organic circuits are addressed. One of the possible approaches is the extraction of the equivalent SPICE parameters by using the particle swarm optimization (PSO) algorithm for the device matching and using the extracted parameters for basic organic circuit simulations in SPICE circuit simulator. Another approach is the use of SEQUEL circuit simulator developed at Indian Institute of Technology, Bombay, Mumbai, India [41]. It facilitates circuit simulation using look-up table (LUT) approach. An LUT is nothing but the device simulation results extracted and interpolated to give a large set of points. The table contains drain current, gate charge, drain charge values against gate and drain voltages which are obtained here by measurements on fabricated OFETs. This LUT is now the device model that can be used with SEQUEL for simulating circuits wherein we can define different widths for different transistors for simulating circuits.

2.1. Fabrication of P3HT Based OFETs and Extraction of Parameters for Simulation Model

Patterned gate OFETs were fabricated on a silicon wafer with 500 nm thick silicon dioxide (SiO_2) acting as the substrate. A thin layer (~80 nm) of aluminium (Al) was deposited by thermal evaporation and patterned for gate electrode by optical photolithography and etching of Al. To realize bottom gate bottom contacts (BGBC) structure a layer of HfO_x (45 nm thickness) was deposited by RF sputtering at room temperature and patterned using lift-off method. Cr/Au layer was patterned for the source-drain electrodes (10 nm/50 nm) and interconnects. P3HT solution (3 mg/ml in Chloroform) was spin-coated (500 rpm for 10 seconds followed by 1000 rpm for 40 seconds) to form organic semiconducting layer. The devices were annealed at 90 °C for 60 mins in ambient before characterization. The schematic of fabricated BGBC configuration OFETs based on P3HT is as shown in Fig. 1.

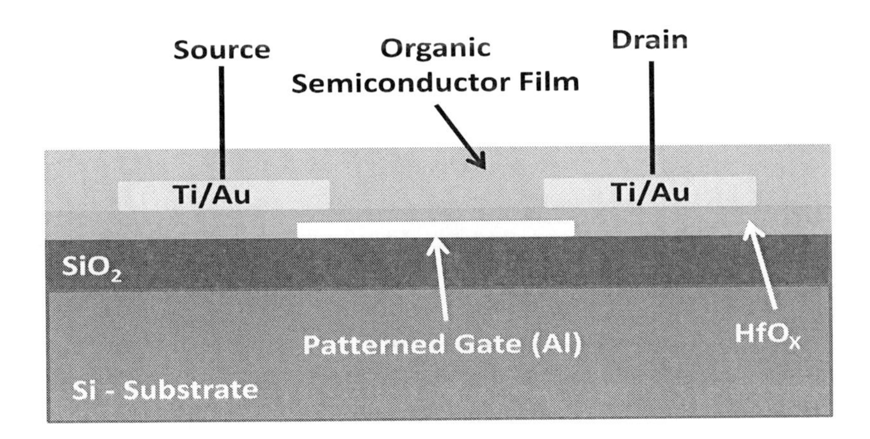

Figure 1. OFET schematic structure in Bottom Gate Bottom contact configuration with Al - patterned gate, HfO_X as high – *k* gate dielectric and Ti/Au as source and drain contacts, using thick SiO_2 grown silicon-wafer as a mechanical support.

The devices were characterized in normal ambient environment on a probe station using a Keithley 4200 SCS setup by measuring the output characteristics (I_{DS} vs. V_{DS} at constant V_{GS}), and the transfer characteristics (I_{DS} vs. V_{GS} at constant V_{DS}) for each device. Fig. 2 and 3 show the output and transfer characteristics of the fabricated OFETs respectively, with HfO_x as gate dielectric allowing lower operating voltages, with V_{GS} varied up to -4V. Table I lists the various parameters for the device with extracted parameters from $I_{DS}^{1/2}$ -V_{GS} curve (Fig. 3) [19] which were used for preparing the simulation model.

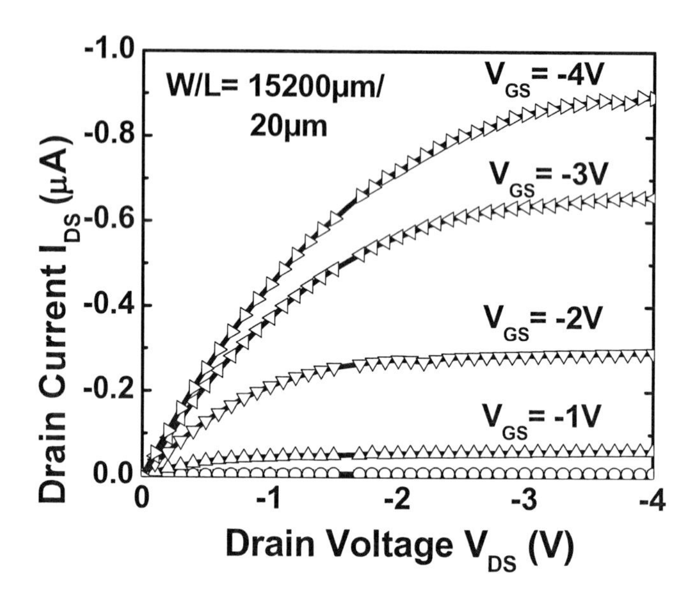

Figure 2. Output (I_{DS}-V_{DS}) characteristics of a typical patterned gate OFET fabricated with HfO_x as gate dielectric for low operating voltages

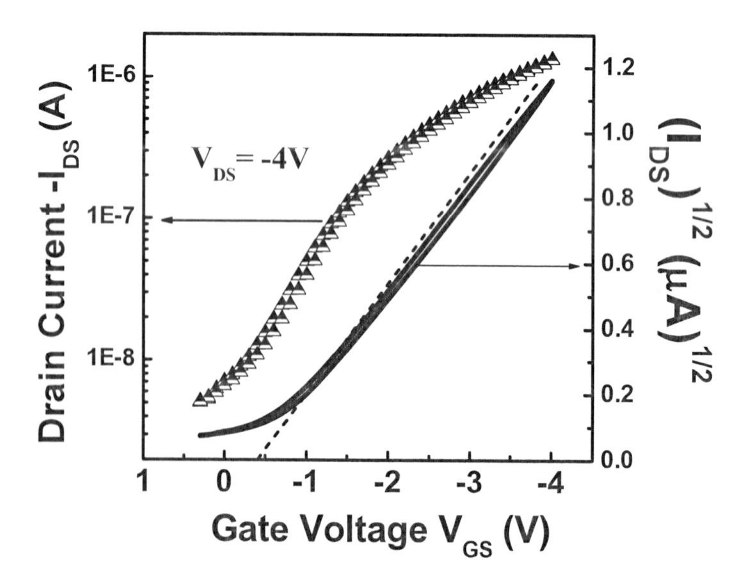

Figure 3. I_{DS}-V_{GS} (triangles) and $I_{DS}^{1/2}$ -V_{GS} (solid line) characteristics for a typical patterned gate OFET fabricated with with HfO_x gate dielectric (W/L=15200μm /20μm) for low operating voltages

The OFETs exhibit low on/off ratios because of two reasons. The first reason is that the roughness of these sputtered/deposited dielectrics on Al is always higher compared to the commonly used thermally grown SiO_2. Secondly, device characterizations in ambient also affects the device performance due to oxygen doping and moisture [38]. Low threshold voltages of these OFETs enable us to design low operating voltage circuits. To simulate various organic electronic circuits it is necessary to develop a model which fits these actual measured characteristics of the OFET and help in successful realization of various organic circuits by varying various design parameters and employing various design techniques.

Table I. Measured parameters for the fabricated device

Parameters	Values
W/L	15200/20 μm
M	2.33 x 10^{-3} cm^2/Vs
t_{ox}	45 nm
I_{ON}/I_{OFF}	200
V_{TH}	-0.7 V

2.2. OFET Simulation Using SPICE Based Approach

Since SPICE has built-in models for the standard silicon semiconductor devices, the user need to specify only the pertinent model parameter values. It has been discussed earlier that well-developed physical models are not available for organic transistors. Looking at this, MOS LEVEL 2 model has been chosen to capture the behavior of organic transistors. This model has 13 parameters, out of which six parameters listed in Table II were observed to be significant to capture the measured organic transistor data with MOS LEVEL model. The PSO algorithm, which does not require any initial guess for the parameters to be extracted, is one of the global optimization techniques shown for extraction of efficient parameters for MOSFET model [42], which has been used in this work. The PSO algorithm is based on real life phenomenon of birds finding their food [43]. In this algorithm, the range of each parameter to be extracted is specified, and in the beginning, many possible solutions for parameters are randomly generated within the range given by the user. Each possible solution is evaluated for its fitness (i.e. closeness to experimental data), and best possible solution is found out, which is called as a leader. The information of parameter values of the leader are used to update the parameter values of other candidates for better solution. In this process, algorithm may reach a globally optimum solution after certain number of iterations. Generally, this globally optimum solution provides a very good matching between experimental data and model results. These equivalent spice parameters which are extracted are used as device model parameters for circuit simulation in SPICE wherein we can define different widths and channel length depending on the required circuit configuration.

The match between the simulated and measured device characteristics was obtained within acceptable limits. Fig. 4 shows the measured vs. simulated characteristics for a particular device [44].

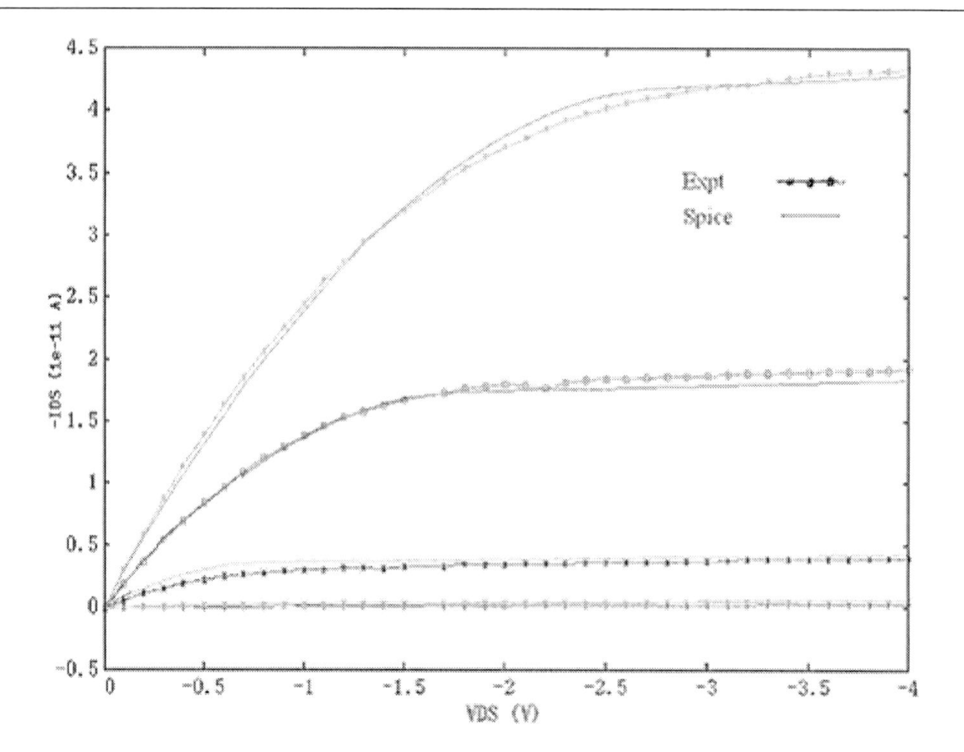

Figure 4. Comparison of $I_{DS} - V_{DS}$ Characteristics measured on a fabricated device vs. SPICE model based simulated $I_{DS} - V_{DS}$ Characteristics for patterned gate OFETs (W=1μ) with high-k gate dielectric.

Organic transistors being three terminal devices are easy to characterize for their DC parameters, but the C-V characterizations are not always straight forward because of the absence of a body terminal. So the possibility of LUT extraction from measured device data is ruled out. It is assumed that the device operates in a quasi-static regime, i.e., the terminal currents and charges depend only on the instantaneous values of the terminal voltages [45]. The charge transport in organic semiconductor thin films (Pentacene) follows a mechanism similar to that in amorphous silicon thin films. This mechanism is dominated by the presence of high number of traps in the semiconductor lattice [46], so amorphous silicon model can be used as a good approximation to organic thin film transistors (TFTs) if silicon material parameters in the model are replaced with those of appropriate organic semiconductor (like pentacene, P3HT, PQT etc.) parameters [47].

2.3. Low Operating Voltage Organic Inverters with P3HT

Based on this model, a p-type organic semiconductor based inverter was simulated for achieving the best performance by varying the width of the OFETs. An inverter with width ratio 4 : 1 for driver to load was found to be the best for achieving an optimum performance. An inverter with similar fabrication procedure to OFET was fabricated with $(W/L)_{M1}$= 24850μm/50μm and $(W/L)_{M2}$= 5050μm/50μm. Interconnects were patterned during gate and S/D electrodes layer fabrication.

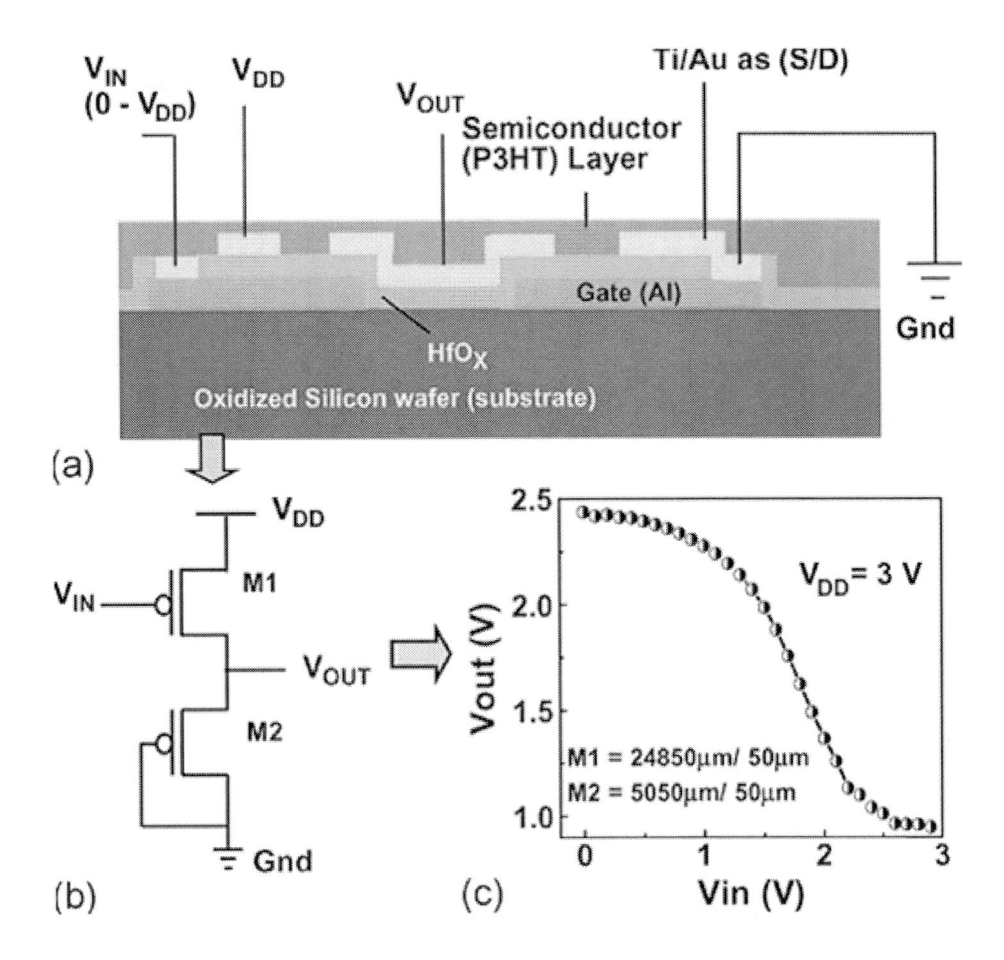

Figure 5. (a) Schematic corss-section of an Organic Inverter, (b) Circuit realizing organic inverter using two p-type OFETs acting as a driver and a load, and (c) DC transfer characteristics of the fabricated organic inverter with high-*k* gate dielectric for low voltage operation [16].

Fig. 5 (a) shows a schematic cross section of the inverter which was fabricated to realize OFET circuit for inverter realization shown in fig. 5 (b). The bias voltage V_{DD} was kept at 3V to get the output characteristics of the inverter and the input voltage V_{IN} was varied from 0 to 3V. The inverter shows a good switching behaviour with a DC gain (A_v) of -1.5 which is lower than the estimated gain value -1.93 through simulation as shown in the fig. 5 (c). The particular inverter shows a output high voltage (V_{OH}) and an output low voltage (V_{OL}) values of 2.4 and 0.9 V, respectively, when operated between 0 and 3 V with V_{DD} = 3V resulting in a low noise margin for the p-type only organic inverter. A good match between the actual device characteristics and characteristics based on the obtained model for simulation suggests that the model can now be used for simulations of bigger circuits based on P3HT based OFETs.

In the organic inverter, although the V_{OH} voltage of all-p-type inverter approaches toward V_{DD}, the output low voltage of these inverters does not approach 0 V due to the threshold voltage drop in the load transistor. To overcome this problem, the bootstrapping technique was suggested for a-Si:H and polycrystalline-Si circuits [48, 49]. This technique was also demonstrated on organic inverters with pentacene [50].

2.4. Improved Output Swing with Solution Processed Bootstrapped Organic Inverters

To improve the output swing, for the first time this design technique was employed with solution processed P3HT for bootstrapped organic inverters incorporating high-k gate dielectric for low operating voltage operation [51]. The device parameters were decided using this new design strategy (bootstrap technique) for better swing at the output based on the simulation results [44]. A bootstrap inverter circuit adds an OFET and a capacitor to the normal inverter with two OFETs. The extra OFET serves as a biasing means for the active load, and the capacitor is used as a bootstrapping feedback element for the active load. The circuit for bootstrapping inverter is shown in Fig. 6 (a). Here, C_p is the parasitic capacitance between the drain and gate electrodes of the M3 transistor and C_b is the bootstrap capacitor fabricated using gate layer electrode and S/D layer electrode separated by a gate dielectric in between. The efficiency of the bootstrapping is dependent on the ratio C_b / (C_b + C_p). The ideal ratio should be nearly 1 for better operation of the circuit. To achieve the ratio near to 1, the area of the electrodes of C_b is kept 100 times larger than the overlapping area of source and gate electrode of M3 transistor. The voltage across C_b capacitor keeps transistor M2 in the active state during the switching of input voltage at the gate of M1, helping the output to decrease to nearly 0 V, resulting in a full swing operation compared to a normal organic inverter.

Fig. 6 (b) shows the DC transfer characteristics of two p-type inverters with W/L_{M1} = 17350 μm / 50 μm and $W/L_{M2\ (\text{and M3 for bootstrapped inverter})}$ = 3550 μm / 50 μm, with a snapshot of the fabricated bootstrapped inverter as in inset. The Fig. 6 (b) includes DC transfer characteristics of a simple organic inverter with identical dimensions for load and driver OFETs comparing the result with a bootstrap inverter with an additional transistor with W/L_{M3} = 3550 μm / 50 μm, and bootstrapping capacitor C_b with area 2070 μm x 1000 μm. The characterization was done on a probe station with three SMU channels, two from Keithley 2602 and one from Keithley 236 and the transient response was measured with the help of a pulse generator HP S1101A and a DSO TDS 1002. The output characteristics shows excellent switching behavior with a dc voltage gain (Av) of -1.7 with V_{OH} and V_{OL} values of 3.3 V and 0.35 V respectively for a DC bias voltage, V_{DD} = 4V. V_{OH} values for simple inverter and bootstrapped inverter are identical which agrees with the theory but an improvement from 1.18 V to 0.35 V can be observed in V_{OH} improving the output swing from ~2 to ~3V.

To compare the dynamic performance of both the inverters, a pulse train of square wave was given as an input and the output was measured with high input impedance buffer circuit connected to the DSO. Transient response of a simple inverter and the bootstrapped inverter are shown in Fig. 6 (c) and (d) respectively. Due to an additional large capacitance in the bootstrapped inverter, dynamic behavior is better with a simple organic inverter while one can observe an improvement in the swing of output characteristics with the bootstrapped inverter.

Thus, experimental results on solution processed OFETs with P3HT and high-k gate dielectric for low operating voltages have opened a wide field of circuit applications for polymer electronics. Moreover, simulation models extracted for an organic semiconductor enable researchers to simulate and realize various organic circuits with improved performance using novel design techniques such as bootstrapped organic inverters.

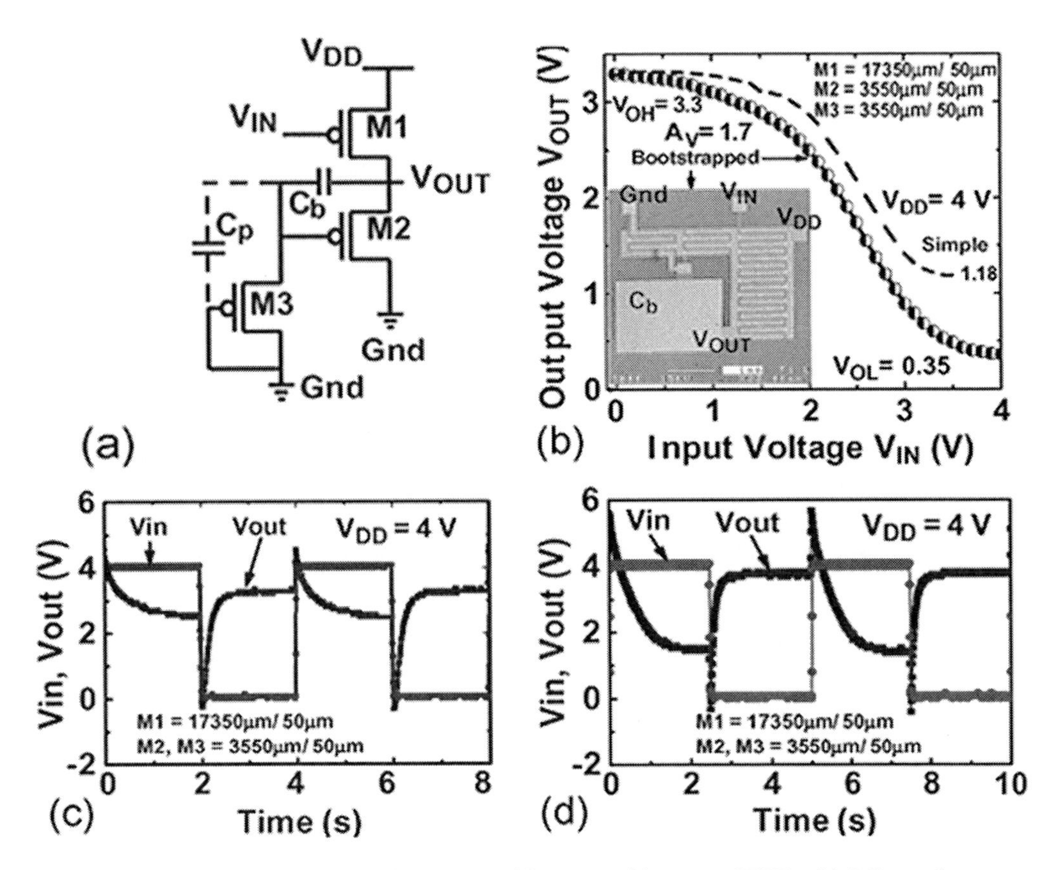

Figure 6. (a) The circuit diagram of a bootstrapped inverter with p-type OFETs, (b) DC transfer characteristics of a fabricated bootstrapped organic inverter showing excellent switching behavior with V_{OH} and V_{OL} values of 3.3 and 0.35 V, respectively with the sweep rate of the input to be 0.1 V/s. The inset figure shows a SEM photograph of the fabricated bootstrap inverter, where, for M2 and M3, the S/D layer is connected to the gate layer by interconnect vias through the dielectric layer, (c) Transient response of the normal inverter, (d) Transient response for the bootstrapped inverter [16].

3. OFET Based Sensors for Explosive Vapor Detection and Radiation Dose Measurements

Organic semiconducting materials, due to their chemical properties, interact with a variety of matters around and show significant change in its properties to find their use in various sensing applications. In this section of the chapter, use of solution processed P3HT based OFET for sensing explosive vapors and determining ionizing radiation are discussed in detail.

3.1. Explosive Vapor Detectors Using Organic Semiconductor Materials and Devices

Vapor phase sensors with good sensitivity and selectivity to explosive vapors are the need of the hour for a variety of security applications. Because of their various advantages organic semiconducting polymers have proved themselves to be the best suited for the sensing applications. The possible use of metalloporphyrins for detection of explosives (mainly 2,4,6-trinitrotoluene (TNT)) [27 - 29], use of P3HT and metalloporphyrins for detection of explosives [30] and toxic gases such as NO_2 [31] have been recently reported.

In this section of the chapter, the focus is on the sensing property of OFET in particular for the explosives vapors. Especially, change in the properties of P3HT has been studied with a blend of other chemicals for various explosive sensing applications. Highly doped n-type silicon wafer was used here as a gate with thermally grown SiO_2 acting as a gate dielectric and a Ti/Au (10nm/60nm) layer was patterned using photolithography lift off process for S/D electrodes to realize BGBC structure. Fig. 7 (c) shows the schematic of device structure while chemical structure of P3HT and Copper Tetraphenylporphyrin (CuTPP) are shown in fig. 7 (a) and fig. 7 (b) respectively.

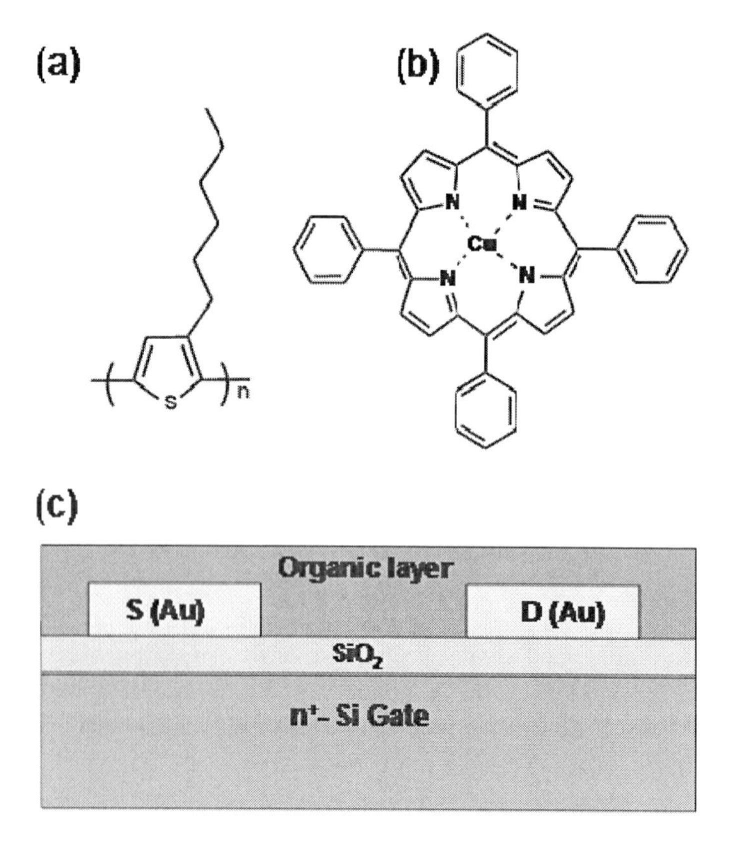

Figure 7. (a) Chemical structure of P3HT, (b) Chemical structure of CuTPP, and (c) The schematic of the structural cross section of an OFET with n^+- silicon wafer acting as a common gate for all OFETs fabricated on the wafer [26].

The OFETs were fabricated with different channel lengths (30μm to 70μm) and widths (17050 μm to 24850μm). The blend of P3HT (from Sigma Aldrich) & CuTPP (synthesized as reported in literature [52]) was dissolved in chloroform (3mg/ml) and a thin layer of organic material was formed by spin coating done at 500 rpm for 10 sec followed by 1000 rpm for 40 s. I-V measurements were done for various gate (0V, -5V, -10V, -15V, & -20V) and drain voltages (0 to -40V). Measurements for each sensor were taken before and after exposure of analyte vapors. OFET with the P3HT as an organic layer was exposed to 1,3,5-trinitro-1,3,5-triazacyclohexane (RDX) and TNT vapors to study the changes in characteristics. When the RDX vapors were exposed to the OFET, almost no variation was observed in the output characteristics shown in Fig. 8 (a). The saturation current was found to be almost unaffected and the change in conductance of the film, measured in the linear region of operation, was also found to be negligible (1-2%). The inability of the non-aromatic RDX molecular interaction with the π- cloud of P3HT could be a possible reason for this negligible response to the RDX vapors. Now when TNT vapors were exposed, an appreciable change in saturation current (15-20%) and conductance (8-10%) were observed. The higher sensitivity to TNT could be due to the strong π - π interactions between the electron-deficient aromatic TNT molecule and the P3HT polymer.

Eq. 1 was used to obtain conductance [53], where the values of field-effect mobility (μ) and threshold voltage (V_T) were obtained by a linear fit between the I_D-V_{GS} data measured at V_{DS}= -2 V and eq. 2, expression for drain current of a transistor in the linear region [54].

$$S = \mu C_{ox} \frac{W}{L} (V_{GS} - V_T) \tag{1}$$

$$I_{DS} = \mu C_{ox} \frac{W}{L} \left[(V_{GS} - V_T) V_{DS} - \frac{1}{2} V_{DS}^2 \right] \tag{2}$$

$$\sigma = Nq\mu \tag{3}$$

where μ is the field-effect mobility, C_{OX} is the capacitance per unit area of the gate dielectric [F/cm^2], V_T is the threshold voltage, W (width) and L (length) are the dimensions of the semiconductor channel defined by the source and drain electrodes of the transistor, σ is the film conductivity, N is the total number of carriers per unit area in the film, and q is the electron charge.

The experiments were repeated with P3HT and CuTPP as the organic layer. Motivation behind the use of CuTPP was a previous study which showed formation of molecular complexes between metalloporphyrins and various acceptors [55]. These films made up of P3HT and CuTPP were exposed to vapors of RDX as well as TNT. Interestingly there was a considerable change in the saturation current observed for both vapors of RDX and TNT. Fig. 8 (b) shows the output characteristics in this case when exposed to RDX vapors. Solid and dotted lines represent the I-V characteristics with and without exposure of RDX vapors respectively.

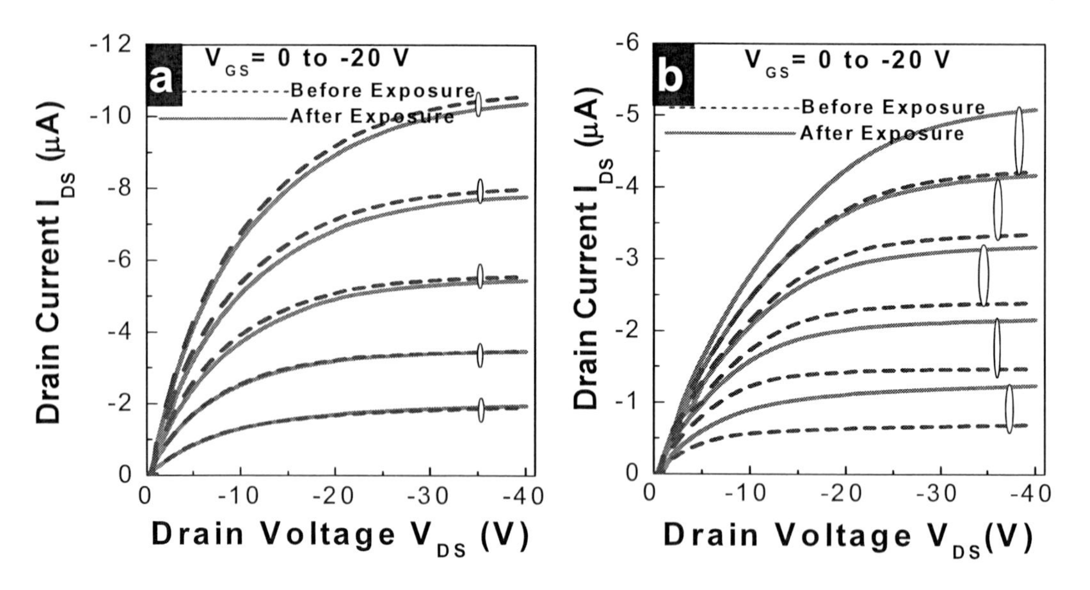

Figure 8. (a) I_{DS}-V_{DS} characteristics for P3HT based OFET for various values of V_{GS}, the dotted curve showing the response without RDX vapors and the solid curve with exposure to RDX vapors and (b) I_{DS}-V_{DS} characteristics for P3HT+CuTPP composite OFET for various values of V_{GS}. The dotted curve shows the response without RDX vapors and the solid curve with exposure to RDX vapors [26].

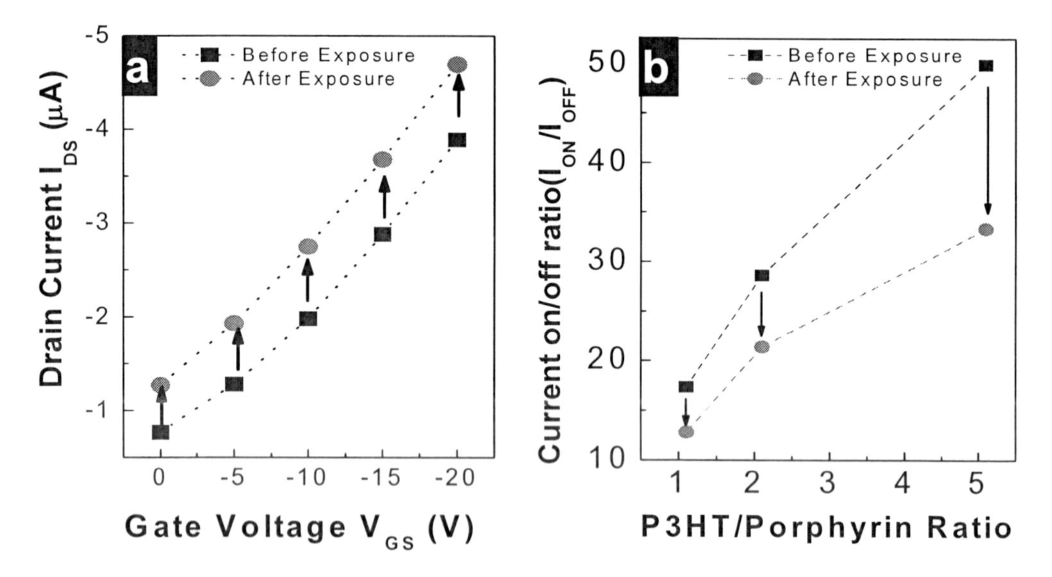

Figure 9. (a) I_{DS}-V_{GS} plot indicating the rise in current after exposing to RDX vapors. (■) shows the current without exposure to RDX while the (●) shows the current with exposure (b) I_{ON}/I_{OFF} ratio plotted for various composite ratios of P3HT+CuTPP, wherein I_{ON}/I_{OFF} as well as the change in I_{ON}/I_{OFF} is found to be maximum for 5:1 ratio of P3HT+CuTPP. (■) points show the values without exposure while (●) shows the values with exposure to RDX vapors [26].

A significant change in the saturated drain current was observed for different gate voltages shown in fig. 9 (a). Significant change (10-15%) was also observed in the conductance of the film. Superior sensor response in this case compared to the case of only P3HT based OFET could be due to the well-known strong tendency to form coordinate bonding between the metalloporphyin molecule and the nitro group [55] as well as the π-stacking between the porphyrins and aromatic rings of the P3HT polymer [56]. A comparatively higher response to the TNT vapors could be due to the additional π- π interaction between the P3HT and TNT molecules. To optimize the ratio of P3HT and CuTPP, various composition ratios were used to perform the experiments. Fig. 9 (b) shows the variation in I_{ON}/I_{OFF} ratio with different P3HT/CuTPP ratio. Here I_{ON}/I_{OFF} ratio with and without exposure to analyte is compared and we found that a P3HT/CuTPP ratio of 5:1 gives the highest I_{ON}/I_{OFF} ratio and hence for further experiments this ratio was used

The P3HT and CuTPP based transistors were exposed to various analytes to address the selectivity issue. We selected nitrobenzene (NB), DNB and TNT – all nitro-based explosive compounds, along with benzoquinone (BQ) and benzophenone (BP) – non-nitro, non-explosive compounds. Characterization and analysis of experimental results were obtained for all the analytes. The significant response was observed only for nitro-based explosive compounds (with the exception of NB). However, the negligible response was found for the other analytes (BP, BQ and NB). As explained in the case of RDX & TNT, strong interactions between metalloporphyrins and the nitro-group could be responsible for the selectivity towards nitro-based compounds. Our experiments were in agreement with the behavior that the metalloporphyrins have large binding constants for nitro-aromatics relative to free-base porphyrins [55]. On the other hand, the poor sensor response to strongly oxidizing BQ could be due to the low binding strength (Kb), leading to low polymer-quinone interactions [52].

3.1.1. Role of Porosity to Enhance the Sensitivity of Sensor

Role of porosity in OFET based on P3HT-metalloporphyrin sensors for nitro based explosive compounds were also studied experimentally, as intuitively better sensitivity can be achieved with a larger surface area or in other words with a higher porosity of the film. Here we have shown that the addition of a specific polymer in binary composite of P3HT and CuTPP leads to an increase in the porosity of the film and hence there is an increase in sensitivity. A binary composite of P3HT/CuTPP was taken as a starting point and the porosity was modified by addition of ADB to obtain a ternary composite. This binary composite material P3HT/CuTPP is shown to have an increased porosity with the addition of ADB which significantly increased the sensitivity towards explosives.

3.1.2. Use of ADB Copolymer to Increase the Porosity of the P3HT/Cutpp Film

The polymer used for increasing the porosity is actually a fluorescent co-polymer of Diethynyl-pentiptycene and Dibenzyl-ProDOT (ADB) synthesized locally for TNT sensing. Fig. 10 shows the excitation and the emission spectra and along with it the repeating unit of the copolymer, ADB. The chloroform was used to record the absorption and emission spectra of the polymer.

For the initial experiments, as a demonstration of the plausible use of the ADB polymer as a sensor for nitro aromatics, a chloroform solution of the polymer was mixed with chloroform solutions of TNT, and MNT (*m*-nitrotoluene) respectively. This resulted in the fluorescence quenching of the polymer. The sensitivity was measured by adding 50 μL of 10^{-5} M TNT solution in chloroform (Fig. 11). ADB was chosen due to its high molecular weight, specificity towards TNT molecules, and rigid aromatic polycyclic structure which could lead to enhancement in the porosity when mixed with P3HT/CuTPP. The diffusion of the nitroaromatic explosive molecules in the bulk of the film could enhance due to the increase in the porosity of the P3HT polymer film by addition of ADB. The standard BET analysis was used to measure the porosity of the film. BET analysis showed that the binary (P3HT/CuTPP) composite film was not porous, but the addition of ADB polymer increases proportionally the porosity of the ternary composite (Table II). As seen from the values reported in Table II, the higher concentration of ADB results in a higher porosity of the ternary composite (Fig. 12 and Fig. 13). The incremental difference in the surface area in terms of the pore width is shown in Fig. 6. The surface area was seen to be increasing with the increase in concentration of ADB polymer in the composite. The surface area was about 0.003 m^2/g with the maximum pore width being 17 Å for ADB of 5 wt%, surface (Fig. 12 (a)) and the it increases to 0.016 m^2/g (Fig. 12 (b)) and 0.2 m^2/g (Fig. 12 (c)) for 10 wt% and 20 wt% of ADB respectively. The other measure of increase in porosity was the absorption of the quantity of N_2, here the experimental quantity absorbed and relative pressure was found to be increasing with increase in ADB wt% for P3HT/CuTPP composite. Fig. 13 clearly shows the quantity of N_2 absorbed in the pores with increase in the relative pressure of the column. This can be attributed to the increase in the pore widths with increase in concentration of ADB polymer and thus these results indicate the increase in porosity with the addition of ADB polymer.

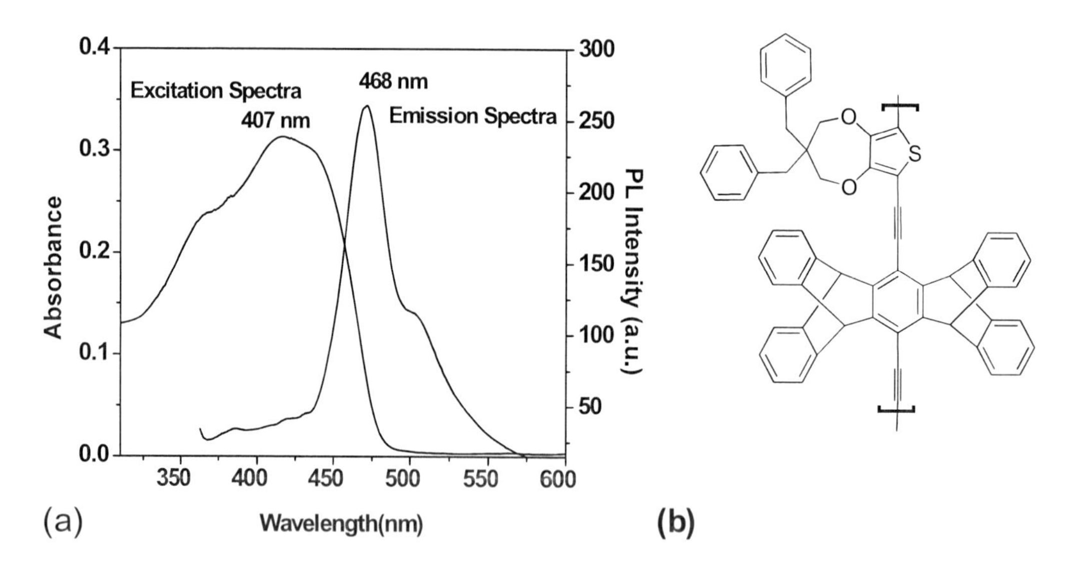

Figure 10. (a) Excitation (407nm) and Emission (468nm) spectra of ADB polymer showing (b) Chemical structure of ADB polymer [30].

Table II. Variation in the BET surface area and pore area with different ADB concentrations

Parameters → Composites ↓	BET Surface Area (m^2/g)	Langmuir Surface Area (m^2/g)	Total Area in Pores (m^2/g) (BET Pore Size ≥ 14.83 A°)
P3HT/CuTPP	------------	-----------	-------------
P3HT/CuTPP/ADB (5 wt %)	0.0085	0.0095	0.004
P3HT/CuTPP/ADB (10 wt %)	0.0467	0.0572	0.035
P3HT/CuTPP/ ADB (20 wt %)	0.2585	0.3803	1.396

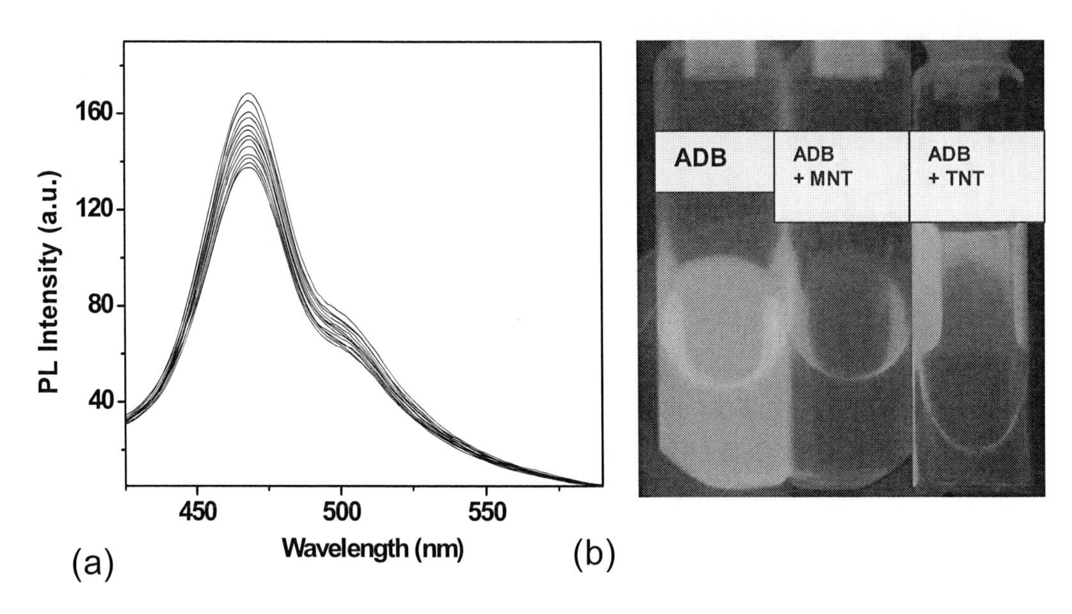

Figure 11. (a) Fluorescence quenching response of ADB on addition of 50μL of 10^{-5} M TNT in chloroform. (b) (1) Fluorescence seen under UV lamp (2) with MNT and (3) with TNT showing the quenching of polymer [30].

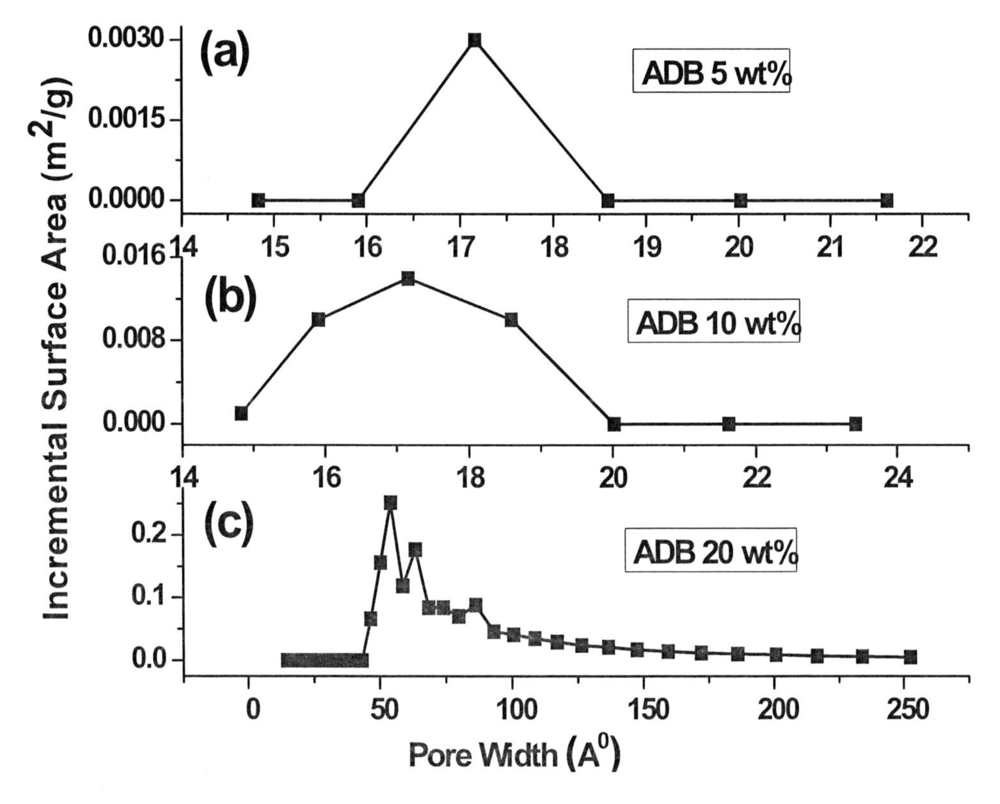

Figure 12. Incremental surface area as a function of Porewidth for different ADB concentration in P3HT/CuTPP Composite. Surface area is more with higher pore width for higher concentration of ADB [30].

The higher concentration of ADB in the composite films deposited as OFET results in a very high porosity and shows reduced conductivity and hence are not suitable for sensor applications. The concentration of ADB was optimized (5 wt %) for basic operation of OFET. The films obtained were characterized by various studies like AFM and EFM. AFM was used to study the surface morphology and surface roughness. As expected, increase in the surface roughness was observed with the addition of ADB polymer to the P3HT/CuTPP film. The rms surface roughness of P3HT/CuTPP film was observed to be 0.72 nm compared to 0.93 nm surface roughness of P3HT/CuTPP/ADB film. This increase in rms roughness, favorable for the sensing applications, was seen to be proportional to the percentage of ADB in the composite. AFM images of the P3HT/CuTPP and P3HT/CuTPP/ADB films are shown in Fig. 14. Fig. 14 (a) and (b) respectively, where a conformal difference can be seen between the two films.

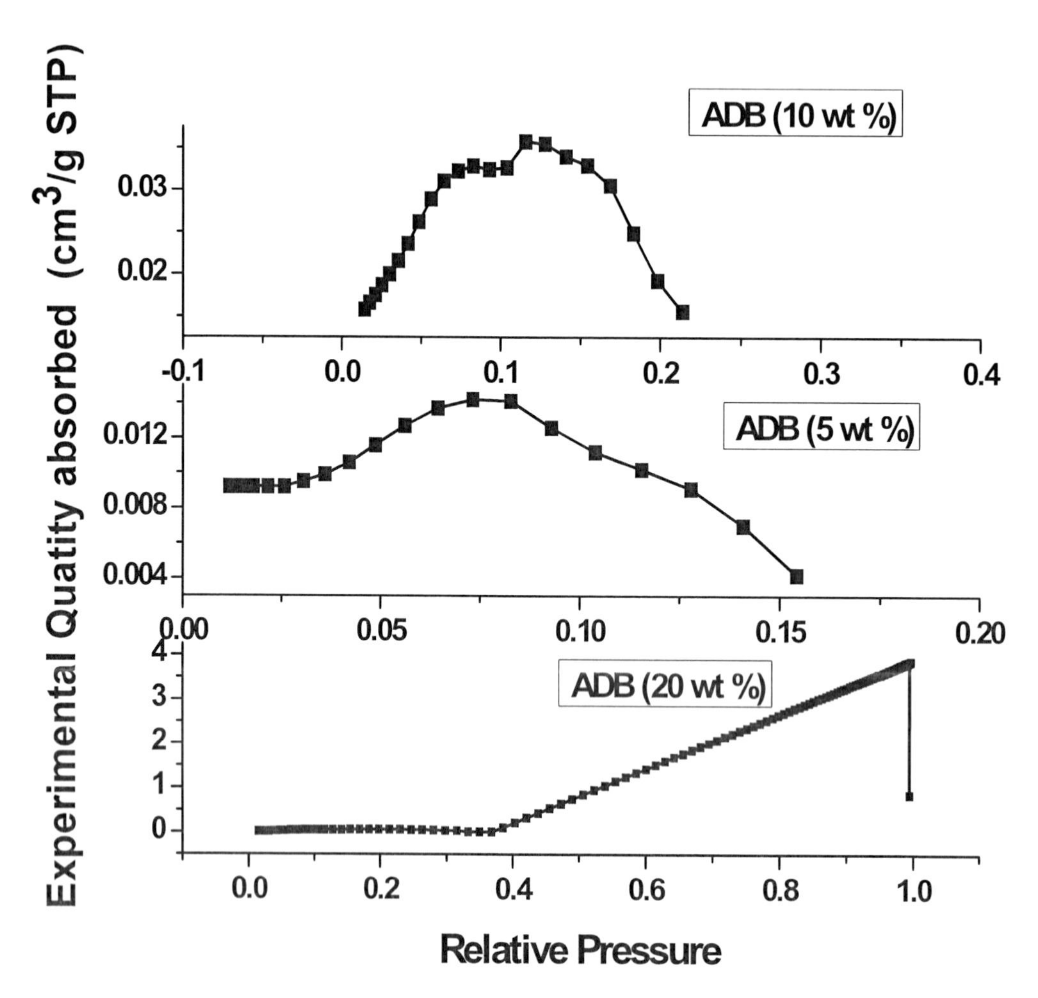

Figuer 13. Experimental quantity absorbed with different ADB concentration in P3HT/CuTPP composite increase in quantity observed for higher concentration of ADB [30].

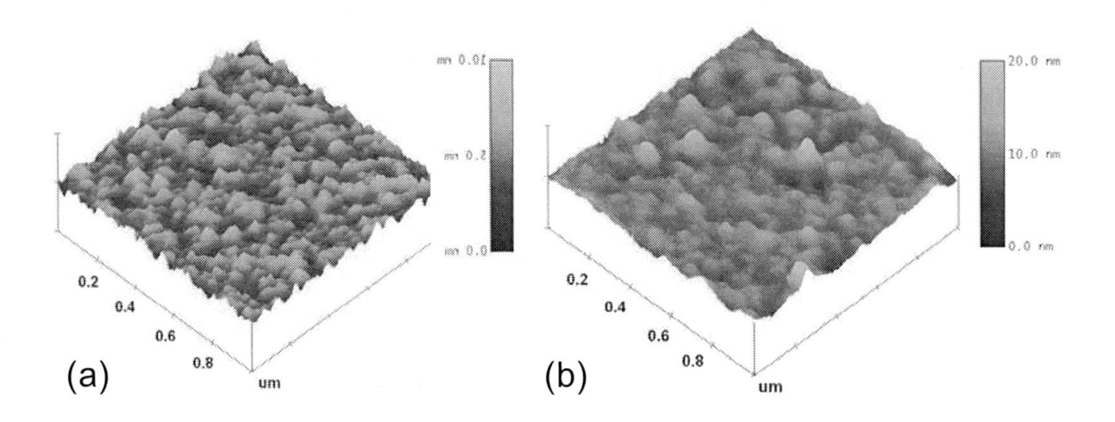

Figure 14. AFM images: (a) 3-D image of P3HT/CuTPP and (b) 3-D image of P3HT/CuTPP/ADB [30].

3.1.3. EFM Characterizations of the Polymer Composite Films

This Electrostatic Force Microscopy (EFM) is a very useful technique to study the film characteristics from the electrical conductance point of view. From the current – voltage (I-V) characteristics of the OFET with both the polymer composites it was observed that the current values were smaller for the ternary composite compared to the binary composite. An EFM study was done to know the reason behind decrease in the initial conductance of the ternary composite film. LiftMode was used to do EFM with the 50nm gap between the substrate (polymer coated) and the platinum tip of AFM cantilever. EFM images were taken with and without a biasing voltage applied in LiftMode. In the EFM image of binary composite shown in Fig. 15 (a), an almost connected lighter region in the form of continuous pathways was seen, confirming the good conductivity of the film. On the other hand, the lighter regions are seen in the form of isolated patches in fig. 15 (b), an EFM image of ternary composite, indicating that the film has discontinuity in the conductivity. EFM study was continued in the biased mode as well; where a bias of 3 V was applied between sample and the tip of cantilever and the resultant EFM images for the binary and ternary films are as shown in Fig. 16 (a) and Fig. 16 (b) respectively. Here, the contrast generated is higher due to the induced field by applying a voltage between the tip and the sample. For further investigation a sectional analysis of the films was carried out and the results obtained, as shown in Fig. 16, are in agreement with this theory. The modulated frequency was determined from the sectional analysis to study the film conductivity. The modulated frequency is proportional to the electric field gradient. Here, in the case of binary composite the modulated frequency was observed to be 0.876 Hz while in the case of ternary composite it was found to be 4.413 Hz. The higher modulating frequency in the case of ternary composite shows the higher electrical field gradients (between darker and lighter contrast regions). Thus, in this case, the increase in localized charge trapping would be responsible for the decrease in the initial conductivity of the composite film after addition of ADB. However, the increase in porosity for this ternary composite was an advantage for the improvement in the sensitivity of the sensor.

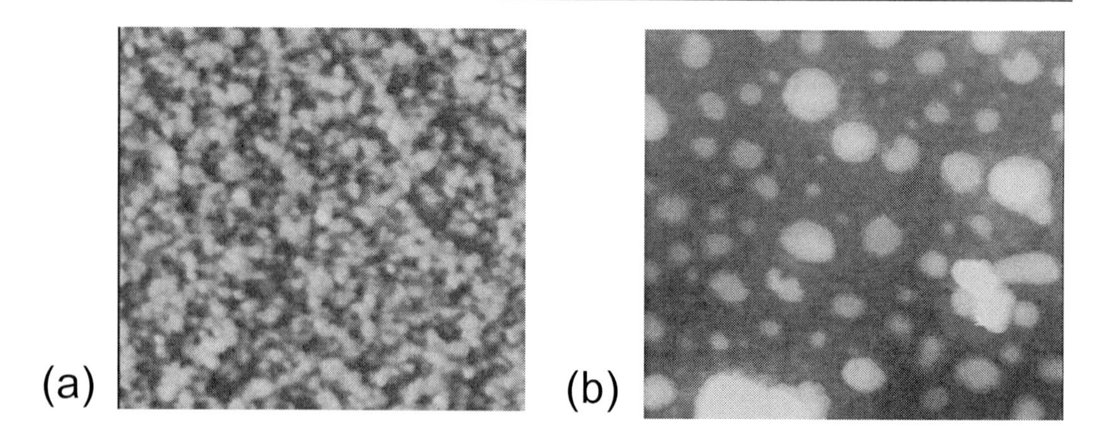

Figuer 15. EFM images for zero bias in LiftMode: of P3HT/CuTPP (a) and (b) of P3HT/CuTPP/ADB[30].

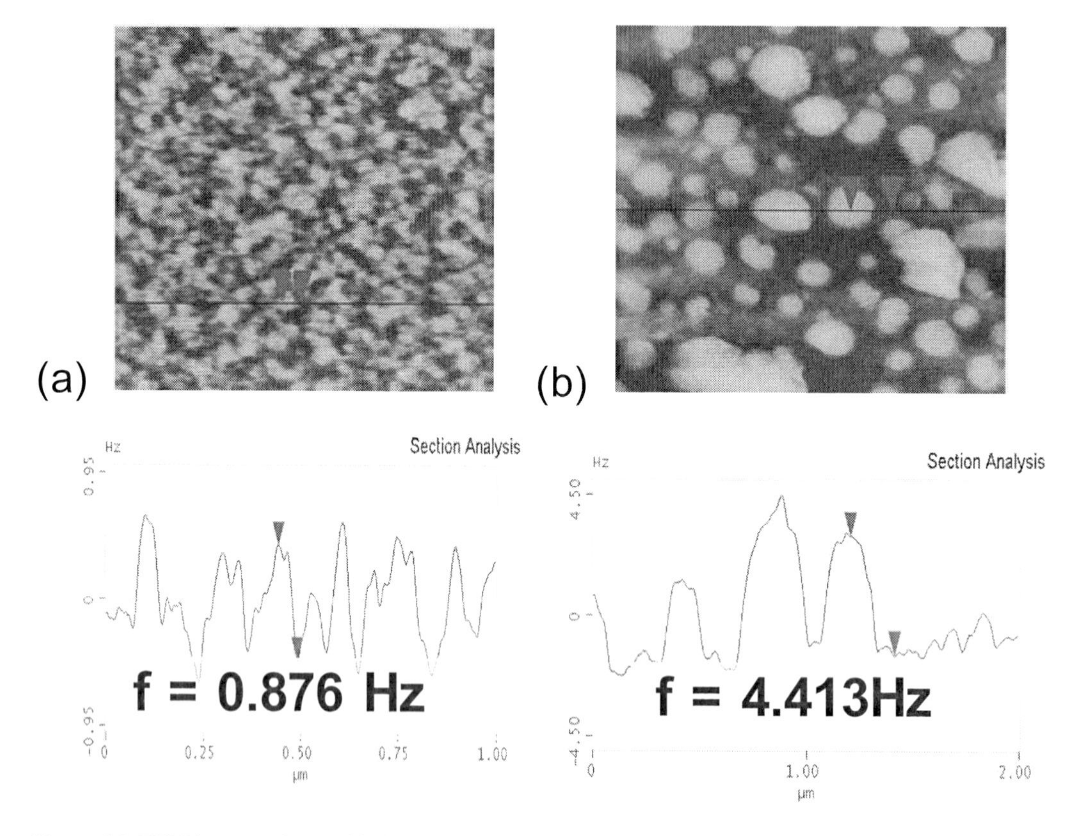

Figure 16. EFM images along with the sectional analysis for bias of 3V : (a) of P3HT/CuTPP and (b) of P3HT/CuTPP/ADB.

3.1.4. I-V Characterizations After Analyte Exposure

For performing experiments to record I-V characteristics of OFETs for both the composites an exactly identical protocol was used while passing the analyte vapors for the same time span. In every graph we have used dotted blue color lines for indicating results before an exposure of analytes and solid lines red in color are used to indicate response after an analyte exposure.

The I-V characteristics of the OFETs of the ternary composites compared with those obtained for the OFETs of the binary composites after exposure to the same range of explosives (TNT, RDX and DNB). There was a considerable increase in the sensitivity observed on addition of 5 wt% of ADB to the binary composites. Moreover, almost negligible change was observed in case of the vapors of other analytes such as NB, BQ and BP (non-explosive compounds). As shown in Fig. 17, I-V characteristics show a higher change in the saturation current after exposure in the case of ternary composite compared to the binary composite. Under similar exposure conditions of TNT, the I-V characteristics of the ternary composite and binary composite are depicted in Fig. 18. The TNT concentration used here was less than 10 ppb while exposure. Similar results were observed also for RDX and DNB and are summarized in Table III.

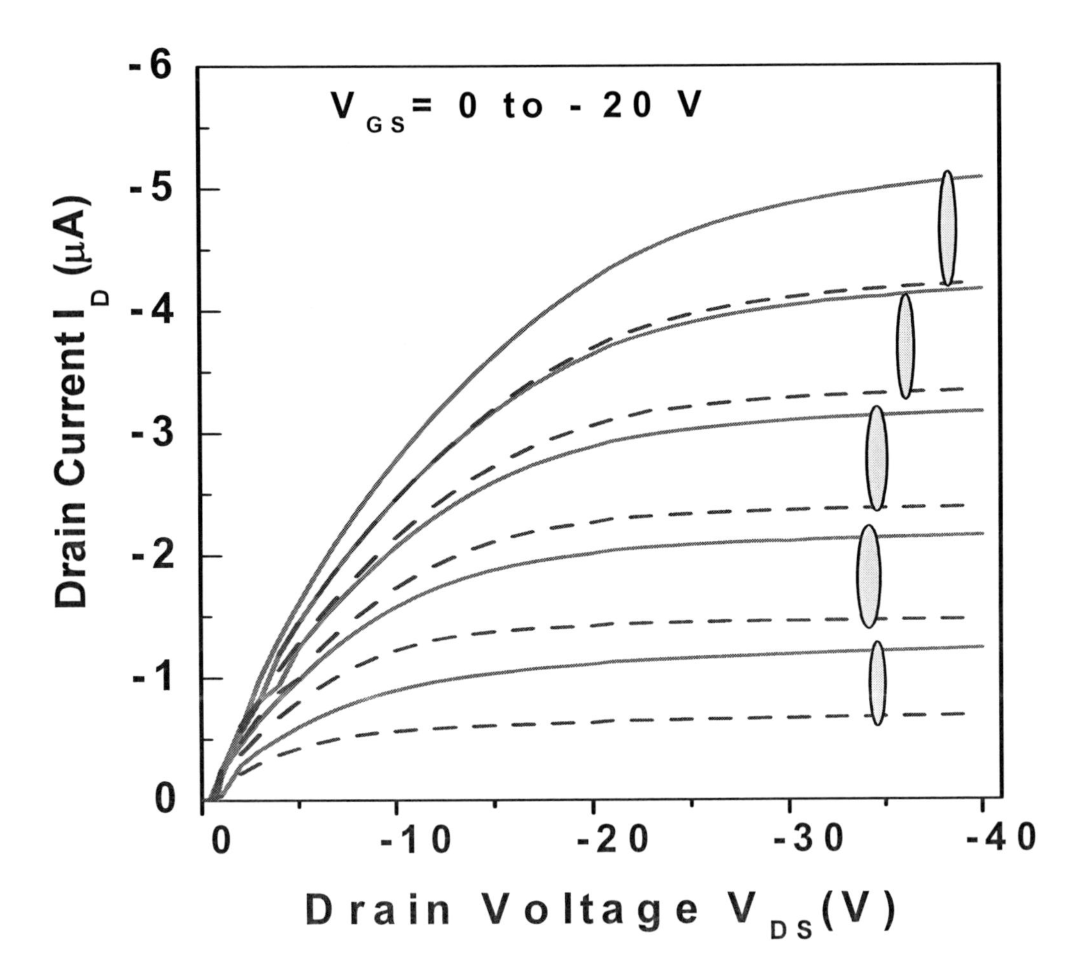

Figure 17. I_{DS}-V_{DS} characteristics for P3HT/CuTPP based OFET for various values of V_{GS}, the dotted curve showing the response without TNT vapors of and the solid curve with TNT vapors [30].

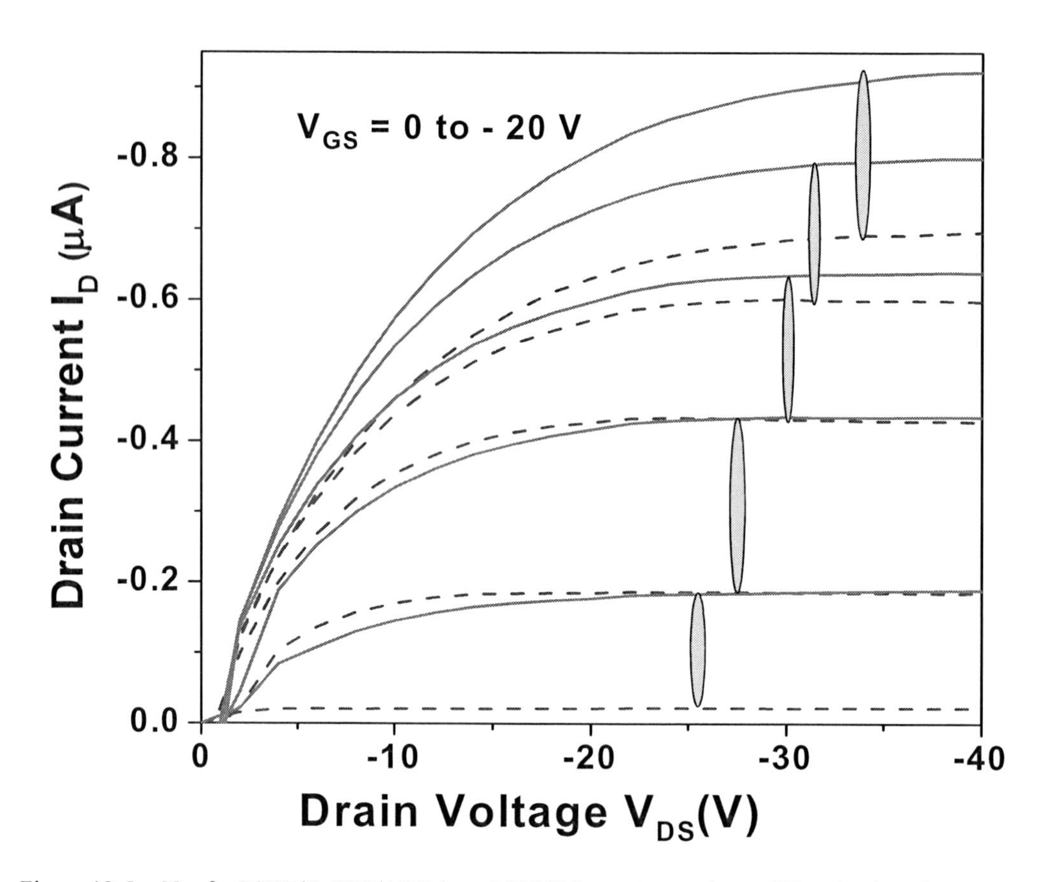

Figure 18. I_{DS}-V_{DS} for P3HT/CuTPP/ADB based OFET for various values of V_{GS}, the dotted curve showing the response without TNT vapors and the solid curve with TNT vapors [30].

Finally the effect of increase in sensitivity due to an addition of ADB is summarized in Fig. 19. The increase in parameters (saturation current and S (Conductance)) was seen for the films after exposure of TNT, RDX and also DNB, while there was a negligible change in the case of NB, BQ and BP. Improved sensor response in this case can be attributed to the porous nature of the film created by the presence of ADB in P3HT/CuTPP. Presence of iptycene moiety in ADB as side chain to the polymeric backbone creates interstitial space around the molecules resulting in an increase in the porosity and the surface area of the thin films [36]. An improved increase in the accessibility of the receptor sites due to the increase in surface area gives rise an increase in the sensitivity. As discussed earlier the response in the case of P3HT/CuTPP could be due to the well-known strong tendency to form a coordinate bonding between the metalloporphyrin molecule and the nitro group [52] as well as the π-stacking between the porphyrins and the aromatic rings of the P3HT polymer [56]. The higher selectivity towards the nitro based compounds could be due to the presence of large binding constants for nitroaromatics in metalloporphyrins as compared to the free-base porphyrins [52]. The poor response to BQ and BP, although they are as good oxidizing agents as TNT and RDX, could be due to their low binding constants (K_b) and hence they lead to low polymer-quinone interactions [52].

Table III. Sensor parameter response (conductance and Ion) to various oxidizing nitro- and non-nitro based compounds. (I) P3HT/CuTPP Composite (discussed in Section 3.1); (II) P3HT/CuTPP/ADB Composite

S.No.	IUPAC Name	Structure	E_{red} (V)	Change in Conductance (S)	Change in I_{ON}
1.	1,3,5-Trinitro-[1,3,5]Triazinane (RDX)	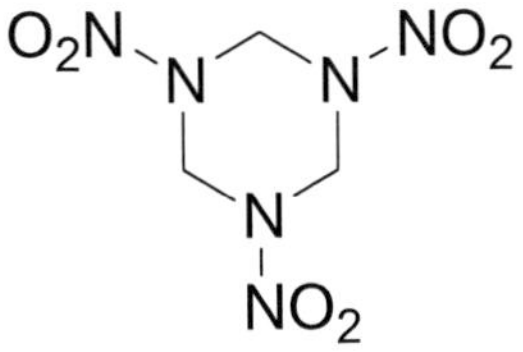 O_2N, NO_2, N, N, N, NO_2	-	9% (I) 14% (II)	20% (I) 26% (II)
2.	2,4,6-Trinitrotoluene (TNT)	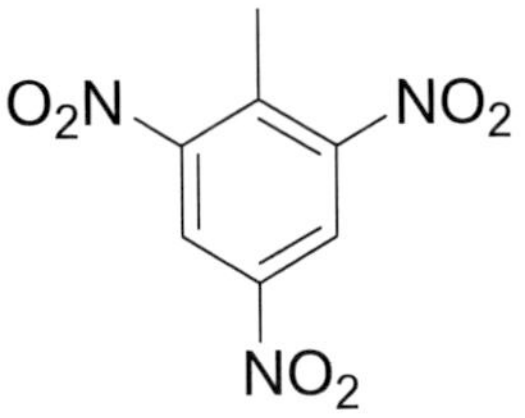 O_2N, NO_2, NO_2	- 0.7	13% (I) 18% (II)	22% (I) 29%(II)
3.	1,3-Dinitrobenzene (DNB)	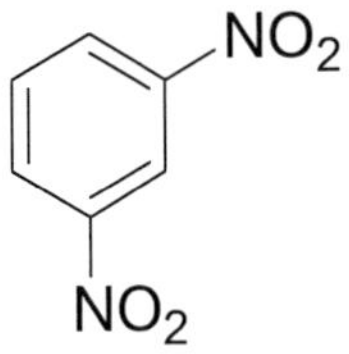 NO_2, NO_2	- 0.9	7% (I) 11% (II)	12% (I) 17% (II)
4.	Nitrobenzene (NB)	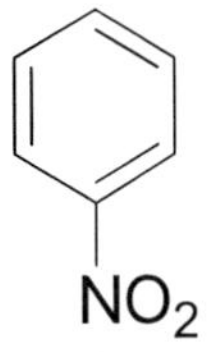 NO_2	-1.15	<1% (I) 2% (II)	3% (I) 4% (II)
5.	1,4-Benzoquinone (BQ)	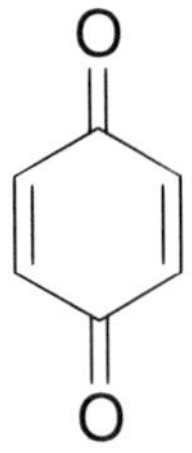 O, O	- 0.5	<1% (I) 1% (II)	2% (I) 3% (II)
6.	Benzophenone (BP)	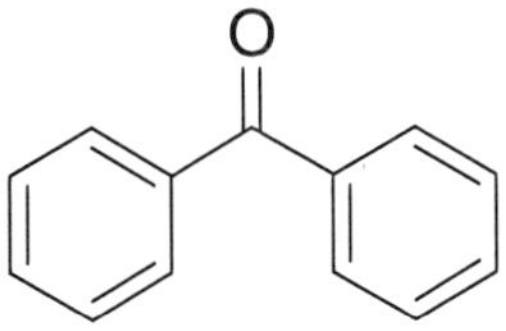 O	-1.6	1% (I) 2% (II)	3% (I) 3% (II)

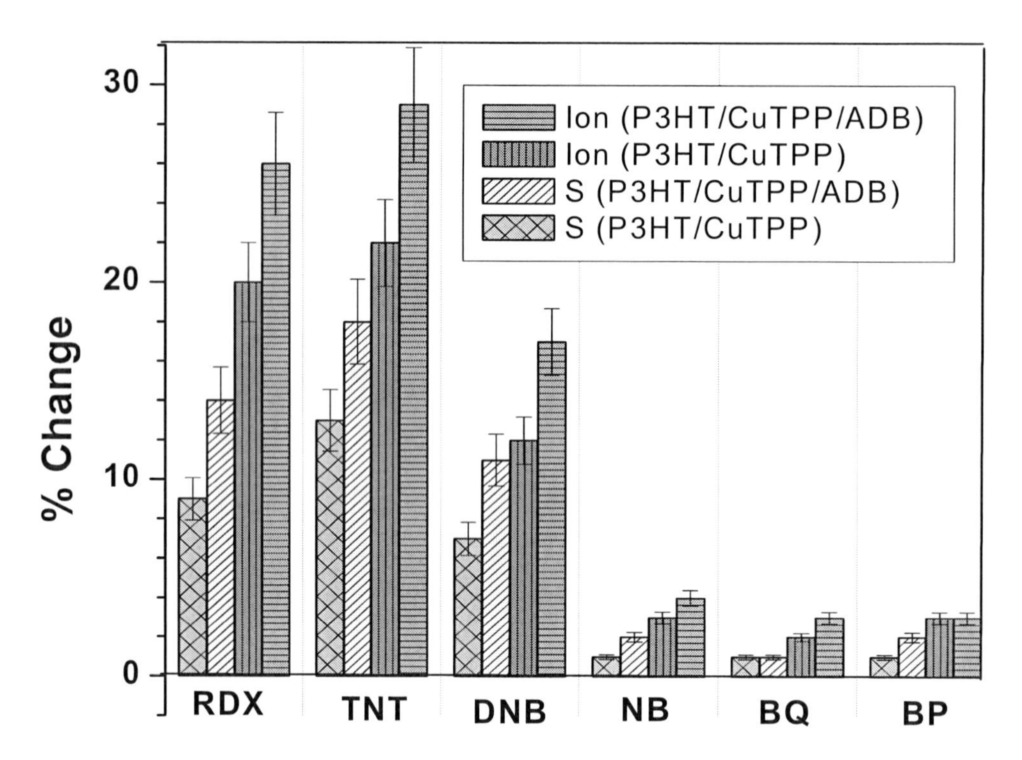

Figure 19. Sensor response (Ion and conductance (S)) comparison for various analytes with and without ADB polymer clearly shows the increase in sensor response with ADB for nitro based explosive compounds and negligible response to other analytes [30].

3.2. Determining Ionizing Radiation Using Organic Semiconductor Materials and Devices

This section of the chapter describes the techniques to determine total dose ionizing radiation using organic semiconductor materials' properties and sensors based on these. To study the effect of ionizing radiation on P3HT, a p-type organic semiconductor, the solution of P3HT prepared in chloroform was exposed to ionizing radiation using ^{60}Co source for radiation and UV-VIS spectrum of the material was studied after different dose of radiation as shown in Fig. 20 (c). The UV-VIS spectrum of the material shows a rising absorption peak at 610 nm wavelength (Fig. 20 (d)) for an increasing dose of ionizing radiation to the P3HT solution along with a peak of absorption at 450 nm wavelength for unexposed P3HT solution. This phenomenon suggests the presence of a new state of the material with less distance between the HOMO and LUMO levels (decreased bandgap) making the P3HT more conductive. The thiophene molecule in the chain of this polymer was oxidized after going to a polaron state and stabilized to a bipolaron and a neutral state [57, 58].

The simplest means of tracking the change in conductivity of the organic semiconductor P3HT is to check the change in resistance of a resistor made up of P3HT. The solution of P3HT prepared in chloroform with 3 mg/ml weight ratio was spin coated at 500 rpm for 15 sec followed by 1000 rpm for 40 sec on interdigitated Ti/Au (10nm/50nm) electrodes fabricated using lift off technique on a thick (~500 nm) layer of SiO_2 grown by thermal oxidation of silicon wafer. The sample was annealed at 90 °C for 90 min in ambient condition. Schematic diagram of the organic resistor is shown in Fig. 20 (a). This organic semiconductor resistor sensor was then irradiated by γ-rays from ^{60}Co radiation source for different doses by varying the time period for irradiation. The change in resistance of the sensor was measured after each total dose of ionizing radiation using Keithley 2602 Source – Measure Unit (SMU) at room temperature in normal atmospheric conditions by varying voltage applied between two electrodes of the resistor and measuring current flowing between them.

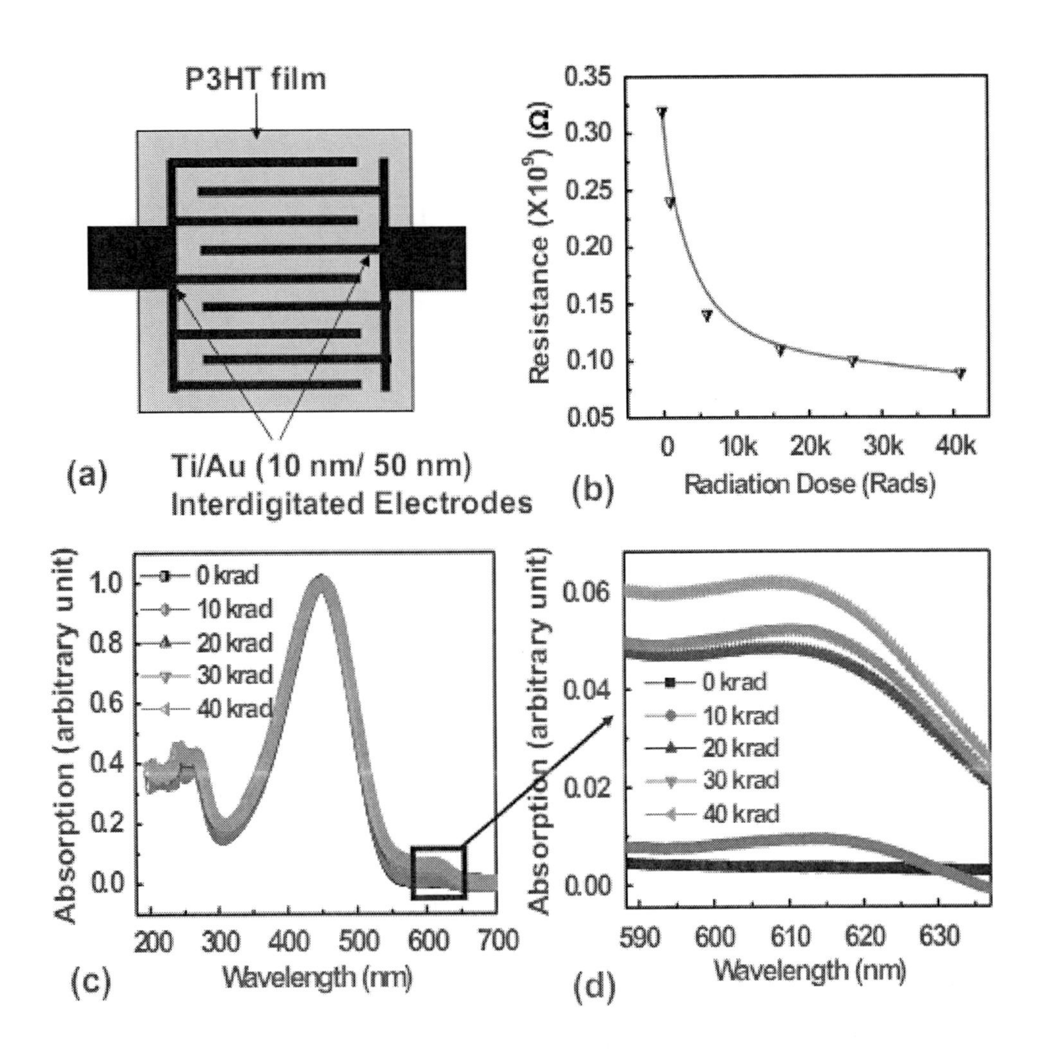

Figure 20. (a) Schematic top-view of an organic semiconductor resistor sensor using P3HT as organic semiconductor, fabricated using two interdigitated electrodes covered by P3HT film, (b) Change in the resistance of the sensor with increasing dose of ionizing radiation, (c) UV-VIS spectrum for the P3HT solution prepared in chloroform and exposed to ^{60}Co radiation for increasing dose, and (d) The oxidation peak in fig. 20 (c), showing increase in oxidizing peak with increasing radiation dose [34].

The resistance of the organic semiconductor resistor sensor was decreased from 320 MΩ to 80 MΩ for un-exposed sensor to the one exposed to 41 krad radiation dose. The response is linear at low radiation dose range (upto 8 krad). The sensitivity of the organic semiconducting resistor sensor was found to be ~20 kΩ /rad for ionizing radiation upto 8 krad. Fig. 20 (b) shows this decrease in resistance of the sensor with increasing dose of ionizing radiation. Change in resistivity of P3HT has proved it to be a potential candidate for determining ionizing radiation employing a simple detection technique. As our results demonstrate, a change in resistance in P3HT is proportional to the dose of ionizing radiation.

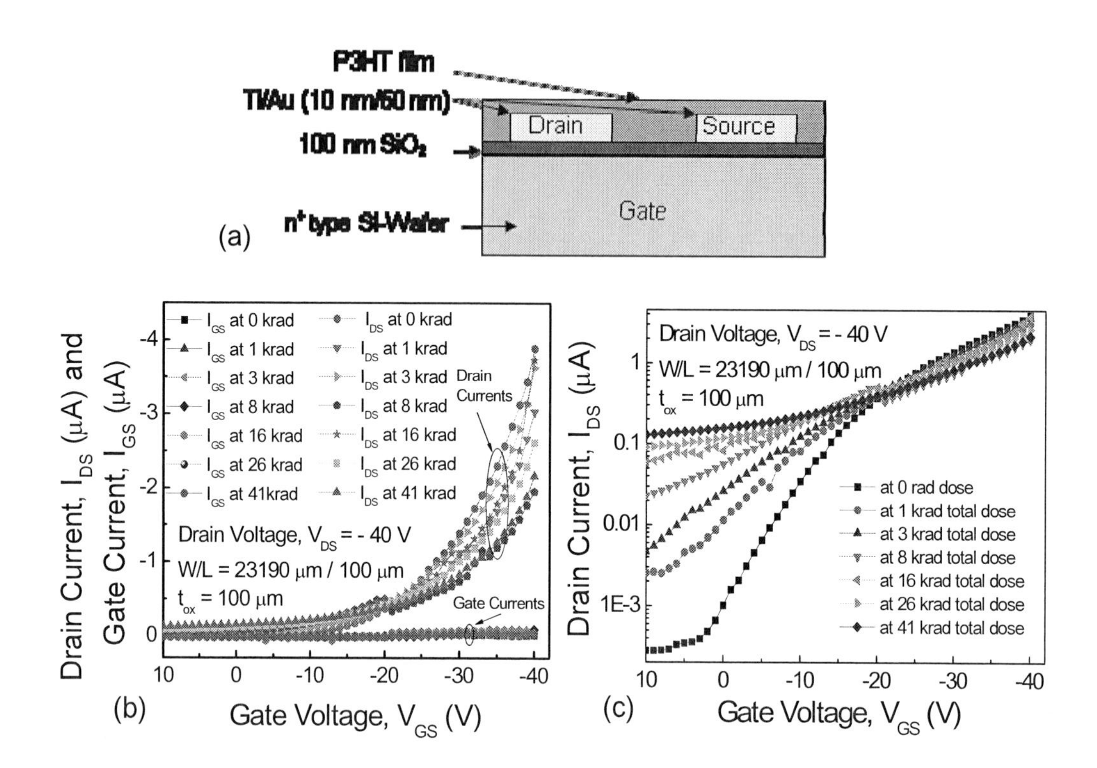

Figure 21. (a) Schematic cross section of a bottom-gate-bottom-contact p-type OFET with P3HT as its active material with thermally grown SiO2 as gate dielectric on n+ silicon wafer acting as gate, (b) Change in I_{DS}-V_{GS} characteristics of the OFET with increasing ionizing radiation dose of ^{60}Co, and (c) Showing significant change in OFF current and sub-threshold degradation (log I_{DS} vs. V_{GS} plot) for the OFET [34].

Taking advantage of this simple detection procedure along with transistor action for improving the sensitivity, a bottom-gate-bottom-contact OFET with P3HT as an organic semiconductor was also exposed to ionizing radiation using ^{60}Co as radiation source for different values of radiation dose by varying the time for exposure. The schematic cross section of the OFET sensor is shown in Fig. 21 (a). To fabricate the OFET sensor, a layer of SiO_2 (100 nm thick) was grown thermally on a highly doped n-type silicon wafer which is acting as a gate for the OFET and using SiO_2 layer as a gate dielectric with a capacitance = 34.5 nF/cm^2. The contacts for source and drain were made by depositing Ti/Au (10 nm / 50 nm) using RF sputtering and patterning an interdigitated structure using photolithography with lift-off technique. To improve the interface between gate dielectric and organic semiconductor (P3HT) the sample was treated with HMDS surface treatment. HMDS was spin coated at 1000 rpm for 10 sec followed by 4000 rpm for 40 sec and heated at 120 °C for

5 min. The P3HT solution prepared in chloroform was then spin coated as described earlier. IV characterization of OFET was done by voltage between gate and source (V_{GS}) was varied from 0 V to -40 V in steps of -1 V keeping voltage between drain and source electrodes (V_{DS}) constant at -40 V and measuring the current flowing through drain terminal (I_{DS}) ($I_{DS}V_{GS}$ characteristics). The I_{DS}- V_{DS} was obtained by varying the V_{DS} from 0 V to -40 V in steps of -1 V at different values of V_{GS} varying from 0 V to – 40 V in steps of -10 V ($I_{DS}V_{DS}$ characteristics) on a probe station with Keithley 2602 and Keithley 236 SMUs at room temperature in the normal atmospheric conditions.

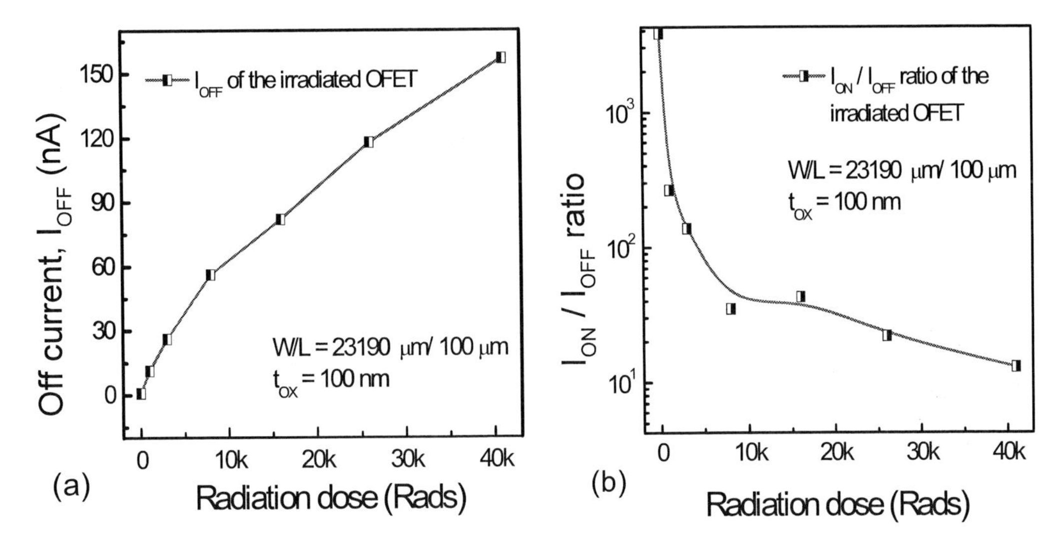

Figure 22. (a) OFF current degradation in the OFET with increasing dose of ionizing radiation. There is a change of about 150 X in the I_{OFF} after a 41 krad ionizing radiation dose using a ^{60}Co radiation source, and (b) Degradation in I_{ON}/I_{OFF} ratio (by 300X) of the OFET for 41 krad dose of ionizing radiation [34].

The OFETs were exposed to ionizing radiation for different values of radiation doses and IV characteristics were measured after each dose for measuring changes due to exposure of the sensor to ionizing radiation. The $I_{DS}V_{DS}$ characteristics of the OFET sensor with a W/L ratio of 23190 μm/100 μm and t_{ox} = 100 nm are shown in Fig. 21 (b) showing the change in drain current and gate current with increasing dose of ionizing radiation. In Fig. 21 (c) the change in drain current is shown on log scale for easy observation of increasing OFF current (I_{OFF}) with increasing total dose of ionizing radiation by varying time of exposure for different dose of radiation. As can be seen, the irradiated sensor shows significant changes in electrical characteristics which can be used to extract various parameters such as I_{OFF}, ON current to OFF current ratio (I_{ON}/I_{OFF}), shift in subthreshold swing, increasing number of interface states, etc. to relate the dose of ionizing radiation with these changing parameters. With no radiation, the sensor shows I_{ON}/I_{OFF} ratio of ~3800 and a mobility of ~1.7×10^{-3} $cm^2 V^{-1}s^{-1}$.

In Fig. 22 (a) the increasing I_{OFF} of the sensor due to increasing conductivity of the organic semiconductor (P3HT) is shown with increasing total dose of ionizing radiation showing an increase of about 150X at 41 krad of radiation dose. The sensitivity of the sensor was found to be ~7 pA/rad for 8 krad of radiation dose considering the linear change in I_{OFF}.

I_{OFF}/I_{ON} for the sensor with increasing dose of radiation is shown in Fig. 22 (b) which degrades rapidly due to a significant increase in the off current as can be seen in Fig. 21 (c).

Degradation in the sub-threshold swing of an OFET is extracted from Fig. 21 (c) for subsequent radiation doses. The sub-threshold swing increases linearly with subsequent radiation doses as shown in Fig. 23 (a). The change (increase) in the number of interface states (ΔN_{it}) can be easily calculated considering the change in sub-threshold swing, ΔS, assuming that the generated interface states are uniformly spread over the bandgap. Eq. 4 shows a relationship between ΔN_{it} and change in subthreshold swing (ΔS).

$$\Delta N_{it} = \frac{C_{ox}\varepsilon_g \, \Delta S}{kT(\ln 10)} \tag{4}$$

Here, band gap (ε_g) for P3HT can be taken as the difference between the highest occupied molecular orbital (HOMO) and lowest unoccupied molecular orbital (LUMO). Reported ε_g values for P3HT vary between 1.9 eV to 2.1 eV [59, 60]. $\varepsilon_g = 2.1$ eV for P3HT is used in this work for calculation of ΔN_{it}. Calculated ΔN_{it} values are plotted in the inset of Fig. 23 (a). To verify that the changes in the electrical characteristics of an OFET due to irradiation are permanent in nature, a time dependent study was done with the OFET sensor for which the sensor was irradiated for 40 krad dose with ^{60}Co radiation. The time dependent annealing characteristics are shown in Fig. 23 (b) indicating that the effects are not transient in nature. Here the slight degradation in the characteristics was due to the effect of moisture and oxygen on P3HT film as the sample was kept open in the atmosphere.

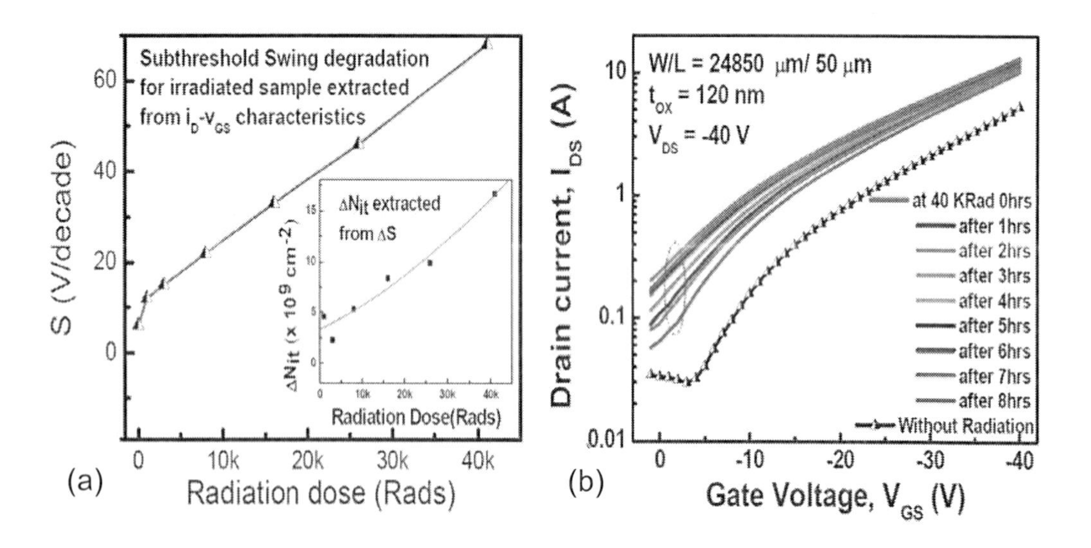

Figure 23. (a) Sub-threshold swing degradation in the OFET (W/L = 23190 μm /100 μm, tox = 100 nm and P3HT as p-type semiconductor) with ^{60}Co radiation dose, the inset figure shows increase in number of interface traps extracted using change in subthreshold swing with increasing ionizing radiation, and (b) annealing (time) study on an OFET (W/L = 24850 μm /50 μm, tox = 120 nm and P3HT as p-type semiconductor) subjected to 40 kRad ionizing radiation dose of ^{60}Co [34].

As has been shown by us earlier, this degradation can be reduced by about 400 % with a proper passivation technique [61]. As the oxygen and humidity degrades the semiconductor, the entire study is performed with a non-irradiated sensor sample, which is kept in the same ambient as the irradiated sample and characterized along with the irradiated sample. From the characterized data it can be observed that the degradation in the characteristics due to oxygen and humidity in the atmosphere is less than 10 % of the change due to irradiation.

Thus, exposure of organic semiconductor material, P3HT, to ionizing radiation increases its conductivity by oxidizing the material which can be detected by various material characterization techniques. At the same time, employing a simple technique like electrical measurements, an organic resistor and an OFET can be used as a sensor for determining ionizing radiation. Change in various parameters extracted from I-V characteristics of an OFET after exposure various dose of ionizing radiation make it a better choice for determining total dose ionizing radiation.

4. Solution Processed Polymer Microcantilever Platform for Biochemical Sensing Applications

Biosensors in general are devices that couple a biochemical recognition element with a transducer. The biochemical element may be biomolecules such as an enzyme, antibody, antigen, living cells, tissues, etc or some other chemicals that can be the selective counterpart for the target molecule to be detected [62]. First biosensor developed was glucose sensor in the early 1960s by Clark and Lyons [63]. Since then the demands in biosensing applications were for fast, easy to use, inexpensive and highly sensitive methods for recognition of target analytes [64]. The demand for these miniaturized high throughput sensors supported by advancements in microfabrication technology lead the biosensor community to develop microdevices for biosensors. One of the most promising candidates was a Micro Electro Mechanical Systems (MEMS) based sensor named "Microcantilevers".

Microcantilevers are simple microfabricated devices making their entry into the MEMS community as atomic force microscopic probes [65]. They have a huge potential as a platform for the developments of many physical [66, 67], chemical [68-70], and biological sensors [71-73]. These are high throughput sensors that can offer an improved sensitivity and an excellent dynamic range. When the microcantilevers are to be used for sensing immunoreaction, these devices are coated with a specific antibody and when they come in contact with the corresponding antigen, the very specific bio-molecular interaction between antigen and antibody leads to a surface stress change or a mass change that can be detected by monitoring the resultant nanomechanical motion of the beam. Microantilever is a promising candidate due to the advantages such as label free detection, feasibility for fabricating multi-element sensors arrays that support high degree of parallelization at lower cost, ease of integration of sensors with signal conditioning circuitry and capability of mass microfabrication leading to a cost effective production. They have two basic modes of operation, static or dynamic. In the former, the static motion or deflection of the microcantilever that arises due to surface stress change is the parameter of interest and in the case of latter, the change in resonance frequency due mass change or stiffness change of microcantilever is the key parameter [73-75]. These two modes have their own advantages and disadvantages and the microcantilever designs need to be optimized differently for respective modes. The bending or resonant frequency changes of microcantilevers can be measured by different optical or electrical transduction schemes. Microcantilever sensors with an optical transduction scheme are expected to offer the highest sensitivity. However, because of the practical limitations of optical transduction schemes, particularly for deployment in the field, electrical transduction methods are preferred for sensing the cantilever deflection. The most common among the electrical transduction scheme is piezoresistive transduction where a microcantilever with an integrated piezoresistor performs electrical transduction of strain by a resultant change in resistance. The differential surface stress generated from biomolecular interaction of the surface of a piezoresistive microcantilever results in a change in resistance of the piezoresistive layer which can be represented using eq. (5) [74].

$$\frac{\Delta R}{R} = -K\left(\frac{1}{\sum_i E_i h_i} + \frac{Z_T . Z_R}{\sum_i E_i h_i \left(Z_{ic}^{\;2} + \frac{1}{3}\left(\frac{h_i}{2}\right)^2 \right)}\right) \cdot \sigma_S \quad (5)$$

where K is the gauge factor, σ_S is the surface stress, Z_T is the position of top layer, Z_R is the position of piezoresistive layer and E_i, h_i and Z_{ic} are the Young's modulus, height and position of the ith layer with respect to neutral axis. From this expression it is clear that the surface stress sensitivity depends on the ratio of gauge factor (K) of the piezoresistive film to the Young's modulus (E) of the structural material. Though the conventional microcantilever materials are normally silicon based, they have lot of disadvantages like lower sensitivity due to higher Young's modulus, complex fabrication processes involving high thermal budget unit processes that leads to increased cost of production etc.

So the demand for highly sensitive and cost effective microcantilever sensors lead to a diversion from silicon based microcantilever structural materials towards solution processable polymers such as SU-8 with its proven properties such as lower Young's modulus, simple and low temperature processing, ability to define high aspect ratio structures and its well known UV and electron beam resist properties.

SU-8 microcantilevers with piezoresistive transduction with different strain sensitive layers such as gold (Au) [76] and polysilicon have been reported earlier. SU-8/Au/SU-8 microcantilevers are less sensitive due to the lower gauge factor of gold. A higher gauge factor material polysilicon, that can be deposited using low temperature deposition processes like hotwire CVD (HWCVD) [36, 77], is a better substitute as piezoresistive layer for improving the sensitivity. However the film should be thin enough such that it does not add to the stiffness of the structure. On the other hand, the lower limit of thickness is decided by the fact that, the thinner polysilicon leads to a decreased signal to noise ratio. So a better option would be to incorporate a compliant piezoresistive film with proper process compatibility with SU-8 processing. In this scenario, a strain sensitive conductive polymer composite made out of SU-8 and conductive fillers would be the right choice. Such polymer composites being solution processable like SU-8 also aid in reducing the cost of fabrication. Such a device concept with cantilevers of thickness 7 μm with carbon black as the filler with a gauge factor in the range of 15 -20 have been reported with 16 wt% of CB in SU-8 [78]. Better sensitivity can be achieved by improving the dispersion of CB in SU-8 and by reducing the thickness of the cantilever structure, which has been detailed in the following sections.

4.1. Material and Methods

The microcantilevers are made of SU-8 obtained from Microchem. SU-8 photoresist is designed to work based on the cationic photo polymerization mechanism, in which cross linking is triggered thermally during the post exposure bake process in the areas where photoacid generation takes place due to the exposure to near UV, laser or electron beam. Thus SU-8 allows simple batch microfabrication using standard lithographic process steps such as spin coating followed by near UV exposure, post exposure bake and development steps.

The polymer composite is prepared by ultrasonic mixing of a high structured Carbon Black (CB) (Conductex 7067 Ultra from Columbian Chemicals) in SU-8 and Nanothinner (Microchem). In order to use SU8/CB as a piezoresistor layer, the SU8/CB composite material should satisfy certain prerequisites such as (i) proper dispersion of CB in SU-8 (ii) the resultant material being spin coatable (iii) photolithographically patternable (iv) conductive and (v) having good adhesion with the gold, which is the electrode material. Experiments were carried out to optimize on these properties.

Carbon black used in these experiments was Conductex 7067 Ultra from Columbian chemical company. Polymer composite samples were prepared with different weight percentage of CB conductive filler in SU-8 2002. The CB wt% is defined as

$$\text{CB wt\%} = \left(\frac{Mass_{CB}}{Mass_{CB} + Density_{SU-8} \times Volume_{SU-8}} \right) \times 100 \tag{6}$$

The composite was prepared by dispersing CB in SU-8 2002 by an ultrasonication step using probe sonicator. Sonication process generates lot of heat which was compensated with the help of an ice bath. Improved dispersion of CB in SU-8 was achieved by addition of Microchem Nano thinner 2000 to SU-8 2002 prior to sonication process that yields a spin coatable polymer composite.

With the addition of CB, the photosensitive nature of SU-8 2002 needs to be retained. It was observed that like SU-8, this material can be photolithograhically patterned but with an increased UV dose [37]. After development in the SU-8 developer solution, CB residues were seen even in the non patterned areas (Fig. 24 (a)). A separate ultrasonication step in iso-propyl alcohol was introduced in order to remove these carbon residues and the sample looked very clean after this step (Fig. 24 (b)).

Polymer composite thin films of varying filler concentrations were prepared and the thicknesses of these composite films were measured using contact stylus profilometer. It was observed that thickness increased with increase in CB filler concentration as shown in figure. 25 which can be attributed to the increase in the viscosity of SU-8/CB spin coatable resist with increased filler loading.

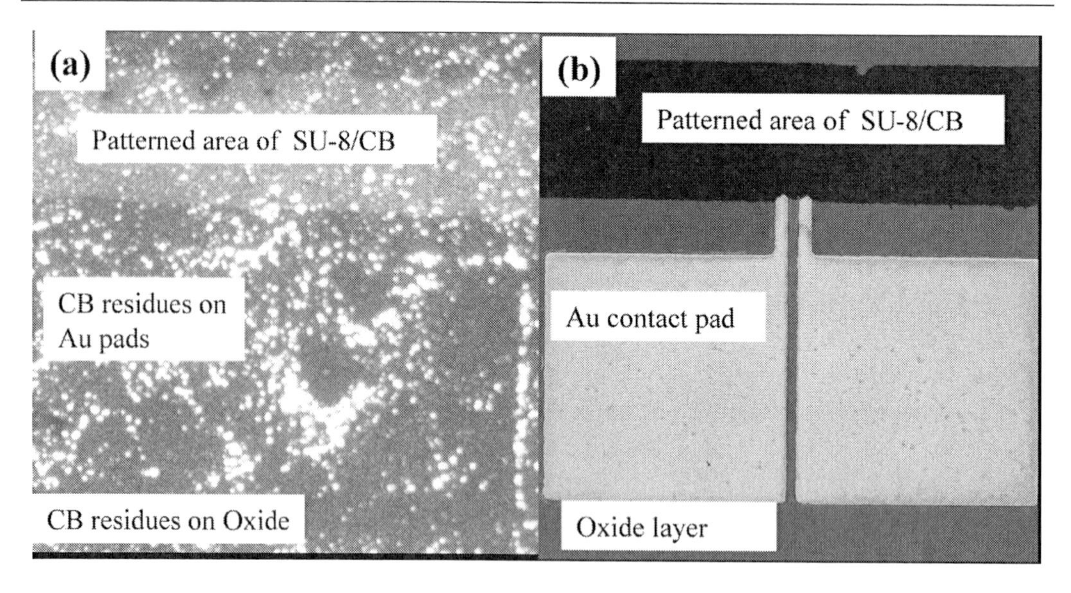

Figure 24. (a) Optical micrograph of photolithographically patterned SU-8/CB film before and (b) after ultrasonication clean step in iso-propyl alcohol [37].

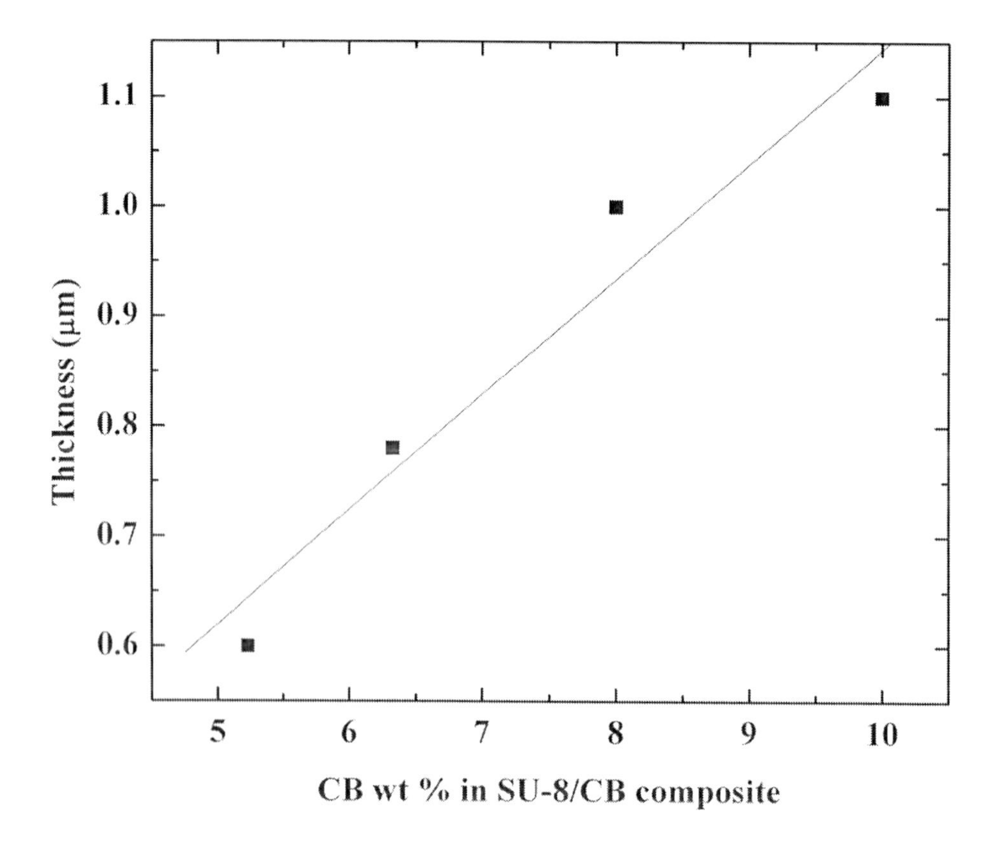

Figure 25. Thickness of SU-8/CB film as a function of CB wt% [37].

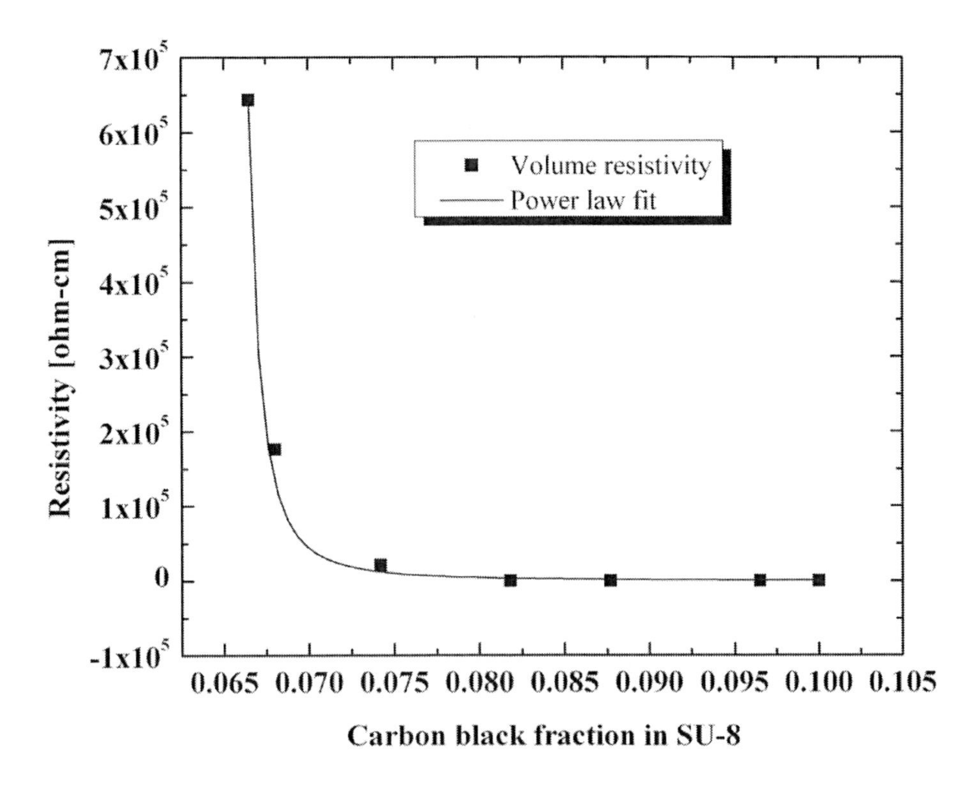

Figure 26. Percolation characteristics of SU-8/CB composites [37].

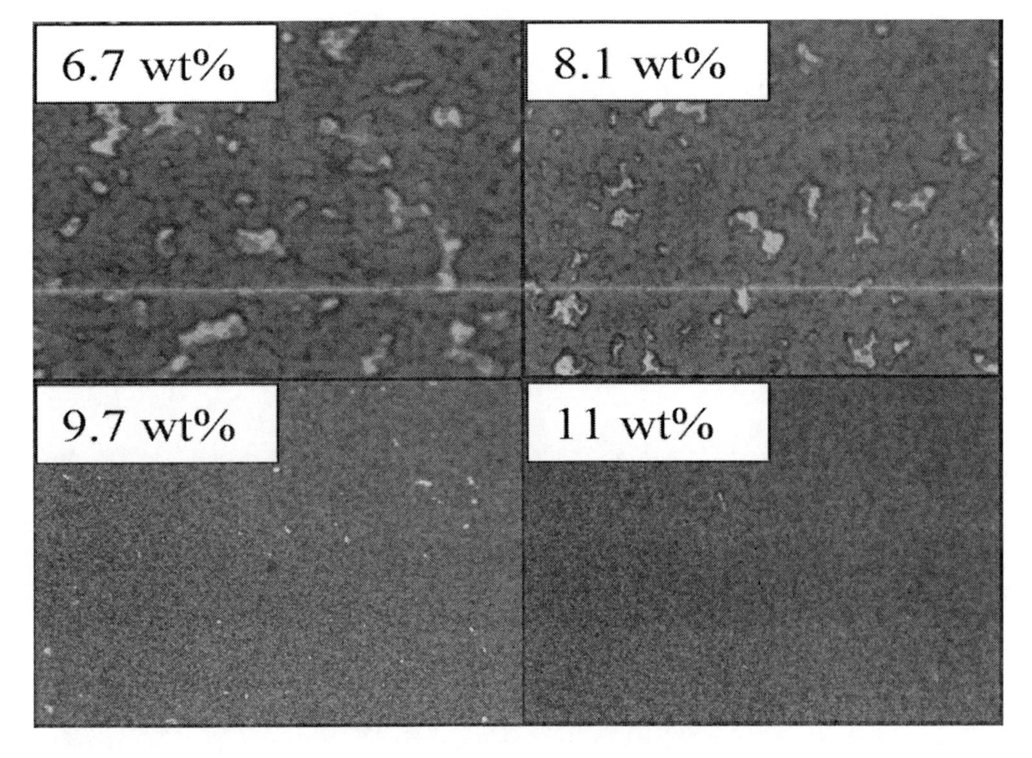

Figure 27. Optical micrograph (20X) of SU-8/CB film on SU-8 for different CB concentration showing the density of conductive filler network.

Next task is to study the nature of conductivity of the composite. For this oxidized silicon wafers were patterned with Cr/Au as contact pads and were dehydrated at 1200C for 30 minutes. SU-8/CB samples with different weight percentages were prepared and spin coated at 3000 rpm for 30 seconds with a spread cycle at 500 rpm for 5 seconds. Then a prebake at 68oC and 90oC was performed followed by a UV exposure in Karl Suss MJB3 standard aligner. Post exposure bake similar to softbake was done in order to complete the UV initiated cross linking. The samples were developed using SU-8 developer and finally ultrasonicated in IPA leaving SU-8/CB strip patterns between the Au electrodes (Fig. 24b.). The resistances of these samples were measured using standard Karl Suss probe station and Keithley 2602 source measuring unit. It was found that the resistivity decreases with increase in CB wt % (Fig. 26) and obeys a power law equation for percolation, given by resistivity, ρ ~ (f - fc)-t where f is the fraction of the conductive filler which in this case being the CB wt%, fc is the critical fraction known as the percolation threshold and t is the critical exponent.

This power law fit on the experimental data yielded a typical percolation threshold of 6.5 wt%. The conduction in these films is due to the formation of carbon black conductive network which becomes denser as the conductive particle concentration increases as shown in Fig. 27.

4.2. Device Fabrication

According to equation (1), the deflection sensitivity increases with decrease in the thickness of the cantilever, with increase in the distance of piezoresistive layer from neutral axis and with decrease in ratio of length of piezoresistor film to the length of the cantilever. The schematic of the cantilever die is given in Fig. 28.

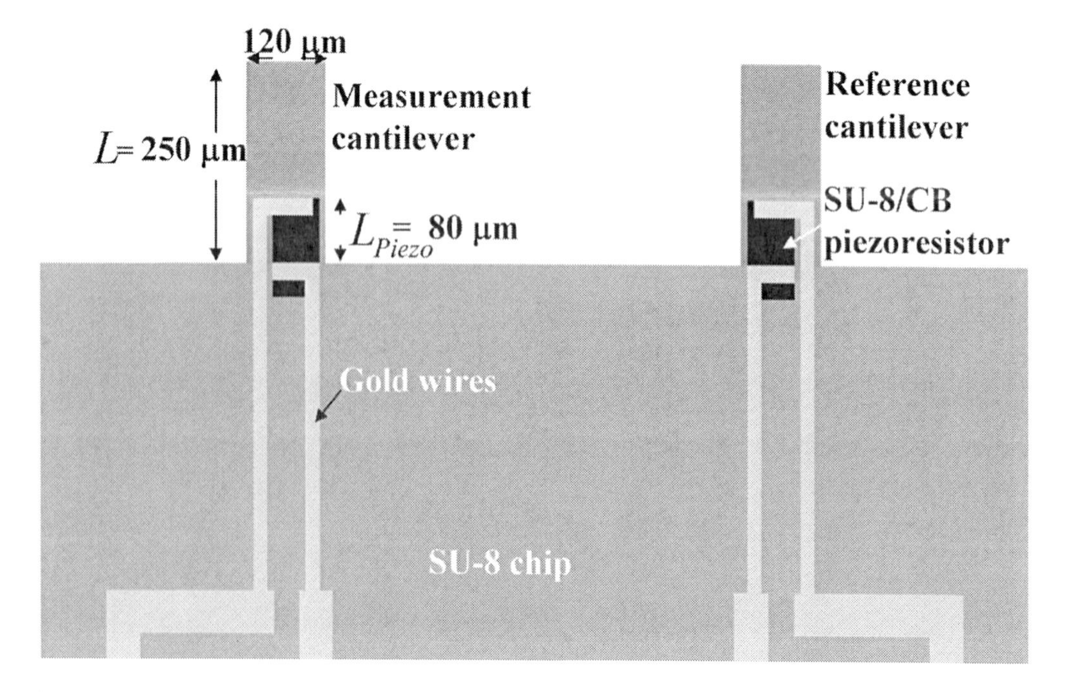

Figure 28. Schematic of the cantilever die.

The fabrication of these microcantilevers involves six levels of optical lithography (Fig. 29). A Silicon wafer that is used as a dummy substrate is piranha cleaned and a sacrificial layer of silicon dioxide is sputtered. Then the first SU-8 layer for cantilever is patterned following the regular SU-8 processing steps. Then the contact pads are formed using 20 nm/250 nm layer of Cr/Au followed by lift-off patterning of Ti/Au electrodes. SU-8/CB composite was then pattered on these gold electrodes followed by the encapsulation layer of SU-8. Next the anchor for cantilever and frames were formed by patterning 160 micron thick SU-8. The frames were finally released from the substrate by etching the sacrificial layer in BHF solution. The floating SU-8 frames in the HF solution are slowly soaked in DI water and the Cr layer on the gold contact pads is removed using Cr etchant. The optical and scanning electron microscopic images of the fabricated devices are shown in Fig. 30 and 31.

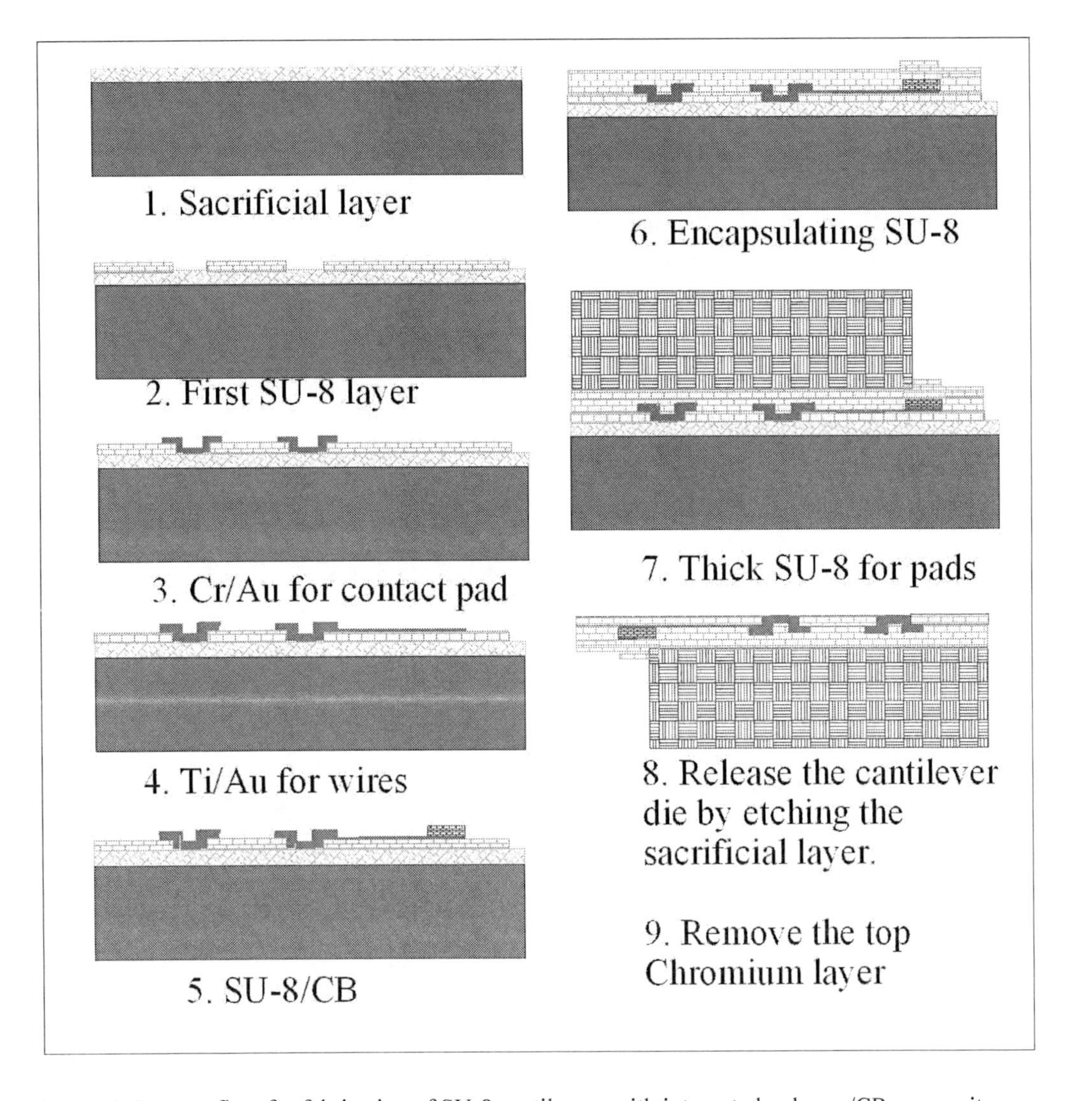

Figure 29. Process flow for fabrication of SU-8 cantilevers with integrated polymer/CB composites.

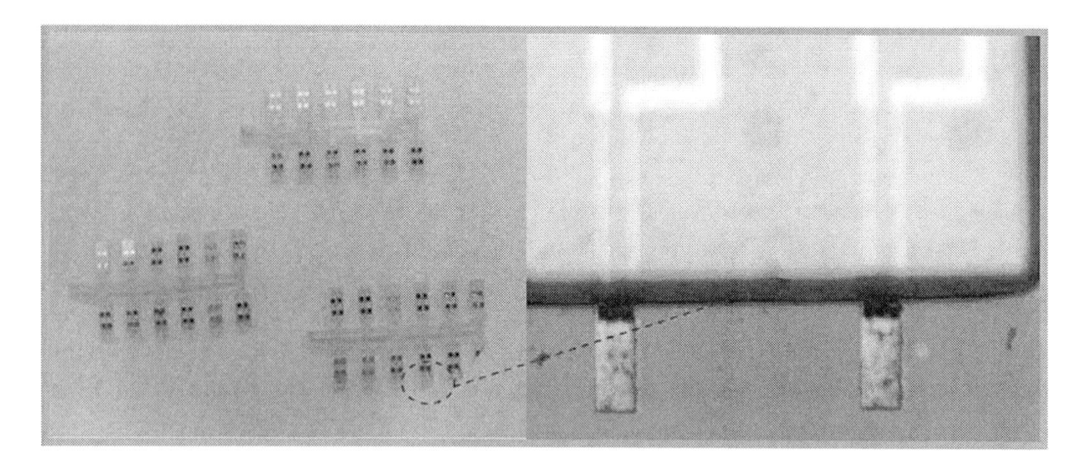

Figure 30. Optical micrographs of the fabricated cantilever die with two cantilevers.

Figure 31. SEM image of the released die showing straight cantilevers.

4.3. Characterization

The fabricated microcantilevers were electromechanically characterized to demonstrate their piezoresistive behavior. Deflection sensitivity which is an important performance parameter for piezoresistive microcantilevers is a measure of the relative change in resistance ($\Delta R/R$) of the embedded piezoresistor as a function of cantilever deflection. The measurement was performed by deflecting the cantilever using a calibrated micromanipulator with a resolution of 10 μm and simultaneous recording of resistance. The measurements were performed using Keithley 4200 source measuing unit with the microcantilevers connected inside a low noise shielded probe station from Suss Microtech. ΔR/R of piezoresistor as a function of deflection of tip of microcantilever is shown in Fig. 32.

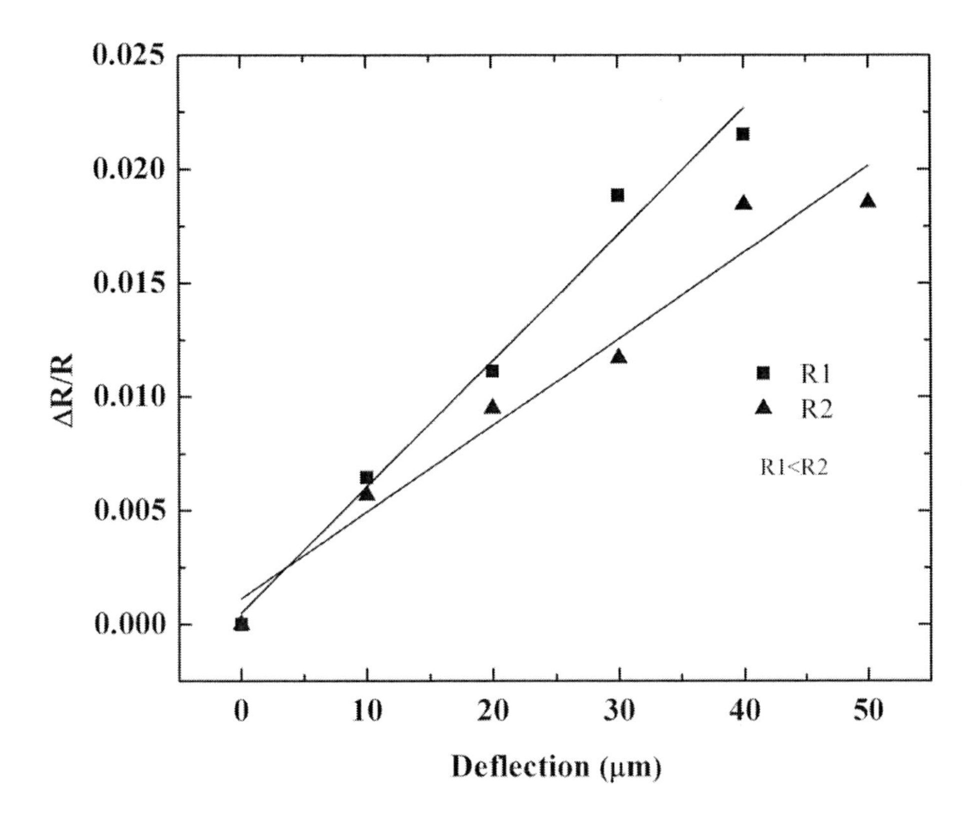

Figure 32. Electromechanical characterization of cantilevers for deflection sensitivity.

It is observed that deflection sensitivity is better for those cantilevers with a higher nominal resistance. The highest deflection sensitivity achieved with these devices is 0.55 ppm/nm which is significantly higher compared to the cantilevers with a Au strain gauge. The surface stress sensitivity for these devices is calculated using equation (2). It is assumed that the Young's modulus of SU-8/CB composite is identical to that of SU-8 (approx. 5 GPa [67]) and this assumption is reasonable since the carbon black loading at these low concentrations would not potentially increase the Young's modulus of SU-8/CB composite. The calculated surface stress sensitivity for these devices is 4.1x10-3 [N/m]-1 which is more than 10 times the sensitivity of the SU-8 cantilever with Au[6]. A typical antibody-antigen interaction surface stress of 5 mN/m [63] is therefore expected to give rise to a ΔR/R of around 20 ppm with the cantilevers presented in this work. Hence the sensitivity of these cantilevers is sufficient for sensing biomolecular interactions.

References

[1] Hideki Shirakawa, Edvin J. Louis, Alan G. MacDiarmid, Chwan K. Chiang and Alan J. Heeger, "Synthesis of Electrically Conducting Organic Polymers: Halogen Derivatives of Polyacetylene, $(CH)_x$," *J. Chem. Soc. Chem. Comm.*, pp. 578, 1977.

[2] Christos D. Dimitrakopoulos and Patrick R. L. Malenfant, "Organic Thin Film Transistors for large area electronics", *Adv. Mater.*, Vol.14, No. 2, pp. 99, 2002.

[3] S. A. Carter, M. Angelopoulos, S. Karg, P. J. Brock, and J. C, Scott, "Polymeric anodes for improved polymer light-emitting diode performance," *Appl. Phys. Lett.*, Vol. 70, pp. 2067, 1997.

[4] "Polymers in Electronics", Saurabh Goyal, V.Ramgopal Rao; Specialty Polymers: Materials and Applications, I.K. International Private Limited, Edited by Dr. Faiz Mohammad, Ed. 2007, *Category : Physical Sciences*, ISBN : 8188237655.

[5] C. Tang, and S. VanSlyke, "Organic Electroluminescent Diodes," *Appl. Phy. Lett.*, 51, pp. 913-915, 1987.

[6] L.J. Rothberg, and A.J. Lovinger, "Status of and prospects for organic electroluminescence," *Journal of Materials Research*, 11, pp. 3174-3187, 1996.

[7] P. E. Burrows, G. Gu, V. Bulovic, Z. Shen, S. R. Forrest, and M. E. Thompson, "Achieving full-color organic light-emitting devices for lightweight, flat-panel displays," *IEEE Transactions on electron devices* 44 (8), pp. 1188-1203, 1997.

[8] M. Shibusawa, M. Kobayashi, J. Hanari, et al, "A 17-inch WXGA full-color OLED display using the polymer ink-jet technology," *IEICE Transactions on Electronics* E86C, pp. 2269-2274, 2003.

[9] B.A. Gregg and M.C. Hanna "Comparing organic to inorganic photovoltaic cells: Theory, experiment, and simulation" *J. of Appl. Phy.*, 93, pp. 3605-3614, 2003.

[10] F.C. Krebs "Fabrication and processing of polymer solar cells: A review of printing and coating techniques" *Solar Energy Materials & Solar Cells*, 93, pp. 394-412, 2009.

[11] D. Braun and A. J. Heeger, "Visible light emission from semiconducting polymer diodes," *Appl. Phys. Lett.* 58, pp.1982-1984, 1991.

[12] Christoph J Brabec, Christoph winder, N. Serder Scriciftci, Jan C. Hummelen, Anantharaman Danabalan, Paul a van Hal, and Rene A. J. Janssen, "A Low-Bandgap Semiconducting Polymer for Photovoltaic Devices and Infrared Emitting Diodes,"*Adv. Funct. Mater.*, Vol. 12, Issue 10, pp. 709-712, 2002.

[13] Kazuhito Tsukagoshi, Jun Tanabe, Iwao yagi, Kunji Shigeto, Keiichi Yanagisawa and Yoshinobu aoyagi, "Organic light-emitting diode driven by organic thin film transistor on plastic substrates,"*J. Appl. Phys.*, 99, id. 064506, 2006.

[14] T. W. Kelley, L.D. Boardman, T. D. Dunbar, D. V. Muyres, M. J. Pellerite, and T. P. Smith, "High-Performance OTFTs Using Surface-Modified Alumina Dielectrics," *J. Phys. Chem. B.*, 107, 5877, 2003.

[15] H. Klauk, M. Halik, U. Zschieschang, F. Eder, D. Rohde, G. Schmid, and C. Dehm, "Flexible organic complementary circuits," *IEEE Trans. on Elect. Dev.*, 52, 618, 2005.

[16] Harshil N. Raval, S. P. Tiwari, Ramesh R. N., S. G. Mhaisalkar, and V. Ramgopal Rao, "Solution-Processed Bootstrapped Organic Inverters based on P3HT with High-k Gate Dielectric Material", *IEEE Electron Device Letters* 30, 484 – 486, 2009.

[17] Ramesh R. Navan, Harshil N. Raval, M. Shojaei Baghini, and V. Ramgopal Rao, *"Low Voltage Patterned Gate Pentacene Organic Circuits with Hafnium oxide High-K Gate Dielectric,"* 7th Organic Semiconductor Conference (OSC-09) London Heathrow Marriott, London, UK, 2009.

[18] Ramesh R. Navan, Rajesh A. Thakker, S. P. Tiwari, M. Shojaei Baghini, M. B. Patil,S. G. Mhaisalkar, and V. Ramgopal Rao, *"DC & Transient Circuit Simulation Methodologies for Organic Electronics"*, Proceedings of the IEEE International Workshop on Electron Devices & Semiconductor Technology, June 1-2, 2009.

[19] S. P. Tiwari, D. Maji, Srinivas P., Ramesh R. N., Harshil N. Raval, V. Ramgopal Rao, *"Patterned gate OFETs with inorganic High-K gate dielectric materials for all P-type Organic circuits,,"* Proceedings of the European Materials Research Society (EMRS) Spring Meeting, 2007.

[20] S.P. Tiwari, D.maji, Ramesh R. N., Harshil N. Raval, Srinivas P., V. Ramgopal Rao, *"Low Voltage Organic circuits with Polythiophene semiconductor and High-K Gate dielectric,"* 10th International Conference on Advanced Materials (ICAM), organized by the International Union of Materials Research Societies (IUMRS), 2007.

[21] S. P. Tiwari, D. Maji, Srinivas P., Harshil N. Raval, Nitin S. Kale and V. Ramgopal Rao, *"Ambipolar Organic Field Effect Transistors with semi conducting blends for CMOS-type Organic circuits,"* Proceedings of the European Materials Research Society (EMRS) Spring Meeting, 2007.

[22] S. P. Tiwari, V. Ramgopal Rao, Huei Shaun Tan, E. B. Namdas, Subodh G Mhaisalkar, *"Pentacene Organic Field Effect Transistors on Flexible substrates with polymer dielectrics,"* Proceedings of the 14 th International Symposium on VLSI Technology, Systems, and Applications (2007 VLSI-TSA) Hsinchu, Taiwan, 2007.

[23] Paragiotis E. Keivanidis et al, "X-Ray Stability and Response of Polymeric Photodiodes for Imaging Applications", *Appl. Phys. Lett.*, Vol. 92, Issue 2, id. 023304, 2008.

[24] T. Agostinelli et al, "A polymer/fullerene based photodetector with extremely low dark current for x-ray medical imaging applications", *Appl. Phys. Lett.*, Vol. 93, Issue 20, id. 203305, 2008.

[25] Heny Wohltjen, *"Organic Semiconductor Vapor Sensing Method",* US patent, US 4, 572, 900, 1986.

[26] Ravishankar S. Dudhe, S. P. Tiwari, Harshil N. Raval, Mrunal A. Khaderbad, Jasmine Sinha, M. Yedukondalu, M. Ravikanth, Anil Kumar, and V. Ramgopal Rao, "Explosive Vapor Sensor Using Poly (3-hexylthiophene) and Cu^{II} tetraphenylporphyrin composite based Organic Field Effect Transistors", *Appl. Phys. Lett.*, Vol. 93, Issue 26, id. 263306, 2008.

[27] Shengyang Tao, Guangtao Li and Hesun Zhu, "Metalloporphyrins as sensing elements for the rapid detection of trace TNT vapor", *J. Mater. Chem*, 16, 4521–4528. 2006.

[28] Shengyang Tao, Zhenyu Shi, Guangtao Li, and Pei Li, "Hierarchically Structured Nanocomposite Films as Highly Sensitive Chemosensory Materials for TNT Detection," *Chem. Phys. Chem.* 7, pp. 1902 – 1905, 2006.

[29] Jose M. Pedrosa, Colin M. Dooling, Tim H. Richardson, Robert K. Hyde, Chris A.Hunter, Marıa T. Martın, and Luis Camacho, "Influence of molecular organization of asymmetrically substituted porphyrins on their response to NO^2 gas,"*American Chemical Society,* 18, 7594-7601, 2002.

[30] Ravishankar S. Dudhe Jasmine Sinha, Anil Kumar and V. Ramgopal Rao, "Polymer composite-based OFET sensor with improved sensitivity towards nitro based explosive vapors", *Sensors and Actuators*, Accepted for publication, 2010. doi:10.1016/j.snb.2010.04.022 .
[31] Mahmudur Rahman, H. James Harmon, "Absorbance change and static quenching of fluorescence of meso-tetra(4-sulfonatophenyl)porphyrin (TPPS) by trinitrotoluene (TNT)," *Spectrochimica Acta Part A: Molecular and Biomolecular Spectroscopy*, Vol. 65, Issues 3-4, pp. 901-906, 2006.
[32] L. Torsi, A.Dobalapur, N. Cioffi, L. Sabbatini and P.G. Zambonin, "Multi-parameter gas sensors based on organic thin-film-transistors," *Sensors and Actuators B Chemical*, Volume 67, Issue 3, pp. 312-316, 2000.
[33] E. A. B. Silva et al, "Low dose ionizing radiation detection using conjugated polymers", *App. Phys. Lett.*, Vol. 86, Issue 13, id. 131902, 2005.
[34] Harshil N. Raval, Shree Prakash Tiwari, Ramesh R. Navan, and V. Ramgopal Rao, "Determining Ionizing Radiation using Sensors Based on Organic Semiconducting Material", *Appl. Phys. Lett.*, Vol. 94, Issue 12, id. 123304, 2009.
[35] Jose M. Lobez and Timothy M. Swager, "Radiation Detection: Resistivity Responses in Functional Poly (Olefin Sulfone)/Carbon Nanotube Composites", *Chem. Int. Ed.,* Vol. 49, Issue 1, pp. 95-98, 2010.
[36] V. Seena, N. S. Kale, S. Nag,M. Joshi, S. Mukherji, V. R. Rao, Developing a polymeric microcantilever platform technology for biosensing applications", *Int.J. Micro and Nano Syst.*, vol.1, pp.65-70, 2009.
[37] V. Seena, A. Rajorya, P. Pant, S. Mukherji and V. R. Rao., "Polymer microcantilever biochemical sensors with integrated polymer composites for electrical detection" , *Solid State Sci.*,vol.11,pp.1606-1611, 2009.
[38] S. Hoshino, M. Yoshida, S. Uemura, T. Kodzasa, N. Takada, T. Kamata and Kiyoshi Yase "Influence of moisture on device characteristics of polythiophene-based field-effect transistors", *J. Appl. Phys. Lett.*, Vol.95, 2004.
[39] Huiping Jia, Srinivas Gowrisanker, Gaurang K. Pant, Robert M. Wallace, and Bruce E. Gnade, "Effect of poly"3-hexylthiophene film thickness on organic thin film transistor properties," *J. Vac. Sci. Technol.* Vol. A 24, pp. 1228-1232, 2006.
[40] P. F. Baude, D. A. Ender, M. A. Haase, T. W. Kelley, D. V. Muyres and S. D. Theiss, "Pentacene-based radio-frequency identification circuitry," *Appl. Phys. Letters,* Vol. 82, No. 22, pp. 3964-66, June 2003.
[41] M. B. Patil, SEQUEL Users Manual, available at *http://www.ee.iitb.ernet.in/microel/faculty/mbp/sequel1.html.*
[42] R. A. Thakker, N. Gandhi, M.B. Patil and K.G. Anil, "Parameter extraction for PSP MOSFET model using Particle swarm optimization," *Proc. Int. Workshop Phy, of Sem. Dev.*, pp. 130-133, 2007.
[43] J Kennedy and R. C Eberhart., "*Particle swarm optimization*", Proc. of IEEE Int. Conf. on Neural Networks, Piscataway, NJ, pp. 1942-1948, 1995.
[44] Ramesh R. Navan, Rajesh A. Thakker, S. P. Tiwari, M. Shojaei Baghini, M. B. Patil,S. G. Mhaisalkar, and V. Ramgopal Rao, "*DC & Transient Circuit Simulation Methodologies for Organic Electronics*", Proceedings of the IEEE International Workshop on Electron Devices & Semiconductor Technology, June 1-2, 2009.

[45] D. Vinay Kumar, Nihar R. Mohapatra, Mahesh B. Patil, and V. Ramgopal Rao, *"Application of Look-up Table Approach to High-K Gate Dielectric MOS Transistor circuits",* Proceedings of the 16th International Conference on VLSI Design, (VLSI03) 2003.

[46] V. I. Arkhipovt, V. A. Kolesnikovi and A. I. Rudenkot, "Dispersive transport of charge carriers in polycrystalline pentacene layer", *J. Phys. D: Appl. Phys.* 17, 1984.

[47] David C. Prentice, *"Simulation of Pentacene Organic Metal-Oxide Field Effect Transistors"*, Division of Research and Advanced Studies of the University of Cincinnati, 2003.

[48] E. S. Schlig and J. L. Sanford, "An SXGA reflective liquid crystal projection light valve incorporating inversion by pixel bootstrapping," *IBM J. Res. Develop.*, vol. 44, no. 6, pp. 909–918, 2000.

[49] L. Lee, S. Al-Sarawi, and D. Abbott, "Dynamic bootstrapped shift register for smart sensor arrays," *Smart Mater. Struct.*, vol. 14, no. 4, pp. 569–574, Aug. 2005.

[50] S. H. Han, S. M. Cho, J. H. Kim, J. W. Choi, J. Jang, and M. H. Oh, "Ring oscillator made of organic thin-film transistors produced by selforganized process on plastic substrate," *Appl. Phys. Lett.*, vol. 89, no. 9, pp. 093 504-1–093 504-3, Aug. 2006.

[51] S. H. Han, S. M. Cho, J. H. Kim, J. W. Choi, J. Jang, and M. H. Oh, "Ring oscillator made of organic thin-film transistors produced by selforganized process on plastic substrate," *Appl. Phys. Lett.*, vol. 89, no. 9, pp. 093 504-1–093 504-3, Aug. 2006.

[52] J. S. Yang, T. M. Swager, "Fluorescent Porous Polymer Films as TNT Chemosensors: Electronic and Structural Effects, " *J. Am. Chem. Soc.* 120 (1998) 11864-11873.

[53] A Assadi, G Gustafsson, M Willander, C Svensson and O. Inganäs, "Determination of field-effect mobility of poly(3-hexylthiophene) upon exposure to NH_3 gas" *Synthetic Metals,* Vol. 37, Issues 1-3, pp. 123-130, 1990.

[54] S. M. Sze, *Semiconductor Devices Physics and Technology* (Wiley, New York, 1981.

[55] T. K. Chandrashekar and V. Krishnan, "Donor properties of metallomacrocyclic tetrapyrrole pigments with sym-trinitrobenzene, " *Inorg. Chem.* 20, pp. 2782, 1981.

[56] S. Tao, G. Li, and H. Zhu, "Metalloporphyrins as sensing elements for the rapid detection of trace TNT vapor, " *J. Mater. Chem.* 16, 4521, 2006.

[57] G. Zotti and S. Zecchin, "Decay of electrochemically injected polarons in polythiophenes. Bipolaron stabilization by structural factors," *Synth. Met.* 87, 115, 1997.

[58] Z. Wei, J. Xu, S. Pu, and Y. Du, "Electrosyntheses of high-quality freestanding poly(fluorene-*co*-3-methylthiophene) films with tunable fluorescence properties," *J. Polym. Sci., Part A: Polym. Chem.* 44, 4904, 2006.

[59] M. Berggren, O. Inganas, G. Gustafsson, M. R. Andersson, T. Hjertberg, and O. Wennerstrom, "Controlling color by voltage in polymer light emitting diodes," *Synth. Met.,* 71, pp 2185-2186, 1995.

[60] Krumm, J. E. Eckert, W. H. Glauert, A. Vllmann, W. fix and W. Clemens, "A polymer transistor circuit, " *IEEE Elect. Dev. Lett.*, Vol. 25, pp. 399 – 401, 2004.

[61] S. P. Tiwari, Srinivas P., Shriram S, Nitin S. Kale, Subodh G Mhaisalkar, V. Ramgopal Rao, "Organic FETs with Hot-wire CVD (HWCVD) Silicon Nitride as a Passivation and Gate Dielectric Layer", *Thin Solid Films*, Vol. 516, Issue 5, pp. 770-772, 2008.

[62] S. P. Mohanty and E. Kougianos, "Biosensors: A tutorial review, " *IEEE Potentials,* 25(2), pp. 35-40, 2006.

[63] L. C. Clark and C. Lyons "Electrode Systems For Continuous Monitoring in Cardiovascular Surgery," *Ann. N. Y. Acad. Sci.* 102 , p. 29, 1962.

[64] L. G. Carrascosa, M. Moreno, M. A′ lvarez, L.M. Lechuga_Nanomechanical biosensors: a new sensing tool, *Trends in Analytical Chemistry,* 2005.

[65] W. P. King, T. W. Kenny, K. E. Goodson, G. L. W. Cross, M. Despont, U. T. Dürig, H. Rothuizen, G. Binnig, and P. Vettiger, "Design of atomic force microscope cantilevers for combined thermomechanical writing and thermal reading in array operation," *J. Microelectromech. Syst.*, vol. 11, pp. 765–774, 2002.

[66] R. Berger, H. P. Lang, Ch. Gerber, J.K. Gimzewski, J.H. Fabian, L. Scandella, E. Meyer and H.-J. Güntherodt, "Micromechanical thermogravimetry" , *Chem. Phys. Lett.,vol.* 294, pp. 363–369, 1998.

[67] L. T. Chen, CY. Lee and W. H. Cheng, "MEMS-based humidity sensor with integrated temperature compensation mechanism", *Sens. and Act. A:Phys.*, vol.147, pp.522-528, 2008.

[68] S. H. Lim, D. Raorane, S. Satyanarayana,, A. Majumdar, "Nano-chemo-mechanical sensor array platform for high-throughput chemical analysis", *Sen. and Act.. B*, pp.466–474, 2006.

[69] L. A. Pinnaduwage, J. E. Hawk, V. Boiadjiev, D. Yi, and T. Thundat, "Use of Microcantilevers for the Monitoring of Molecular Binding to Self-Assembled Monolayers " *Langmuir*,vol.19, pp.7841-7844, 2003.

[70] L. Senesac and T. Thundat " Nanosensors for Trace Explosive Detection", *Materials Today*, vol. 11, pp.28-33, Mar. 2008.

[71] L. G. Carrascosa, M. Moreno, M. Álvarez and L. M. Lechuga "Nanomechanical biosensors: a new sensing tool", *Trends in Anal. Chem.*,vol. 25, pp.196-206, 2006.

[72] Y. Arntz, J. D. Seelig, H. P. Lang, J. Zhang, P. Hunziker, J. P. Ramseyer, E. Meyer, M. Hegner and C. Gerber , "Label-free protein assay based on a nanomechanical cantilever array," *Nanotechnology*, vol. 14, no. 1, pp. 86–90, Jan. 2003.

[73] C. Ziegler, "Cantilever-based biosensors", *Anal Bioanal Chem*, vol. 379, pp. 946–959, 2004.

[74] J. Fritz J, M. K. Baller, H. P. Lang, H. Rothuizen, P. Vettiger, E. Meyer, H. Güntherodt, C. Gerber, J. K. Gimzewski, "Translating biomolecular recognition into nanomechanics," *Science*, vol. 288, no. 5464, pp. 316–318, Apr. 2000.

[75] S. Hosaka, T. Chiyoma, A. Ikeuchi, H. Okano, H. Sone, T. Izumi , "Possibility of a femtogram mass biosensor using a self-sensing cantilever" , *Current Applied Physics* vol.6 ,pp.384–388, 2006.

[76] J. Thaysen, A. D. Yalcinkaya, P. Vettiger and A. Menon, "Polymer-based stress sensor with integrated readout ", *J. Phys. D: Appl. Phys.* vol. 35 ,pp. 2698–2703, 2002.

[77] N. S. Kale, S. Nag, R. Pinto, V. Ramgopal Rao,"Fabrication and Characterization of a Polymeric Microcantilever with an Encapsulated Hotwire CVD Polysilicon Piezoresistor", *J. Microelectromech. Syst.*, vol.18, 2009.

[78] L. Gammelgaard, P. A. Rasmussen, M. Calleja, P. Vettiger, and A. Boisen, "Microfabricated photoplastic cantilever with integrated photoplastic/carbon based piezoresistive strain sensor," *App. Phys.Lett.,* vol.88, 113508, 2006.

In: Organic Semiconductors
Editor: Maria A. Velasquez
ISBN 978-1-61209-391-8

Chapter 4

QUANTUM-CHEMICAL DESIGN OF NEW ORGANIC SEMICONDUCTORS: MOLECULAR DESIGN GUIDELINES FOR FULLERENE DERIVATIVES

Ken Tokunaga*
General Education Department, Kogakuin University, Tokyo, Japan

Abstract

Fullerene C_{60} and its derivatives are recently expected as very useful organic semiconductors for organic devices such as the organic field-effect transistor (OFET). In this Chapter, effect of chemical addition on carrier transport properties of C_{60} is theoretically estimated and systematically discussed by the density functional theory (DFT) calculation, taking fullerene derivative $C_{60}X_n$ as examples. Based on the Marcus theory, carrier transport properties are related to the reorganization energy. Hole-transport property of C_{60} derivatives is strongly dependent on the position and the number of added groups, but is almost independent of chemical properties of added groups. On the other hand, electron-transport property of C_{60} is little influenced by the chemical addition. These results are discussed from viewpoints of geometric and electronic structures. It is found that the values of reorganization energies are almost proportional to the degree of geometrical relaxation upon the carrier injection. Delocalization of frontier molecular orbitals on C_{60} sphere results in small reorganization energy and fast carrier transport. From these analyses, specific guidelines for efficient design of useful organic semiconductors are proposed.

Keywords: Organic Devices, Fullerene, Carrier Transport, Marcus Theory, Reorganization Energy, Electronic Coupling, Density Functional Theory

1. Introduction

Organic semiconductors have been used as essential parts of organic light-emitting devices (OLEDs) and organic field-effect transistors (OFETs) which are expected to be used in next-generation technologies [1] because of the lightness and flexibility. Carrier mobility [2] in

*E-mail address: tokunaga@cc.kogakuin.ac.jp

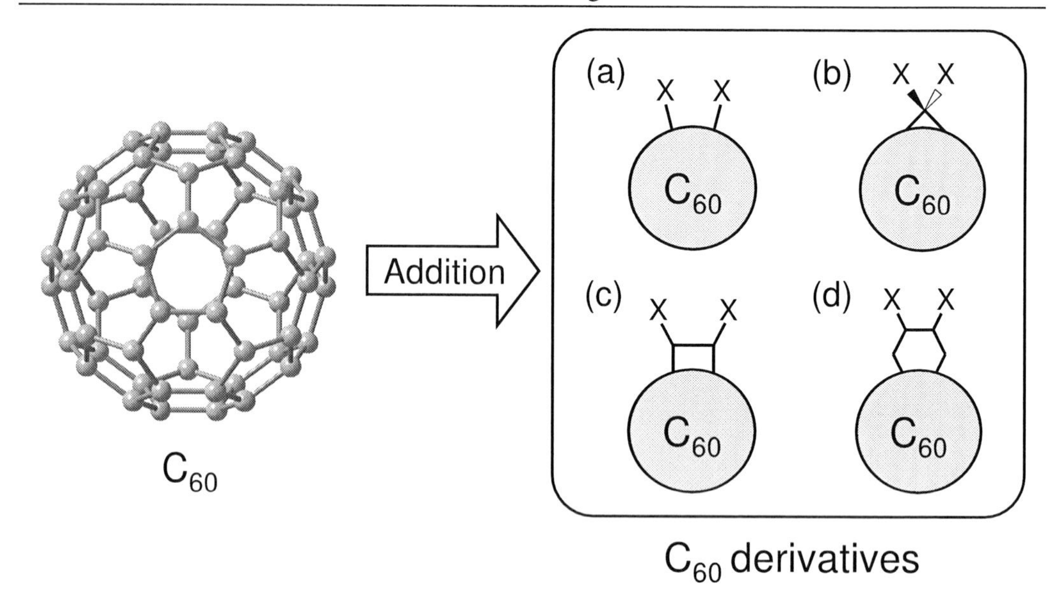

Figure 1. Fullerene C_{60} and some examples of C_{60} derivatives.

organic semiconductor is one of the most important properties in the performance of these devices. Therefore, research and development of new materials with chemical and thermal stability has been recently very active [3, 4, 5, 6]. However, the wide variety of organic materials, which is generally a major advantage of these materials, has been hindering the systematic research and development of new materials. Thus, the establishment of design guidelines for new materials is a matter of great urgency.

Theoretical investigations can give reliable guidelines for the development of such new organic semiconductors. Up to now, much effort has been made to theoretically understand the relationship between the structure and the hole-transport properties of the materials [7, 8, 9, 10]. We have also been studying quantum-chemical design of organic semiconductors based on fullerenes from both scientific and technological viewpoints [11, 12, 13, 14, 15, 16]. C_{60} derivative (Fig. 1) is very interesting from the viewpoint of practical use. Some types of C_{60} derivatives are shown in the figure, and C_{60} derivatives of type (a) and type (b) have been studied in our research groups [11, 12, 13, 14, 15, 16].

It is well known that C_{60} is chemically and thermally stable, and its method of synthesis is also established. Carrier mobility of the amorphous silicon is about 1 $cm^2V^{-1}s^{-1}$ so that the mobility above this value is desirable for organic semiconductors. Hole mobility in C_{60} film is in the 10^{-4} $cm^2V^{-1}s^{-1}$ range [17] so that C_{60} has not been used as hole-transport material (p-channel semiconductor). However, there is a possibility that the chemical addition to C_{60} results in the considerable modification of C_{60} materials. This possibility originates from the fact that C_{60} has electronic degeneracy in its cationic state to due its high symmetry I_h [18, 19] but C_{60} derivatives do not because of symmetry reduction by addition. Thus, the main purpose of our previous studies [11, 12, 15, 16] was to improve the hole mobility of C_{60} by chemical method, that is the hydrogenation. On the other hand, electron mobility in C_{60} film is about 1 $cm^2V^{-1}s^{-1}$ [17, 20, 21] so that C_{60} is one of the most useful electron-transport material. However, in order to realize low-cost production and large-area devices, it is necessary for C_{60} material to have solution-processable form such as [6,6]-phenyl C_{61}-butylic acid methyl ester (PCBM) [22]. Therefore, in the organic electronics, electronic properties of C_{60} derivatives rather than those of the original C_{60} are of much interest and importance. Therefore, analysis of both hole and electron transport properties of C_{60} derivatives is very important for the practical use of C_{60} materials.

In this Chapter, the effect of chemical addition of X to C_{60} on carrier transport properties is systematically discussed from the viewpoint of reorganization energy (λ) of Marcus theory [23]. We focus on the C_{60} derivatives with type (a) in Fig. 1, $C_{60}X_n$, where X is the added group, n is the numbers of added X. There are many isomers for $C_{60}X_n$ so that the position of added X is also an object of discussion. Dependence of carrier-transport property on the kind or chemical nature of added group (X) are discussed from the results of $C_{60}X_2$ (X=H, C_3H_6COOH, and C_4H_8SH). Dependence on the number of added groups are discussed from the results of $C_{60}H_n$ (n=2, 4, and 6). From these discussion, guidelines for effective design of high-performance carrier-transport materials is proposed.

This Chapter is organized as follows: In Section 2, the definition of reorganization energy (λ) and computational details of λ are presented. Structures of $C_{60}X_n$ studied in this Chapter are shown in Section 3. In Section 4, calculated results of λ of both hole and electron transport are shown. From the systematic discussion of these results, dependence of λ on the number (n), the position, and the kind (X) of added groups is revealed in Section 5. Summarizing these discussions, simple guidelines for efficient design of useful C_{60} semiconductors are proposed in the last Section.

2. Carrier Transport Mechanism

2.1. Hopping Mechanism

Carrier mobility of single-crystal C_{60} is around $1\,\mathrm{cm^2V^{-1}s^{-1}}$ at a maximum. Materials having mobility of $0.1-1\,\mathrm{cm^2V^{-1}s^{-1}}$ are categorized into the the boundary region between the band transport and the hopping mechanism [24, 25, 26]. In this work, only the hopping mechanism is considered because the mobilities of C_{60} derivatives are as large as or smaller than $0.1\,\mathrm{cm^2V^{-1}s^{-1}}$ [22, 27, 28, 29].

Schematic picture of carrier hopping mechanism in organic solid is shown in Fig. 2. C_{60} and $C_{60}X_n$ molecules are represented by spheres in Fig. 2 (a). In hopping mechanism treatment, two neighboring molecules are picked up from solid. Then a carrier hopping between these two molecules, that is carrier-transfer reaction, is considered. Repeating such a hopping between neighboring molecules, a carrier travels from the one edge (electrode) to the other edge (electrode) of the organic solid. Initial state and final state of a carrier-transfer reaction is represented as $\mathrm{M}^n(\mathrm{A})\cdots\mathrm{M}(\mathrm{B})$ and $\mathrm{M}(\mathrm{A})\cdots\mathrm{M}^n(\mathrm{B})$, respectively, where n means one hole (h) or one electron (e). Schematic energy diagram of the reaction is shown in Fig. 2 (b). The system is fluctuating around the bottom of potential curve by the molecular vibrations in its initial state. When the system reaches transition state (TS) at which energies of the initial state and that of the final state are same, the system jumps from the initial state to the final state in the rate k. This reaction is written as

$$\mathrm{M}^n(\mathrm{A})\cdots\mathrm{M}(\mathrm{B}) \xrightarrow{k} \mathrm{M}(\mathrm{A})\cdots\mathrm{M}^n(\mathrm{B}). \tag{1}$$

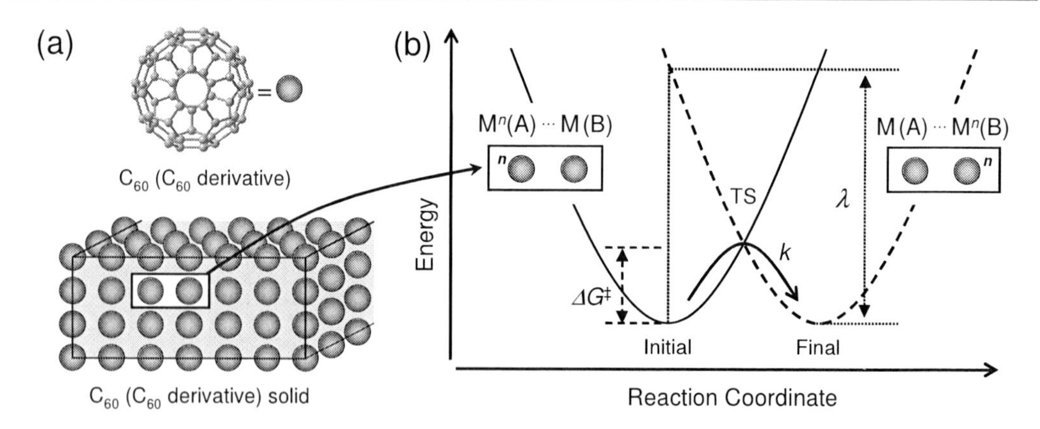

Figure 2. (a) Schematic pictures of C_{60} (C_{60} derivatives) solid. Carrier hopping between two molecules is focused. (b) Schematic reaction diagram of carrier-transfer reaction $M^n(A) \cdots M(B) \rightarrow M(A) \cdots M^n(B)$, where n means hole (h) or electron (e). $\Delta G^{\ddagger}$, λ, and k are potential barrier, reorganization energy, and carrier-transfer rate constant of the reaction.

Localized carrier on one molecule (A) jumps to the neighboring molecule (B). From the Marcus theory [23], the carrier-transfer rate constant k of a homogeneous carrier-transfer reaction can be estimated from two parameters, reorganization energy (λ) and the electronic coupling element (H) between adjacent molecules:

$$k = \frac{4\pi^2}{h} \frac{H^2}{\sqrt{4\pi\lambda k_{\mathrm{B}} T}} \exp\left(-\frac{\lambda}{4k_{\mathrm{B}}T}\right), \tag{2}$$

where k_{B} is the Boltzmann constant, h is the Planck constant, and T is the temperature of the system. We can see that a small λ, large H, and high temperature T result in an effective a fast carrier hopping.

2.2. Reorganization Energy

From the Fig. 2(b), we can see that the reorganization energy (λ) is the difference between "energy of final state with stable nuclear configuration in *final* state" and "energy of final state with stable nuclear configuration in *initial* state". For the calculation of λ, potential energy diagrams of hole-transfer reaction (a) and electron-transfer reaction (b) are shown in Fig. 3. The values λ is obtained by the following procedure: First, the geometries of neutral C_{60} and $C_{60}X_n$ are fully optimized, giving a nuclear coordinate Q_0 and energy E_0^0 of Figs. 3 (a) and (b). At Q_0 of Fig. 3 (a), single-point energy calculations of cations give E_0^+. Next, the structures are fully optimized in their cationic states, giving Q_+ and E_+^+. Single-point energy calculations of the neutral states with geometry Q_+ give E_+^0. Reorganization energy of hole-transfer reaction (λ_{h}) is defined as the summation of λ_{h}^+ and λ_{h}^0 defined as

$$\lambda_{\mathrm{h}}^+ = E_0^+ - E_+^+, \tag{3}$$

$$\lambda_{\mathrm{h}}^0 = E_+^0 - E_0^0, \tag{4}$$

$$\lambda_{\mathrm{h}} = \lambda_{\mathrm{h}}^+ + \lambda_{\mathrm{h}}^0. \tag{5}$$

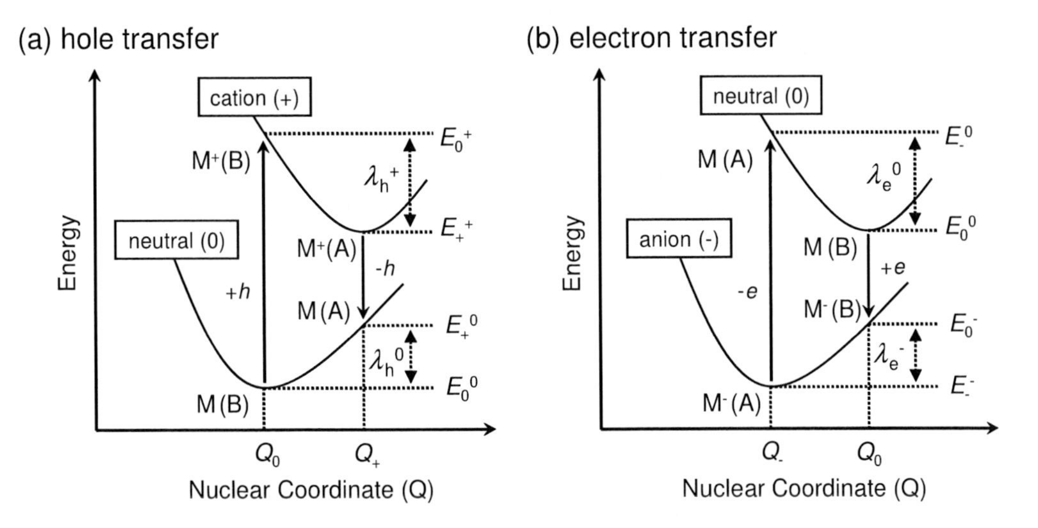

Figure 3. Schematic potential energy surfaces of molecules related to (a) hole-transfer reaction and (b) electron-transfer reaction. Q_0, Q_+, and Q_- mean the nuclear coordinates of stable structure in neutral, cationic, and anionic states. Subscripts on the right hand side of E mean the geometrical structure of the molecule and superscripts on the right hand side of E mean the charge of the molecule.

λ_h^+ and λ_h^0 are relaxation energies in cationic and neutral state shown in Fig. 3 (a), respectively. Similarly, at Q_0 of Fig. 3 (b), single-point energy calculations of anion give E_0^-. Next, the structures are fully optimized in their anionic states, giving Q_- and E_-^-. Single-point energy calculations of the neutral states with geometry Q_- give E_-^0. Reorganization energy of electron-transfer reaction (λ_e) is defined as the summation of λ_e^- and λ_e^0 defined as

$$\lambda_\mathrm{e}^- = E_0^- - E_-^-, \tag{6}$$

$$\lambda_\mathrm{e}^0 = E_-^0 - E_0^0, \tag{7}$$

$$\lambda_\mathrm{e} = \lambda_\mathrm{e}^- + \lambda_\mathrm{e}^0. \tag{8}$$

λ_e^- and λ_e^0 are relaxation energies in anionic and neutral state shown in Fig. 3 (b), respectively.

In this Chapter, calculation and analysis of λ_h and λ_e are focused and electronic coupling element H is set constant for all $\mathrm{C_{60}}$ and $\mathrm{C_{60}}X_n$. This is because that H is dependent on the distance and orientation of the two molecules so that its calculation and analysis is difficult [30]. Therefore, we can not know the numerical values of rate constants k. Instead of numerical values, ratio of rate constants between $\mathrm{C_{60}}$ and $\mathrm{C_{60}}X_n$ is calculated as

$$k_\mathrm{h} = \left(\frac{H_\mathrm{h}^{\mathrm{C_{60}}X_n}}{H_\mathrm{h}^{\mathrm{C_{60}}}}\right)^2 \cdot \sqrt{\frac{\lambda_\mathrm{h}^{\mathrm{C_{60}}}}{\lambda_\mathrm{h}^{\mathrm{C_{60}}X_n}}} \cdot \exp\left(-\frac{\lambda_\mathrm{h}^{\mathrm{C_{60}}X_n} - \lambda_\mathrm{h}^{\mathrm{C_{60}}}}{4k_\mathrm{B}T}\right) \tag{9}$$

$$\approx \sqrt{\frac{\lambda_\mathrm{h}^{\mathrm{C_{60}}}}{\lambda_\mathrm{h}^{\mathrm{C_{60}}X_n}}} \cdot \exp\left(-\frac{\lambda_\mathrm{h}^{\mathrm{C_{60}}X_n} - \lambda_\mathrm{h}^{\mathrm{C_{60}}}}{4k_\mathrm{B}T}\right) \tag{10}$$

$$k_e = \left(\frac{H_e^{C_{60}X_n}}{H_e^{C_{60}}}\right)^2 \cdot \sqrt{\frac{\lambda_e^{C_{60}}}{\lambda_e^{C_{60}X_n}}} \exp\left(-\frac{\lambda_e^{C_{60}X_n} - \lambda_e^{C_{60}}}{4k_B T}\right) \tag{11}$$

$$\approx \sqrt{\frac{\lambda_e^{C_{60}}}{\lambda_e^{C_{60}X_n}}} \cdot \exp\left(-\frac{\lambda_e^{C_{60}X_n} - \lambda_e^{C_{60}}}{4k_B T}\right) \tag{12}$$

on the supposition that $H_h^{C_{60}X_n} \approx H_h^{C_{60}}$, $H_e^{C_{60}X_n} \approx H_e^{C_{60}}$, and T=300 K, where H_h and H_e are the electronic coupling elements of hole- and electron-transfer reactions. Hereafter, λ and k are considered as carrier-transport property and carrier mobility, respectively.

All calculations (geometrical optimizations and self-consistent field (SCF) energy calculations) necessary to obtain the values of energies in Fig. 3 were performed by the quantum-chemical method, that is the density functional theory (DFT) using B3LYP functional. For the calculation of λ_h, 6-311G(d) basis set is adopted. And, for the calculation of λ_e, 6-311++G(d,p) basis set is adopted for $C_{60}H_n$(n=2, 4, and 6) and 6-31+G(d) basis set is adopted for $C_{60}X_2$(X=C_3H_6COOH and C_4H_8SH). All neutral (ionic) systems are calculated in singlet (doublet) states. Calculations were performed by using the GAUSSIAN 03 [31] program package.

3. Fullerene Derivatives

3.1. Hydrogenated Fullerenes $C_{60}H_n$ (n=2, 4, and 6)

We consider three kinds of hydrogenated fullerenes: $C_{60}H_2$, $C_{60}H_4$, and $C_{60}H_6$. The purpose of picking up these derivatives is to clarify the dependence of λ on the number (n) of the addition groups. For each $C_{60}H_n$, many isomers are considered in order to clarify the dependence of λ on the position of the addition groups.

3.1.1. $C_{60}H_2$

There are a total of 23 isomers for $C_{60}H_2$ [32]. 11 isomers of Fig. 4(a) with a small formation energy were selected to consider the possibility of synthesis. Ground state of these isomers is singlet state. Other isomers which are not calculated in this chapter have triplet ground states in semi-empirical calculations [32]. The initial hydrogen atom of $C_{60}H_2$ is already shown in the figure. Second hydrogen atom is added to one of the carbon atoms labeled 1-11. These isomers are named **1** - **11** in bold face. As predicted by Matsuzawa *et al.* [32] by quantum-chemical calculations, two isomers, **1** from 1,2-addition and **5** from 1,4-addition, have been synthesized [33, 34]. Furthermore, isomer **1** has been synthesized by many different methods [35].

3.1.2. $C_{60}H_4$

Although there are totally 4190 isomers for $C_{60}H_4$ [36], we consider 8 isomers originated from two H_2 additions to [6,6]-ring fusions (see Fig. 4 (b)) [37]. In other words, these isomers are resulted from 1,2-addition of $C_{60}H_2$-**1**. In Fig. 4 (b), second H_2 pair is added to one of the carbon atom pairs labeled 1-8 and these isomers are named **1** - **8**. Experimentally,

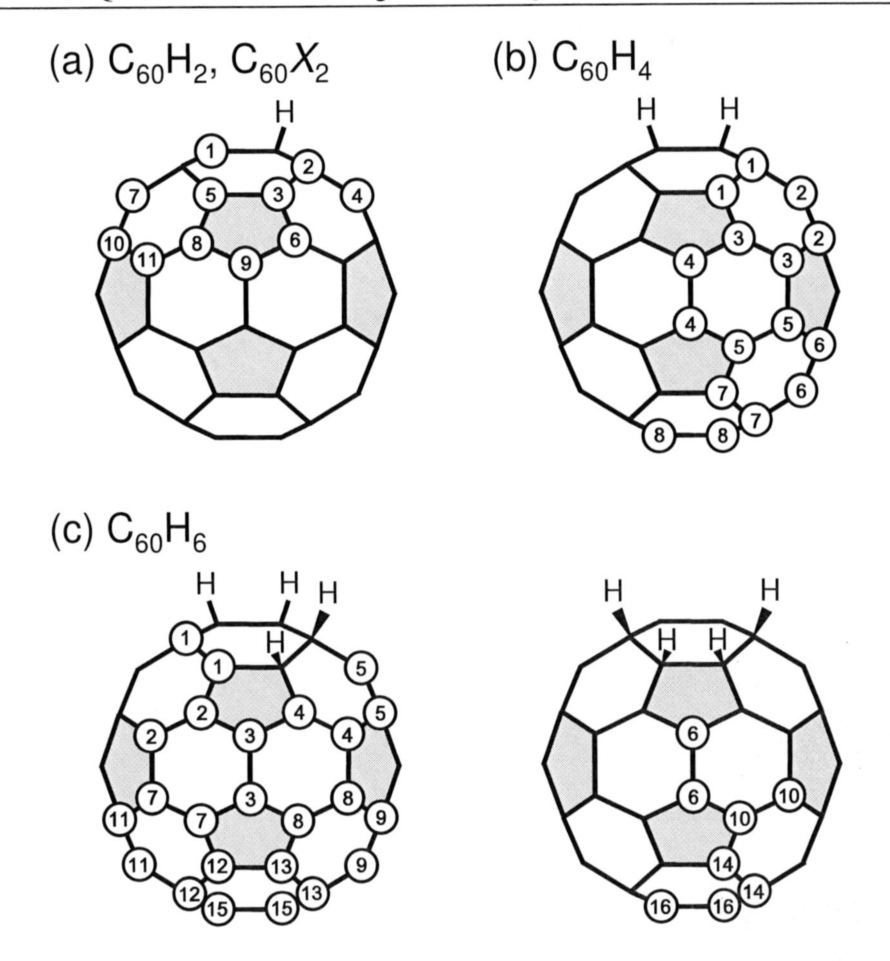

Figure 4. Selected isomers of (a) $C_{60}H_2$ and $C_{60}X_2$, (b) $C_{60}H_4$, and (c) $C_{60}H_6$ (2 figures), where X=C_3H_6COOH and C_4H_8SH. Hydrogen atoms and X are bonding to the labeled carbon atoms.

some of 8 isomers were synthesized, and 4 isomers (**1**, **4**, **6**, and **8**) among them were actually identified [33, 38, 39, 40, 41, 42, 43].

3.1.3. $C_{60}H_6$

Among totally 418470 isomers of $C_{60}H_6$ [36], we consider only 16 isomers in Fig. 4 (c) originated from one H_2 addition to $C_{60}H_4$-**1** at [6,6]-ring fusion. One hydrogen pair is newly added to one of the carbon atom pairs labeled 1-16 and these isomers are named **1**-**16**.

3.2. Fullerene Derivatives with Functional Groups $C_{60}X_2$(X=C_3H_6COOH, and C_4H_8SH)

The purpose of considering $C_{60}X_2$ is to clarify the dependence of λ and k on the kind and chemical nature of addition groups (X). X is selected as C_3H_6COOH and C_4H_8SH. COOH and SH groups are often introduced to C_{60} aiming at the formation of Langmuir-Blodgett [44, 45] and self-assembled monolayer films [46, 47, 48] which can be applied to the device construction. Same as $C_{60}H_2$, there are a total of 23 isomers for $C_{60}X_2$ and 11 isomers named **1-11** shown in Fig. 4 (a) were selected.

4. Reorganization Energies and Rate Constants

4.1. C_{60}

The electronic structure of C_{60}^+ has been considered both experimentally and theoretically due to their electronic degeneracy [49, 50]. C_{60}^+ has nine electrons in five-fold degenerate h_g orbitals. Thus, C_{60}^+ has H_g degenerate electronic state and symmetry lowering due to the Jahn-Teller effect [18] stabilizes C_{60}^+. The most stable structure of C_{60}^+ was calculated as having D_{5d} symmetry, and its λ_h^+ is 95 meV. It should be noted that the calculations of degenerate electronic state based on a single Slater determinant unfortunately result in the splitting of degeneracy [51]. However, energy splitting by DFT calculation is so small that the calculated energy can be approximately considered to be the correct energy of the degenerate state of C_{60}^+. The result of λ_h^+ is qualitatively consistent with previous works based on the static Jahn-Teller effect [52, 53, 54] in which the values of λ_h^+ were calculated as 351 [52], 35 [53], and 71 meV [54]. The value of λ_h^0 of $C_{60}^+(D_{5d})$ was calculated as 74 meV. Thus, the reorganization energy of the hole transport λ_h is 169 meV.

Reorganization energy of electron transport λ_e in C_{60} has been calculated in connection with the superconductivity. C_{60}^- has one electron in three-fold degenerate t_{1u} orbitals. Thus, C_{60}^- has T_{1u} degenerate electronic state and symmetry lowering due to the Jahn-Teller effect stabilizes C_{60}^-. The most stable structure of C_{60}^- was calculated as having D_{3d} symmetry, and its λ_e^- is 72 (76) meV by 6-311++G** (6-31+G*) basis set. The λ_e^0 of $C_{60}^-(D_{3d})$ was calculated as 57 (56) meV. Thus, the value of λ_e is calculated as 129 (132) meV. These values are much smaller than the experimental values (294 meV) obtained by the photoemission spectra [55], but show good agreement with previous theoretical reports (112 meV [56], 120 meV [57], 150 meV [58], and 178 meV [59]).

It should be noted that the values of λ_h^+ and λ_e^- are much larger than those of λ_h^0 and λ_e^0, respectively. This result comes from the fact that the geometrical relaxation in ionic state is very large due to Jahn-Teller distortion [18]. Potential curve of the ionic states is expected to have larger curvature around the minima than the neutral state.

4.2. $C_{60}H_n$ (n=2, 4, and 6)

4.2.1. $C_{60}H_2$

The reorganization energies (λ) and rate constants (k) of all 11 isomers are listed in Table 1. For almost all isomers, λ_h^+ is almost equal to λ_h^0, and λ_e^- is almost equal to λ_e^0. This means

Table 1. Reorganization energies (λ_h, λ_e) and rate constants (k_h, k_e) of carrier transport in of $C_{60}H_2$. Values of the original C_{60} and averaged values over eleven $C_{60}H_2$ isomers are also shown.

$C_{60}H_2$	hole transport				electron transport			
Isomers	λ_h^+	λ_h^0	λ_h	k_h	λ_e^-	λ_e^0	λ_e	k_e
(C_{60})	95	74	169	1.00	72	57	129	1.00
1	66	67	133	1.59	67	71	137	0.90
2	119	109	228	0.48	95	102	197	0.42
3	85	129	214	0.57	147	157	303	0.12
4	77	66	143	1.39	82	76	158	0.68
5	66	76	142	1.41	82	85	167	0.61
6	49	52	101	2.48	86	86	172	0.57
7	89	85	175	0.93	103	82	185	0.49
8	61	77	138	1.49	82	96	178	0.53
9	80	79	160	1.12	99	95	193	0.44
10	131	126	257	0.34	93	95	188	0.47
11	104	123	227	0.49	82	84	166	0.62
Average	84	90	174	0.93	92	93	186	0.48

that the potential curves of ionic and neutral states have almost the same curvature around the minima because the hydrogenation removes the electronic degeneracy in ionic states of C_{60}.

In average, λ_h^+ of $C_{60}H_2$ is smaller than that of C_{60} by 11 meV and λ_h^0 of $C_{60}H_2$ is larger than that of C_{60} by 16 meV. Averaged value of λ_h of $C_{60}H_2$ (174 meV) is almost as large as that of C_{60} (169 meV). However, only the addition of two H atoms lead to the larger difference of λ_h, 101 - 257 meV. It is interesting that six isomers of $C_{60}H_2$ (**1**, **4**, **5**, **6**, **8**, and **9**) have a smaller λ_h than C_{60}. In particular, isomer **6** has the smallest λ_h (101 meV) which is over 40% less than that of C_{60}. Also, the k_h of **6** is about 2.5 times as large as that of C_{60}. From the viewpoint of practical use, it should be noted that the λ_h of the two synthesized isomers, **1** and **5**, are 133 and 142 meV, respectively, which are about 20% smaller than that of C_{60}. The k_h of these isomers are about 1.5 times as large as that of C_{60}.

On the other hand, in average, λ_e^- of $C_{60}H_2$ is larger than that of C_{60} by 20 meV and λ_e^0 of $C_{60}H_2$ is larger than that of C_{60} by 36 meV. Therefore, averaged value of λ_e of $C_{60}H_2$ (186 meV) is much larger as that of C_{60} (129 meV). All isomers has larger λ_e than C_{60}, and isomer **3** has quite large value (303 meV). $C_{60}H_2$ in average has k_e half as large as that of C_{60}. This result is consistent with the fact that C_{60} derivatives generally have smaller electron mobility than the original C_{60}.

These results indicate that hydrogenation can be an effective method for modifying the hole-transport property of C_{60}, but is ineffective for modifying the electron-transport property of C_{60}. Two synthesized isomers of $C_{60}H_2$, **1** and **5**, have potential utility as hole-transport materials.

4.2.2. $C_{60}H_4$

Table 2. Reorganization energies (λ_h, λ_e) and rate constants (k_h, k_e) of carrier transport in $C_{60}H_4$. Values of the original C_{60} and averaged values over eight $C_{60}H_4$ isomers are also shown.

$C_{60}H_4$ Isomers	hole transport				electron transport			
	λ_h^+	λ_h^0	λ_h	k_h	λ_e^-	λ_e^0	λ_e	k_e
(C_{60})	95	74	169	1.00	72	57	129	1.00
1	41	41	83	3.28	69	67	135	0.92
2	87	96	183	0.83	73	72	145	0.81
3	69	73	142	1.41	78	71	150	0.76
4	75	63	138	1.49	82	59	142	0.85
5	62	64	126	1.74	75	74	149	0.77
6	73	77	150	1.26	71	72	144	0.82
7	46	46	92	2.83	66	66	132	0.96
8	63	63	126	1.74	66	66	132	0.96
Average	65	68	134	1.57	72	69	141	0.85

The calculated values of λ and k are listed in Table 2. Similar to the results of $C_{60}H_2$, neutral and ionic states have almost the same reorganization energies. In average, λ_h^+ of $C_{60}H_4$ is smaller than that of C_{60} by 30 meV and λ_h^0 of $C_{60}H_4$ is smaller than that of C_{60} by 6 meV. Averaged values of λ_h of $C_{60}H_4$ (134 meV) is much smaller than that of C_{60} (169 meV) and is almost as large as that of $C_{60}H_2$-**1** (133 meV). Addition of two H atoms to $C_{60}H_2$ results in the large difference in λ_h, 83 - 183 meV. Seven of eight isomers have smaller λ_h than C_{60}. Remarkably, the major product **1** has the smallest λ_h (83 meV) which is over 50% less than that of C_{60}. Also, k_h of **1** is 3.28 times as large as that of C_{60}, and more than twice as large as that of the synthesized $C_{60}H_2$-**1** [11]. Isomer **7** with the second smallest λ_h has k_h which is 2.83 times as large as that of C_{60}. Other identified isomers **4**, **6**, and **8** also have small λ_h (138, 150, and 126 meV), and k_h of these isomers are 1.49, 1.26, and 1.74 times larger than that of C_{60}.

In average, λ_e^- of $C_{60}H_4$ is same to that of C_{60}, and λ_e^0 of $C_{60}H_4$ (69 meV) is larger than that of C_{60} by 12 meV. Averaged values of λ_e of $C_{60}H_4$ (141 meV) is larger than that of C_{60} (129 meV) and is almost as large as that of $C_{60}H_2$-**1** (137 meV). Contrast to the values of λ_h, values of λ_e are in range of 132 - 150 meV and are not influenced by the further hydrogenation of $C_{60}H_2$.

It is found that the further hydrogenation of $C_{60}H_2$ is very effective for improving the hole-transport property. Synthesized isomers of $C_{60}H_4$, especially the major product **1**, have potential utility as useful hole-transport material. On the other hand, hydrogenation from $C_{60}H_2$ to $C_{60}H_4$ has little effect on the reorganization energy of electron transfer.

Table 3. Reorganization energies (λ_h, λ_e) and rate constants (k_h, k_e) of carrier transport in $C_{60}H_6$. Values of the original C_{60} and averaged values over sixteen $C_{60}H_6$ isomers are also shown.

$C_{60}H_6$	hole transport				electron transport			
Isomers	λ_h^+	λ_h^0	λ_h	k_h	λ_e^-	λ_e^0	λ_e	k_e
(C_{60})	95	74	169	1.00	72	57	129	1.00
1	36	35	71	3.94	65	62	127	1.02
2	46	50	95	2.69	71	72	143	0.83
3	44	61	105	2.33	74	73	147	0.78
4	38	45	84	3.22	72	72	144	0.82
5	78	104	182	0.85	69	73	142	0.84
6	48	41	89	2.97	92	87	179	0.53
7	67	74	141	1.43	71	77	148	0.78
8	86	74	160	1.11	70	77	147	0.79
9	74	90	164	1.06	78	74	153	0.73
10	87	89	176	0.91	73	75	148	0.78
11	49	40	89	2.99	82	84	166	0.61
12	74	75	150	1.27	72	73	145	0.81
13	48	63	111	2.14	75	76	151	0.75
14	54	66	120	1.91	73	75	148	0.78
15	51	46	97	2.63	68	68	136	0.91
16	45	45	91	2.89	65	66	131	0.98
Average	61	66	127	1.73	74	75	150	0.76

4.2.3. $C_{60}H_6$

The calculated values of λ and k of $C_{60}H_6$ isomers are listed in Table 3. Neutral and ionic states have almost same reorganization energies. In average, λ_h^+ of $C_{60}H_6$ is much smaller than that of C_{60} by 34 meV and λ_h^0 of $C_{60}H_6$ is smaller than that of C_{60} by 8 meV. Averaged values of λ_h of $C_{60}H_6$ (127 meV) is much smaller than that of C_{60} (169 meV) but is much larger than that of $C_{60}H_4$-**1** (83 meV). Further addition of two H atoms to $C_{60}H_4$ leads to the large difference in λ_h, 71 - 182 meV. 14 of 16 isomers have smaller λ_h than C_{60}. Isomer **1** has the smallest λ_h (71 meV) which is about 60% less than that of C_{60}. Also, k_h of **1** is 3.94 times as large as that of C_{60}. Isomer **4** with the second smallest λ_h has k_h which is 3.22 times as large as that of C_{60}.

In average, λ_e^- of $C_{60}H_6$ is a little larger than that of C_{60} by 2 meV and λ_e^0 of $C_{60}H_6$ (69 meV) is larger than that of C_{60} by 18 meV. Averaged value of λ_e of $C_{60}H_6$ (150 meV) is larger than those of C_{60} (129 meV) and $C_{60}H_4$-**1** (135 meV). Contrast to the values of λ_h, values of λ_e are in narrow range, 127 - 179 meV.

It is found that the further hydrogenation of $C_{60}H_4$ can reduce λ_h for some isomers. Isomer **1** has very small λ_h. It should be noted, however, that these $C_{60}H_6$ isomers are originated from the hydrogenation of $C_{60}H_4$-**1** and averaged values of λ_h (127 meV) is larger than that of $C_{60}H_4$-**1** (83 meV). This result means that the hydrogenation effect of decreasing λ_h is saturated at $C_{60}H_4$. On the other hand, similar to the results of $C_{60}H_2$ and $C_{60}H_4$, hydrogenation from $C_{60}H_4$ to $C_{60}H_6$ has little effect on the reorganization energy of electron transport.

From the systematic discussion through $C_{60}H_n$ (n=2, 4, and 6), it was found that hydrogenation has large effect on the improvement of hole-transport property (λ_h), but has little effect on the electron-transport property (λ_e). The smallest values of λ_h decrease as the number of hydrogen atoms increases: C_{60} (169 meV) → $C_{60}H_2$-**6** (101 meV) → $C_{60}H_4$-**1** (83 meV) → $C_{60}H_6$-**1** (71 meV). Averaged value of λ_h change as C_{60} (169 meV) → $C_{60}H_2$ (174 meV) → $C_{60}H_4$ (134 meV) → $C_{60}H_6$ (127 meV). Averaged value of $C_{60}H_4$ (134 meV) is almost as large as that of $C_{60}H_2$-**1** (133 meV), but averaged value of $C_{60}H_6$ (127 meV) is much larger than that of $C_{60}H_4$-**1** (83 meV). In electron transport, smallest values change as C_{60} (129 meV) → $C_{60}H_2$-**1** (137 meV) → $C_{60}H_4$-**7**, **8** (132 meV) → $C_{60}H_6$-**1** (127 meV). Average values changes C_{60} (129 meV) → $C_{60}H_2$ (186 meV) → $C_{60}H_4$ (141 meV) → $C_{60}H_6$ (150 meV). Therefore, hydrogenation has small influence on λ_e and tends to increase λ_e.

4.3. $C_{60}X_2$ (X=C_3H_6COOH and C_4H_8SH)

4.3.1. $C_{60}(C_3H_6COOH)_2$

Table 4. Reorganization energies (λ_h, λ_e) and rate constants (k_h, k_e) of carrier transport in $C_{60}(C_3H_6COOH)_2$. Values of the original C_{60} and averaged values over eleven $C_{60}(C_3H_6COOH)_2$ isomers are also shown.

$C_{60}(C_3H_6COOH)_2$	hole transport				electron transport			
Isomers	λ_h^+	λ_h^0	λ_h	k_h	λ_e^-	λ_e^0	λ_e	k_e
(C_{60})	95	74	169	1.00	76	56	132	1.00
1	76	77	153	1.22	73	74	147	0.82
2	124	130	254	0.36	111	116	227	0.30
3	96	137	233	0.46	149	156	305	0.12
4	80	90	170	0.98	92	84	176	0.57
5	86	98	184	0.82	95	104	199	0.42
6	70	69	138	1.48	109	117	227	0.30
7	99	96	195	0.72	101	99	200	0.42
8	92	96	188	0.78	116	149	265	0.19
9	93	94	187	0.79	98	122	220	0.33
10	138	141	279	0.27	100	106	206	0.39
11	128	132	260	0.33	97	112	208	0.38
Average	99	105	204	0.65	104	113	216	0.34

The calculated values of λ and k of $C_{60}(C_3H_6COOH)_2$ isomers are listed in Table 4. Be careful that the values of λ_e in Tables 4 and 5 are calculated using different basis set,

Table 5. Reorganization energies (λ_h, λ_e) and rate constants (k_h, k_e) of carrier transport in $C_{60}(C_4H_8SH)_2$. Values of the original C_{60} and averaged values over eleven $C_{60}(C_4H_8SH)_2$ isomers are also shown.

$C_{60}(C_4H_8SH)_2$	hole transport				electron transport			
Isomers	λ_h^+	λ_h^0	λ_h	k_h	λ_e^-	λ_e^0	λ_e	k_e
(C_{60})	95	74	169	1.00	76	56	132	1.00
1	56	59	114	2.06	67	67	133	0.98
2	113	117	230	0.47	99	103	202	0.41
3	88	129	217	0.55	143	148	291	0.14
4	71	86	157	1.15	81	78	159	0.70
5	56	63	118	1.94	85	91	176	0.57
6	54	53	106	2.30	85	80	165	0.65
7	92	94	186	0.81	95	94	189	0.48
8	77	75	152	1.24	89	95	184	0.51
9	85	84	169	1.00	95	94	189	0.48
10	131	133	264	0.32	92	98	190	0.48
11	111	118	229	0.48	85	87	172	0.60
Average	85	92	177	0.90	92	94	186	0.50

6-31+G(*d*). In average, λ_h^+ of $C_{60}(C_3H_6COOH)_2$ is larger than that of C_{60} by 4 meV and λ_h^0 of $C_{60}(C_3H_6COOH)_2$ is larger than that of C_{60} by 31 meV. Averaged values of λ_h of $C_{60}(C_3H_6COOH)_2$ (204 meV) is much larger than that of C_{60} (169 meV). Addition of two C_3H_6COOH groups leads to the large difference in λ_h, 138 - 279 meV. Only two isomers, **1** and **6**, have a smaller λ_h than C_{60}. Isomer **6** has the smallest λ_h (138 meV) and its k_h is about 1.5 times as large as that of C_{60}.

On the other hand, λ_e^- of $C_{60}(C_3H_6COOH)_2$ is larger than that of C_{60} by 28 meV, and λ_e^0 of $C_{60}(C_3H_6COOH)_2$ is larger than that of C_{60} by 57 meV. Therefore, averaged values of λ_e of $C_{60}(C_3H_6COOH)_2$ (216 meV) is much larger as that of C_{60} (132 meV). All isomers has larger λ_e than C_{60} and isomer **3** has very large value (305 meV). Generally, $C_{60}(C_3H_6COOH)_2$ has much smaller k_e than C_{60}. These results indicate that addition of C_3H_6COOH groups can modify the hole-transport property λ_h of C_{60} for only two isomers. However, all isomers have larger λ_e than C_{60}.

4.3.2. $C_{60}(C_4H_8SH)_2$

The calculated values of λ and k of $C_{60}(C_4H_8SH)_2$ isomers are listed in Table 5. In average, λ_h^+ of $C_{60}(C_4H_8SH)_2$ is larger than that of C_{60} by 10 meV and λ_h^0 of $C_{60}(C_4H_8SH)_2$ is larger than that of C_{60} by 18 meV. Averaged values of λ_h of $C_{60}(C_4H_8SH)_2$ (177 meV) is almost as large as that of C_{60} (169 meV). Addition of two C_4H_8SH groups leads to the large difference in λ_h, 106 - 264 meV. Five isomers of $C_{60}(C_4H_8SH)_2$ (**1**, **4**, **5**, **6**, and **8**) have a smaller λ_h than C_{60}. Isomer **6** has the smallest λ_h (106 meV) and its k_h is about 2.3 times as large as that of C_{60}.

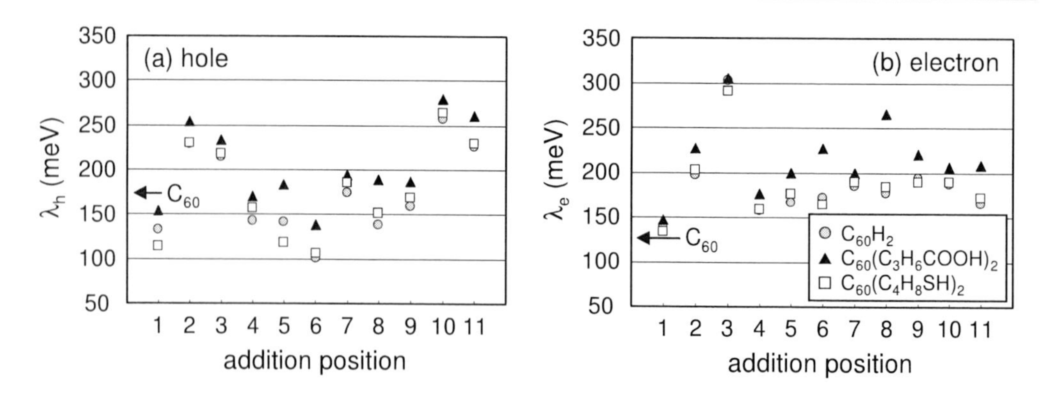

Figure 5. Relationship between reorganization energy (λ), addition position, and the kind of added group (X) about (a) hole transport and (b) electron transport in $C_{60}X_2$.

On the other hand, λ_e^- of $C_{60}(C_4H_8SH)_2$ is larger than that of C_{60} by 16 meV and λ_e^0 of $C_{60}(C_4H_8SH)_2$ is larger than that of C_{60} by 38 meV, in average. Therefore, averaged values of λ_e of $C_{60}(C_4H_8SH)_2$ (186 meV) is also much larger as that of C_{60} (132 meV). All isomers have larger λ_e than C_{60}, and isomer **3** has very large value (291 meV). $C_{60}(C_4H_8SH)_2$ has much smaller k_e than that of C_{60}. These results indicate that addition of C_4H_8SH groups can modify the hole-transport property λ_e of C_{60} for five isomers. All isomers have larger λ_e than C_{60}.

5. Analysis for Molecular Design

5.1. Chemical Nature of Added Groups

Figures 5 (a) and (b) shows the relationship between reorganization energy (λ), addition position, and kind of added group (X) about hole- and electron-transport properties of $C_{60}X_2$ (X=H, C_3H_6COOH, and C_4H_8SH). The values of λ are strongly dependent on the position of X but are almost independent of the kind of X. It should be noted, however, that $C_{60}(C_3H_6COOH)_2$ and $C_{60}(C_4H_8SH)_2$ generally have larger λ than $C_{60}H_2$ for both hole and electron transport. And, $C_{60}(C_3H_6COOH)_2$ have a little larger λ than $C_{60}(C_4H_8SH)_2$. Especially, **6** and **8** of $C_{60}(C_3H_6COOH)_2$ have much larger λ_e than those of $C_{60}H_2$ and $C_{60}(C_4H_8SH)_2$ in Fig. 5 (b).

Reorganization energy is a stabilization energy by geometric relaxation originated from the change in electronic structure [51, 60, 62]. Strong force to nuclei generally result in large λ and geometric relaxation. Here, we define a parameter Δr which characterizes the geometric relaxation as

$$\Delta r = \sum_i |\Delta r_i|, \tag{13}$$

where Δr_i is a difference between ith bond lengths of neutral and ionic states. The summation is taken over all bonds. Figures 6 (a) and (b) shows the relationship between λ and Δr

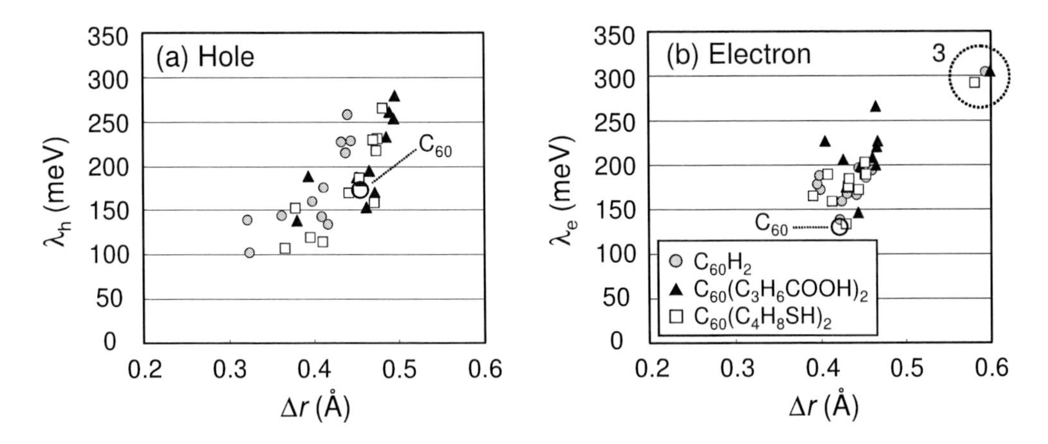

Figure 6. Relationship between reorganization energy (λ) and geometrical relaxation (Δr) about (a) hole transport and (b) electron transport in $C_{60}X_2$ (X=H, C_3H_6COOH, and C_4H_8SH). The values of C_{60} are shown by the larger open circle.

of hole and electron transport in $C_{60}X_2$. About hole transport of Fig. 6 (a), there is almost linear relationship between Δr and λ_h. Isomers of $C_{60}(C_3H_6COOH)_2$ tend to have large Δr and λ_h. On the other hand, isomers of $C_{60}H_2$ tend to have small Δr and λ_h. Isomers of $C_{60}(C_4H_8SH)_2$ have moderate Δr and λ_h. Δr and λ_h of C_{60} are almost as large as averaged values of all $C_{60}X_2$ isomers. About electron transport of Fig. 6 (b), it is found that three **3** isomers of $C_{60}H_2$, $C_{60}(C_3H_6COOH)_2$, and $C_{60}(C_4H_8SH)_2$ with the largest λ_e have the largest Δr regardless of the kind of X. For other isomers, similar to the case of hole transport, $C_{60}(C_3H_6COOH)_2$ isomers have a little larger Δr and λ_e and $C_{60}H_2$ isomers have a little smaller Δr and λ_e. However, these isomers have almost same λ_e and Δr since electron transport property is influenced only a little by an addition.

In order to clarify the reason why $C_{60}(C_3H_6COOH)_2$ isomers have larger λ_h and λ_e in Fig. 5, λ_e and Δr of $C_{60}X_2$ (X=C_4H_9, $C_5H_{10}COOH$, and $C_7H_{14}COOH$) were calculated for isomers **6** and **8**. From the results shown in Fig. 7, it is found that $C_{60}H_2$ and $C_{60}(C_4H_9)_2$ have almost same λ_e and Δr. Addition of alkyl chains itself does not have influence on λ_e and Δr. Therefore, the deviation of $C_{60}(C_3H_6COOH)_2$ from $C_{60}H_2$ in λ_e is due to the interaction between C_{60} and COOH group. This interpretation is assisted by the results of $C_{60}(C_5H_{10}COOH)_2$ and $C_{60}(C_7H_{14}COOH)_2$ in Fig. 7. As the -CH_2- chain length of $(CH_2)_m$COOH becomes longer from m=3 to m=7, the values of λ_e and Δr approach those of $C_{60}H_2$ since the interaction between C_{60} and COOH group decreases. SH group does not interact with C_{60} because atomic orbital energies of S atom do not coincide with those of C atoms. Therefore, the dependence of λ_e in m will not be observed for SH-terminated molecules. These discussions on λ_e will also be applied to the discussion of λ_h

5.2. Number of Added Groups

Figures 8 (a) and (b) show the relation between n, Δr, and λ for hole- and electron-transport properties of $C_{60}H_n$. About hole transport property shown in Fig. 8 (a) , compared to the

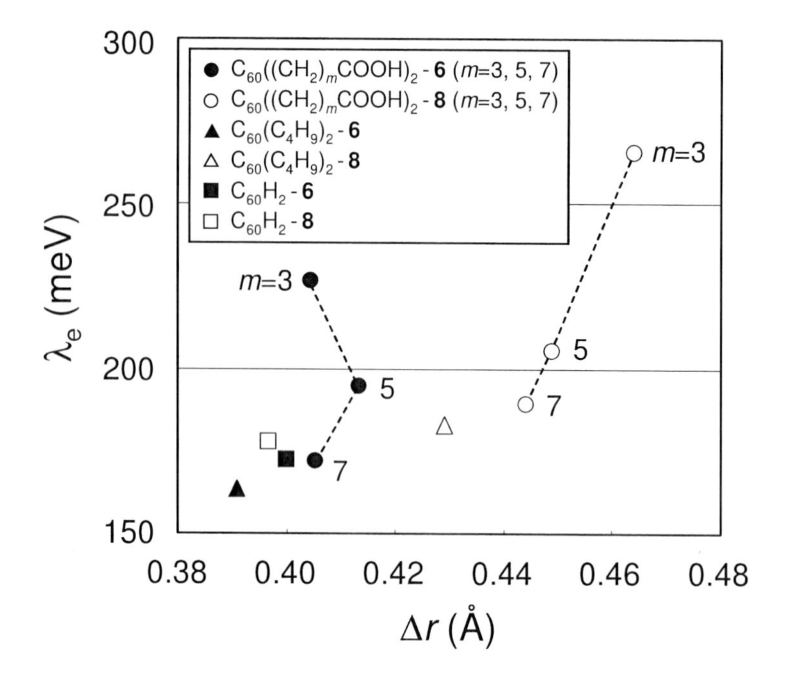

Figure 7. Reorganization energy (λ_e) and geometrical relaxation (Δr) of **6** and **8** isomers of $C_{60}X_2$ (X=H, C_4H_9, C_3H_6COOH, $C_5H_{10}COOH$, and $C_7H_{14}COOH$). Filled marks show the values of isomers **6** and open marks show the values of isomers **8**.

results of $C_{60}X_2$ (Fig. 6 (a)), we can see the clear linear-relationship between Δr and λ_h. $C_{60}H_2$ isomers have large Δr and λ_h, and $C_{60}H_6$ isomers have small Δr and λ_h. Increase of hydrogen atoms generally result in the decrease of Δr and λ_h. For all isomers, Δr of $C_{60}H_n$ is smaller than that of C_{60}.

About electron transport property shown in Fig. 8 (b), most $C_{60}H_n$ isomers have the same values of Δr and λ_e because hydrogenation has little effect on the electron-transport property of C_{60}. Only $C_{60}H_2$-**3** isomer has a quite large Δr and λ_e. Therefore, for also electron transport, we can confirm the linear relationship between Δr and λ_e.

5.3. Molecular Orbital Pattern

Values of reorganization energies are different between isomers. This difference comes from the difference in electronic structures. Devos and Lannoo [63] have reported that larger π-conjugated systems have smaller λ. Also, very recently, it was reported that λ is related to the pattern of the frontier orbital [60, 61, 62]. From these works, it is predicted that there is a close relationship between λ and the size of the π-conjugation region of the frontier orbital.

It is well known that the electronic properties of the highest occupied molecular orbitals (HOMO) and the lowest unoccupied molecular orbitals (LUMO) have close relation to hole-transport property and electron-transport property, respectively. Up to now, many researchers [64, 65, 66] have reported the relationship between the HOMO and λ_h (or k_h).

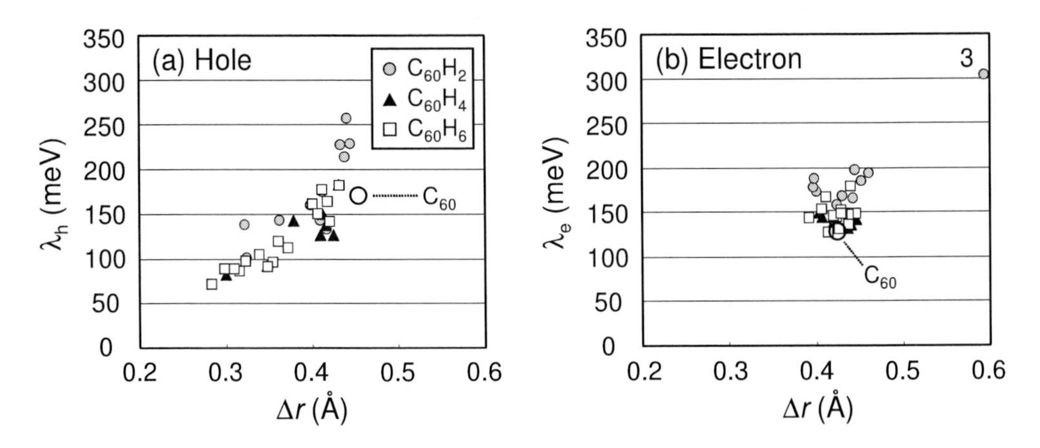

Figure 8. Relationship between reorganization energy (λ) and geometrical relaxation (Δ_r) about (a) hole transport and (b) electron transport in $C_{60}H_n$ (n=2, 4, and 6). The values of C_{60} are shown by the larger open circle.

Kitamura and Yokoyama [64] reported that the uniformity of the positive charge distribution leads to the high mobility k_h in several hydrazone compounds. In contrast, Aratani *et al.* [65] found that when HOMO distribution is localized on triphenylamine part of triphenylamine derivatives, the mobility k_h increase. Recently, Lin *et al.* [66] theoretically reported that λ_h in the hole conduction is determined mainly by the moiety which contributes predominantly to its HOMO. Although a universal relationship between the HOMO and λ_h (or k_h) is still not clear, it should be noted that there is a certain relation which can be applied to the similar compounds. Therefore, for the establishment of design guidelines, it is meaningful to find the relation about the fullerene derivatives.

In Fig. 9 (a), HOMOs of isomers with the smallest λ_h and the largest λ_h are drawn by GAUSSVIEW [67]. LUMOs with the largest λ_e are shown in Fig. 9 (b). However, LUMOs of $C_{60}H_4$ and $C_{60}H_6$ with the largest λ_e and LUMOs of all $C_{60}X_n$ with the smallest λ_e are not shown because these isomers have almost the same λ_e as can be seen in Figs. 6 (b) and 8 (b) and it is difficult to find the clear difference in molecular orbital distribution.

Firstly, we focus on the hole-transport property of $C_{60}H_2$, in Fig. 9 (a). HOMOs of **6** (smallest λ_h) and **10** (largest λ_h) calculated in the neutral state are shown in the left side of Fig. 9 (a). It is clear that the HOMO of **6** is more delocalized on the whole molecule than that of **10**. Isomer **6** with the delocalized HOMO has a large π-conjugation region similar to C_{60}. On the other hand, isomer **10** with the localized HOMO has a small π-conjugation similar to corannulene $C_{20}H_{10}$ [68] on the upper region of the molecule. Therefore, the isomers of $C_{60}H_2$ with a larger π-conjugation region tend to have a smaller λ_h and larger k_h.

About $C_{60}X_2$ (X=C_3H_6COOH and C_4H_8SH), HOMOs of isomer **6** with the smallest λ_h shown in the top of Fig. 9 (a). It is found that the distribution of these HOMOs are similar to that of $C_{60}H_2$-**6**. We can also see that HOMOs of $C_{60}X_2$-**10** with the largest λ_h have similar distribution to the HOMO of $C_{60}H_2$-**10**. Distribution of HOMOs of $C_{60}H_4$-**1** and $C_{60}H_4$-**2** is a little different from those of $C_{60}H_2$-**6** and $C_{60}H_2$-**10** since hydrogenation separates the π-conjugated system of $C_{60}H_2$. HOMO of $C_{60}H_4$-**1** is distributed on the whole molecule, but that of $C_{60}H_4$-**2** is localized on the central and vertical region of the molecule. The distribution of HOMO of $C_{60}H_6$-**1** (smallest λ_h) is much similar to that of $C_{60}H_4$-**1** (smallest λ_h), and the distribution of HOMO of $C_{60}H_6$-**5** (largest λ_h) to that of $C_{60}H_4$-**2** (largest λ_h). It should be noted that HOMOs of $C_{60}H_4$-**1** and $C_{60}H_6$-**1** are more delocalized than that of $C_{60}H_2$-**6** so that $C_{60}H_4$-**1** and $C_{60}H_6$-**1** have smaller λ_h than $C_{60}H_2$-**6**. For all $C_{60}H_n$, delocalized HOMO gives small λ_h (large k_h) and localized HOMO gives large λ_h (small k_h).

For the consideration of the electron transport properties, LUMO of the neutral $C_{60}X_2$-**3** (X=H, C_3H_6COOH, and C_4H_8SH) with the largest λ_e are shown in Fig. 9 (b). It is found that these orbital are localized on the upper region of the molecule and LUMOs of $C_{60}(C_3H_6COOH)_2$-**3** and $C_{60}(_4H_8SH)_2$-**3** have very similar distribution to that of $C_{60}H_2$-**3**.

From these results, we can see that the isomers with delocalized frontier orbitals tend to have small λ_h (large k_h), and the isomers with localized frontier orbitals tend to have have large λ_h (small k_h) and large λ_e (small k_e). This tendency coincides with the intuitive picture of carrier hopping. Delocalized orbitals can easily pass the carrier from one molecule to the other molecule. $C_{60}X_2$ (X=C_3H_6COOH and C_4H_8SH) have λ which is almost as large as that of $C_{60}H_2$ because these molecules have similar frontier orbitals to $C_{60}H_2$.

6. Summary and Molecular Design Guidelines

Systematic analyses of reorganization energies of C_{60} derivatives, $C_{60}X_2$ (X=H, C_3H_6COOH, C_4H_8SH) and $C_{60}H_n$(n=2, 4, 6), give us very important knowledge for efficient design of useful C_{60} materials:

- Carrier-transport properties of C_{60} derivatives are quite different from those of C_{60} [11, 12, 13],
- Chemical addition can improves hole mobility of C_{60} for some isomers [11, 12]. Inversely, electron mobility of C_{60} is generally not influenced or decreased by the chemical addition [13].
- Values of reorganization energies of both hole-transport and electron-transport are almost independent of the kind (chemical nature) of addition group X, but are strongly dependent on the position of addition group [13]. Therefore, reorganization energies of other kinds of $C_{60}X_2$ will be approximately estimated from the results of $C_{60}H_2$ [13].
- Hole and electron mobilities are closely related to the distribution patterns of HOMO and LUMO, respectively [11, 12, 13]. Delocalized orbital gives large carrier mobility, and localized molecular orbital small carrier mobility.

The results obtained in this work will also be applied to other kinds of $C_{60}X_2$, and give us a guideline for efficient design of novel materials based on C_{60} derivatives in both experimental and theoretical approaches. For example, in the experimental viewpoint, we can freely select X which is proper for thin-film formation and is easily synthesized, without

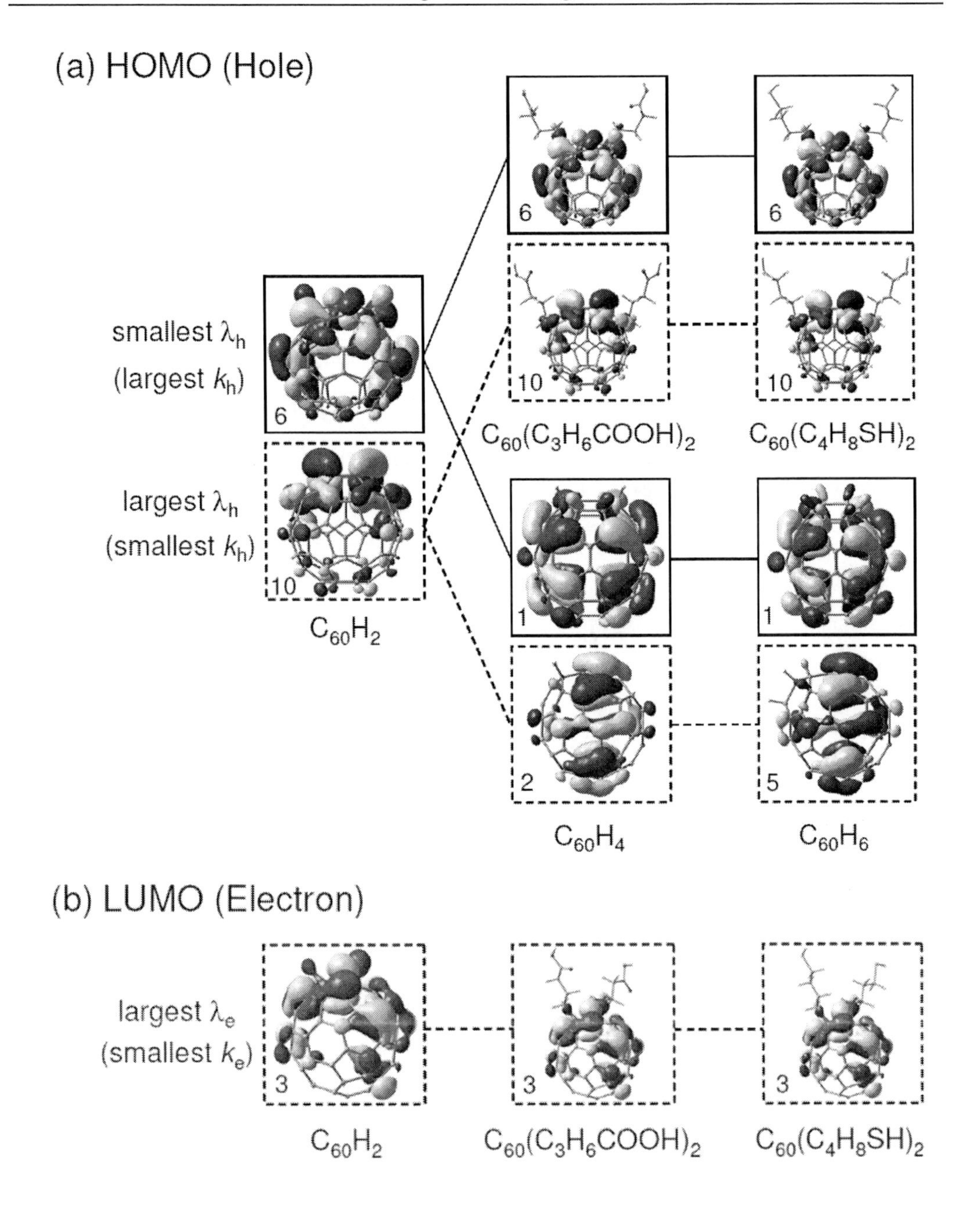

Figure 9. (a) HOMO of $C_{60}X_n$ with the smallest λ_h and the largest λ_h. (b) LUMO of $C_{60}X_2$ with the largest λ_e. Inserted number means the number of the isomer.

considering the influence of X on the electronic properties. In the theoretical viewpoint, this result enables us to save computational time and resources in the material design because various kinds of $C_{60}X_2$ can be simplified to $C_{60}H_2$ in the prediction and discussion of these properties.

We did not calculate the reorganization energies of $C_{60}X_4$ and $C_{60}X_6$ with COOH and SH termination. However, it is expected that these values will be almost same to those of $C_{60}H_4$ and $C_{60}H_6$. Lastly, it should be noted that the values of electronic coupling elements H were not discussed at all in this chapter. Calculation, analysis, and control of H will be also very important for the complete establishment of molecular design guideline of C_{60} derivatives.

Acknowledgements

The author would like to thank Dr. Hiroshi Kawabata of Hiroshima University and Dr. Shigekazu Ohmori of National Institute of Advanced Industrial Science and Technology (AIST) for collaboration in this research theme and helpful comments to this chapter. Theoretical calculations were mainly carried out using the computer facilities at Research Center for Computational Science (Okazaki, Japan) and Research Institute for Information Technology, Kyushu University (Fukuoka, Japan).

References

[1] C. W. Tang and S. A. VanSlyke, *Appl. Phys. Lett.* **51**, 913 (1987).

[2] H. H. Fong, K. C. Lun, and S. K. So, *Chem. Phys. Lett.* **353**, 407 (2002).

[3] J. -C. Brédas, D. Beljonne, V. Coropceanu, and J. Cornil, *Chem. Rev.* **104**, 4971 (2004).

[4] A. R. Murphy and J. M. J. Fréchet, *Chem. Rev.* **107**, 1066 (2007).

[5] V. Coropceanu, J. Cornil, D. A. da Silva Filho, Y. Olivier, R. Silbey, and J. L. Brédas, *Chem. Rev.* **107**, 926 (2007).

[6] T. Mori, *J. Phys.: Condens. Matter,* **20**, 184010 (2008).

[7] K. Sakanoue, M. Motoda, M. Sugimoto, and S. Sakaki, *J. Phys. Chem. A* **103**, 5551 (1999).

[8] J. C. Sancho-García, G. Horowitz, J. L. Brédas, and J. Cornil, *J. Chem. Phys.* **119**, 12563 (2003).

[9] G. R. Hutchison, M. A. Ratner, and T. J. Marks, *J. Am. Chem. Soc.* **127**, 16866 (2005).

[10] J.-H. Pan, H.-L. Chiu, L. Chen, and B.-C. Wang, *Comput. Mater. Sci.* **38**, 105 (2006).

[11] K. Tokunaga, S. Ohmori, H. Kawabata, and K. Matsushige *Jpn. J. Appl. Phys.* **47**, 1089 (2008).

[12] K. Tokunaga, H. Kawabata, and K. Matsushige, *Jpn. J. Appl. Phys.* **47**, 3638 (2008).

[13] K. Tokunaga, *Chem. Phys. Lett.* **476**, 253 (2009).

[14] K. Tokunaga, S. Ohmori, and H. Kawabata, *Thin Solid Films* **518**, 477 (2009).

[15] K. Tokunaga, S. Ohmori, and H. Kawabata, *Jpn. J. Appl. Phys. accepted.*

[16] K. Tokunaga, S. Ohmori, and H. Kawabata, *Mol. Cryst. Liq. Cryst. accepted.*

[17] R. Könenkamp, G. Priebe, and B. Pietzak, *Phys. Rev. B* **60**, 11804 (1999).

[18] H. A. Jahn and E. Teller, *Proc. R. Soc. London, Ser. A* **161**, 220 (1937).

[19] A. Ceulemans and P. W. Fowler, *J. Chem. Phys.* **93**, 1221 (1990).

[20] T. B. Singh, M. Marjanović, G. J. Matt, S. Günes, N. S. Sariciftci, A. M. Ramil, A. Andreev, H. Sitter, R. Schwödiauer, and S. Bauer, *Org. Electron.* **6**, 105 (2005).

[21] M. Kitamura, Y. Kuzumoto, M. Kamura, S. Aomori, and Y. Arakawa, *Appl. Phys. Lett.* **91**, 183514 (2007).

[22] T. B. Singh, M. Marjanović, P. Stadler, M. Auinger, G. J. Matt, S. Günes, N. S. Sariciftci, R. Schwödiauer, and S. Bauer, *J. Appl. Phys.* **97**, 083714 (2005).

[23] R. A. Marcus, *J. Chem. Phys.* **24**, 966 (1965).

[24] G. Horowitz, *Adv. Mater.* **10**, 365 (1998).

[25] S. F. Nelsona, Y. Y. Lin, D. J. Gundlach, and T. N. Jackson, *Appl. Phys. Lett.* **72**, 1854 (1998).

[26] C. D. Dimitrakopoulos and D. J. Mascaro, *IBM J. Res. Dev.* **45**, 11 (2001).

[27] C. Waldauf, P. Schilinsky, M. Perisutti, J. Hauch, and C. J. Brabec, *Adv. Mater.* **15**, 2084 (2003).

[28] M. Chikamatsu, S. Nagamatsu, Y. Yoshida, K. Saito, K. Yase, and K. Kikuchi, *Appl. Phys. Lett.* **87**, 203504 (2005).

[29] S. P. Tiwari, E. B. Namdas, V. R. Rao, D. Fichou, and S. G. Mhaisalkar, *IEEE Electron Dev. Lett.* **28**, 880 (2007).

[30] Q. Wu and T. Van Voorhis, *J. Chem. Phys.* **125**, 164105 (2006).

[31] M. J. Frisch, G. W. Trucks, H. B. Schlegel, G. E. Scuseria, M. A. Robb, J. R. Cheeseman, J. A. Montgomery, Jr., T. Vreven, K. N. Kudin, J. C. Burant, J. M. Millam, S. S. Iyengar, J. Tomasi, V. Barone, B. Mennucci, M. Cossi, G. Scalmani, N. Rega, G. A. Petersson, H. Nakatsuji, M. Hada, M. Ehara, K. Toyota, R. Fukuda, J. Hasegawa, M. Ishida, T. Nakajima, Y. Honda, O. Kitao, H. Nakai, M. Klene, X. Li, J. E. Knox, H. P. Hratchian, J. B. Cross, C. Adamo, J. Jaramillo, R. Gomperts, R. E. Stratmann, O. Yazyev, A. J. Austin, R. Cammi, C. Pomelli, J. W. Ochterski, P. Y. Ayala, K. Morokuma, G. A. Voth, P. Salvador, J. J. Dannenberg, V. G. Zakrzewski, S. Dapprich, A. D. Daniels, M. C. Strain, O. Farkas, D. K. Malick, A. D. Rabuck, K. Raghavachari, J. B. Foresman, J. V. Ortiz, Q. Cui, A. G. Baboul, S. Clifford, J. Cioslowski, B. B. Stefanov, G. Liu, A. Liashenko, P. Piskorz, I. Komaromi, R. L. Martin, D. J. Fox, T. Keith, M. A. Al-Laham, C. Y. Peng, A. Nanayakkara, M. Challacombe, P. M. W.

Gill, B. Johnson, W. Chen, M. W. Wong, C. Gonzalez, and J. A. Pople, *Gaussian 03,* Revision C. 02 (Gaussian, Inc., Pittsburgh, PA, 2003).

[32] N. Matsuzawa, D. A. Dixon, and T. Fukunaga, *J. Phys. Chem.* **96**, 7594 (1992).

[33] A. G. Avent, A. D. Darwish, D. K. Heimbach, H. W. Kroto, M. F. Meidine, J. P. Parsons, C. Remars, R. Roers, O. Ohashi, R. Taylor, and D. R. M. Walton, *J. Chem. Soc., Perkin Trans. 2* **96**, 15 (1994).

[34] C. C. Henderson and P. A. Cahill, *Science* **259**, 1885 (1993).

[35] J. Nossal, R. K. Saini, L. B. Alemany, M. Meier, and W. E. Billups, *Eur. J. Org. Chem.*, **4167** (2001).

[36] Y. Shao and Y. Jiang, *Chem. Phys. Lett.* **242**, 191 (1995).

[37] P. A. Cahill and C. M. Rohlfing, *Tetrahedron* **52**, 5247 (1996).

[38] C. C. Henderson, C. McMichael Rohlfing, K. T. Gillen, and P. A. Cahill, *Science* **264**, 397 (1994).

[39] M. S. Meiser, P. S. Corbin, V. K. Vance, M. Clayton, M. Mollman, and M. Poplawska, *Tetrahedron Lett.* **35**, 5789 (1994).

[40] W. E. Billups, W. Luo, A. Gonzalez, D. Arguello, L. B. Alemany, T. Marriott, M. Saunders, H. A. Jiménez-Vázquez, and A. Khong, *Tetrahedron Lett.* **38**, 171 (1997).

[41] R. J. Cross, H. A. Jiménez-Vázquez, Q. Li, M. Saunders, D. I. Schuster, S. R. Wilson, and H. Zhao, *J. Am. Chem. Soc.* **118**, 11454 (1996).

[42] R. G. Bergosh, M. S. Meiser, J. A. L. Cooke, H. P. Spielmann, and B. R. Weedon, *J. Org. Chem.* **62**, 7667 (1997).

[43] L. B. Alemany, A. Gonzalez, W. E. Billups, M. R. Willcott, E. Ezell, and E. Gozansky, *J. Org. Chem.* **62**, 5771 (1997).

[44] M. Matsumoto, H. Tachibana, R. Azumi, M. Tanaka, T. Nakamura, G. Yunome, M. Abe, S. Yamago, and E. Nakamura, *Langmuir* **11**, 660 (1995).

[45] S. Ravaine, C. Mingotud, and P. Delhaès, *Thin Solid Films* **284-285**, 76 (1996).

[46] K. Chen, W. B. Caldwell, and C. A. Mirkin, *J. Am. Chem. Soc.* **115**, 1193 (1993).

[47] Y. Shirai, L. Cheng, B. Chen, and J. M. Tour, *J. Am. Chem. Soc.* **128**, 13479 (2006).

[48] T. X. Wei, Y. R. Shi, J. Zhai, L. B. Gan, C. H. Huang, T. T. Liu, L. M. Ying, G. B. Luo, and X. S. Zhao, *Chem. Phys. Lett.* **319**, 7 (2000).

[49] S. E. Canton, A. J. Yencha, E. Kukk, J. D. Bozek, M. C. A. Lopes, G. Snell, and N. Berrah, *Phys. Rev. Lett.* **89**, 045502 (2002).

[50] N. Manini, P. Gattari, and E. Tosatti, *Phys. Rev. Lett.* **91**, 196402 (2003).

[51] T. Sato, K. Tokunaga, and K. Tanaka, *J. Chem. Phys.* **124**, 024314 (2006).

[52] R. D. Bendale, J. F. Stanton, and M. C. Zerner, *Chem. Phys. Lett.* **194**, 467 (1992).

[53] A. D. Boese and G. E. Scuseria, *Chem. Phys. Lett.* **294**, 233 (1998).

[54] N. Manini, A. D. Corso, M. Fabrizio, and E. Tosatti, *Philos. Mag. B* **81**, 793 (2001).

[55] O. Gunnarsson, H. Handschuh, P. S. Benchthold, B. Kessler, and G. Ganteför, *Phys. Rev. Lett.* **74**, 1875 (1995).

[56] G. Duškesas and S. Larsson, *Theor. Chem. Acc.* **97**, 110 (1997).

[57] S. Larsson, A. Klimkāns, L. Rodriguez-Monge, and G. Duškesas, *J. Mol. Struct. (Theochem)* **425**, 155 (1998).

[58] A. A. Voityuk and M. Duran, *J. Phys. Chem. C* **112**, 1672 (2008).

[59] N. Koga and K. Morokuma, *Chem. Phys. Lett.* **196**, 191 (1992).

[60] K. Tokunaga, T. Sato, and K. Tanaka, *J. Chem. Phys.* **124**, 154303 (2006).

[61] K. Tokunaga, T. Sato, and K. Tanaka, *J. Mol. Struct.* **838**, 116 (2007).

[62] T. Sato, K. Tokunaga, and K. Tanaka, *J. Phys. Chem. A* **112**, 758 (2008).

[63] A. Devos and M. Lannoo, *Phys. Rev. B* **58**, 8236 (1998).

[64] T. Kitamura and M. Yokoyama, *J. Appl. Phys.* **69**, 821 (1991).

[65] S. Aratani, T. Kawashima, and A. Kakuta, *Jpn. J. Appl. Phys.* **30**, L1656 (1991).

[66] B. C. Lin, C. P. Cheng, and Z. P. M. Lao, *J. Phys. Chem. A* **107**, 5241 (2003).

[67] *GaussView 3.0,* (Gaussian, Inc., Pittsburgh, PA, 2003).

[68] T. Sato, H. Tanaka, A. Yamamoto, Y. Kuzumoto, and K. Tokunaga, *Chem. Phys.* **287**, 91 (2003).

In: Organic Semiconductors
Editors: Maria A. Velasquez
ISBN: 978-1-61209-391-8

Chapter 5

ORGANIC FIELD-EFFECT TRANSISTORS: TETRATHIAFULVALENE DERIVATIVES AS HIGHLY PROMISING ORGANIC SEMICONDUCTORS

M. Mas-Torrent[1], P. Hadley[2], S. T. Bromley[3], J. Veciana[1] and C. Rovira[1]

[1]Institut de Ciència de Materials de Barcelona, Campus UAB, Bellaterra, Spain
[2]Kavli Institute of NanoScience, Delft University of Technology, CJ Delft, The Netherlands
[3]Departament de Química Física, Universitat de Barcelona, Barcelona, Spain

The processing characteristics of organic semiconductors make them potentially useful for electronic applications where low-cost, large area coverage, and structural flexibility are required.[1] Contrary to amorphous silicon, which is widely used in solar cells and flat screen displays, organic materials offer the benefits that they can be deposited on plastic substrates at low temperature by employing solution-based printing techniques. These deposition techniques would, therefore, reduce the manufacturing costs dramatically. The challenge now lies on finding organic semiconductors which are processable, stable and, at the same time, exhibit high enough mobilities (μ>0.1 cm^2/Vs) and ON/OFF current ratios (>10^6) to be used for applications in modern microelectronics.

Transistors play a central role in many electronic circuits, where they usually function as either a switch or an amplifier. A field-effect transistor can be described as a three terminal device in which the current through a semiconductor linked to two terminals (namely source and drain) is controlled at the third terminal (called gate) by a voltage. A typical organic field-effect transistor (OFET) configuration is shown schematically in Figure 1. The voltage applied to the gate (V_G) induces an electric field through the dielectric and causes the formation of an accumulation layer of charges at the interface of the semiconductor deposited above. Then, by applying a source-drain voltage (V_{SD}) it is possible to measure current between the source and the drain (I_{SD}).

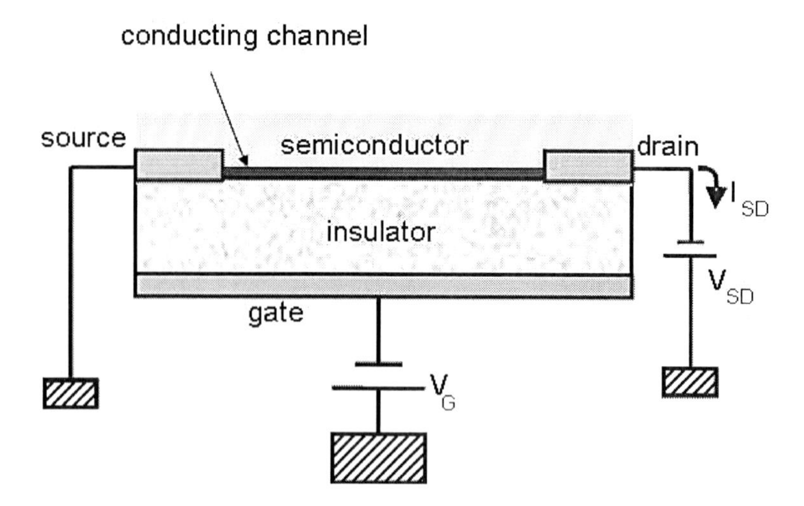

Figure 1. Schematic drawing of an OFET. This device is biased the way a p-type semiconducting OFET would be. The source is grounded and a negative voltage is applied to the drain and the gate to induce holes in the semiconducting channel.

Depending on how strong the interactions between the molecules of the semiconductor are, the electron wavefunctions that correspond to these energy levels can either be localized or extended. Strong molecular interactions lead to extended electron states with the allowed energy levels arranged into bands of allowed energies separated by gaps of forbidden energies. Weak molecular interactions lead to localized states and hopping conduction. Band conductors, with extended electron states, can be good conductors or good insulators depending on the location of the Fermi energy. Electron states much more than k_BT above the Fermi energy are unoccupied. Here k_B is Boltzmann's constant and T is the absolute temperature. States much less than k_BT below the Fermi energy are occupied and states within k_BT of the Fermi energy are partially occupied. The Fermi energy, E_F, is defined implicitly by the relationships,

$$n = \int_{-\infty}^{\infty} g(E) f(E) dE \quad \textbf{and} \quad f(E) = \frac{1}{e^{\frac{E-E_F}{k_BT}} + 1}$$

Here E is the energy, $f(E)$ is the Fermi function, n is the electron density, and $g(E)$ is the energy density of states per unit volume. If the electron density and distribution of energy levels are known, the Fermi energy can be found by iterating these equations. A material with partially filled bands is a metal and a material with the Fermi energy in a forbidden band gap is an insulator if the difference in energy from the Fermi energy to the allowed bands is much greater than the energy of thermal fluctuations k_BT. Semiconductors that are easier to oxidize than to reduce are typically p-type semiconductors where the Fermi energy is in the bandgap but within k_BT above the top of the valence band. In this case, current will flow predominantly due to the motion of positively charged holes. Semiconductors that are easier to reduce than to oxidize are typically n-type semiconductors where the Fermi energy is within k_BT beneath the bottom of the conduction band. In this case, current will flow predominantly due to the motion of negatively charged electrons. Figure 2 shows a schematic energy diagram for all these types of materials.

The Fermi energy can be shifted either chemically or electrostatically. For instance, adding a dopant that increases the number of electron states below the Fermi energy will decrease the Fermi energy. Oxidation shifts the Fermi energy down and reduction shifts the Fermi energy up. In a transistor, a voltage applied to the gate will shift the Fermi energy electrostatically and can thereby modulate the conductivity of the channel. An applied gate voltage can make an originally conducting channel an insulator (depletion mode) or an originally insulating channel conducting (enhancement mode). A gate voltage can even convert an n-type semiconductor into a p-type semiconductor (inversion).

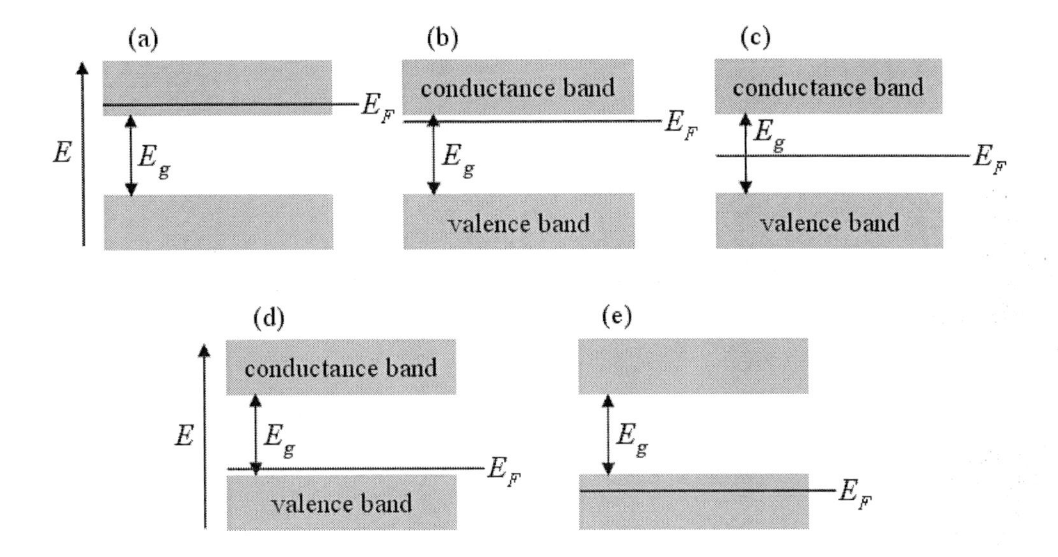

Figure 2. The conductivity of a material depends strongly on the position of the Fermi energy, E_F. The gray shaded areas indicate energies where electron states exist. A forbidden energy gap E_g is indicated for all five cases. (a) A metal with a small work function. (b) An n-type semiconductor. (c) An insulator. (d) A p-type semiconductor. (e) A metal with a large work function.

Whenever two materials come into contact, a transfer of charge takes place to compensate for the difference in the work functions. Electrons can lower their energy by moving to a material with a larger work function. They will do this until the electric field at the interface increases enough to hold them back and the Fermi energy is the same in both materials. Such electric fields typically exist at the source and drain contacts of a transistor. These fields locally shift the energy of the electron states with respect to the Fermi energy. This phenomenon is known as band bending. Consequently, depending on how the bands are bent, the semiconducting region near a contact can be a better conductor or a worse conductor than the rest of the channel. Energy diagrams for two metal/semiconductor contacts are illustrated in Figure 3. In Figure 3a, the Fermi energy is in the middle of the allowed energy band of the metal on the left and it is just above the HOMO states on the right. This means that the semiconductor is p-type. The electron energy states in the semiconductor are bent down at the contacts indicating that the work function of the metal (i.e. the minimum energy required to remove an electron from the Fermi level to the vacuum level) is lower than the work function of the semiconductor. The states near the contacts are further away from the Fermi level making this region less conducting than the middle of the channel. This region of poor conductivity is known as a Schottky barrier. Figure 3b shows the same p-type semiconductor in contact with a high work-function metal. This bends the bands up in the semiconductor forming good ohmic contact to the p-semiconductor. Similarly a low work-function metal makes ohmic contact to an n-type semiconductor while a high work-function metal forms a Schottky barrier at an interface with an n-type semiconductor.

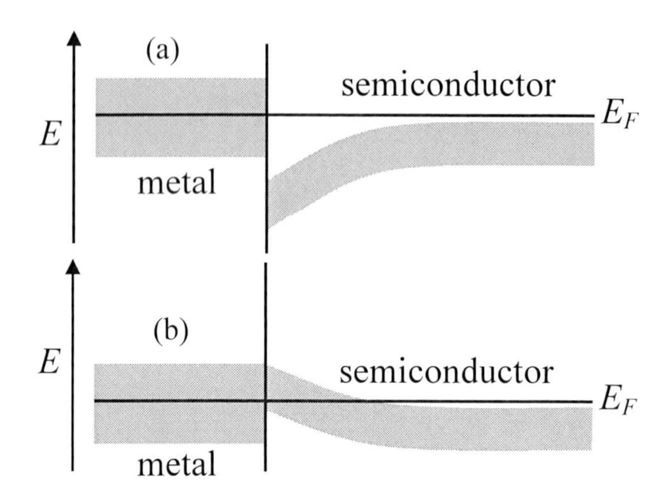

Figure 3. Band bending at interfaces due to differences in work functions. The gray regions indicate the energies at which electron states exist. Just the valence band of the semiconductor is drawn for simplicity. (a) A contact formed between a metal with a low work function and a p-type semiconductor. A poorly conducting Schottky barrier appears at the contact. (b) A contact between a high work-function metal and a p-type semiconductor. This results in ohmic contact.

The current that flows through a transistor depends on all these shifts caused by bringing different materials in contact as well as the shifts caused by applying gate and source voltages. Fortunately there is a relatively simple model that describes the current. The standard model for a Metal Oxide Semiconductor Field Effect Transistor (MOSFET) was originally constructed to describe inorganic semiconductors such as silicon but is commonly applied to OFETs. The equations for the MOSFET model are given in Figure 4. There are two regimes in this model, the linear regime, where paradoxically the current is described by a parabola, and the saturation regime where the source-drain current is independent of the source-drain voltage. An important parameter in this model is the mobility. This parameter can differ by orders of magnitude depending on the materials used for the semiconducting channel. Poorly conducting organic semiconductors have mobilities of $\mu = 10^{-4}$ cm^2/Vs or lower, good quality spin-on organic materials have mobilities of $\mu \sim 10^{-2}$ cm^2/Vs and well ordered organic materials have mobilities in the range $\mu \sim$ 1-10 cm^2/Vs. Amorphous silicon has a mobility of $\mu \sim 1$ cm^2/Vs, and crystalline silicon has mobilities on the order of $\mu \sim 10^3$ cm^2/Vs. Exceptionally clean systems of inorganic semiconductors can have mobilities of 10^7 cm^2/Vs.

There are two main families of organic semiconductors (Figure 5) that can be used as a component in OFETs: i) conjugated polymers (e.g. polyphenylene, polythiophene, poyphenylenevinylene) and ii) small conjugated molecules with low molecular weight (e.g. pentacene, oligothiophene, phthalocyanine).

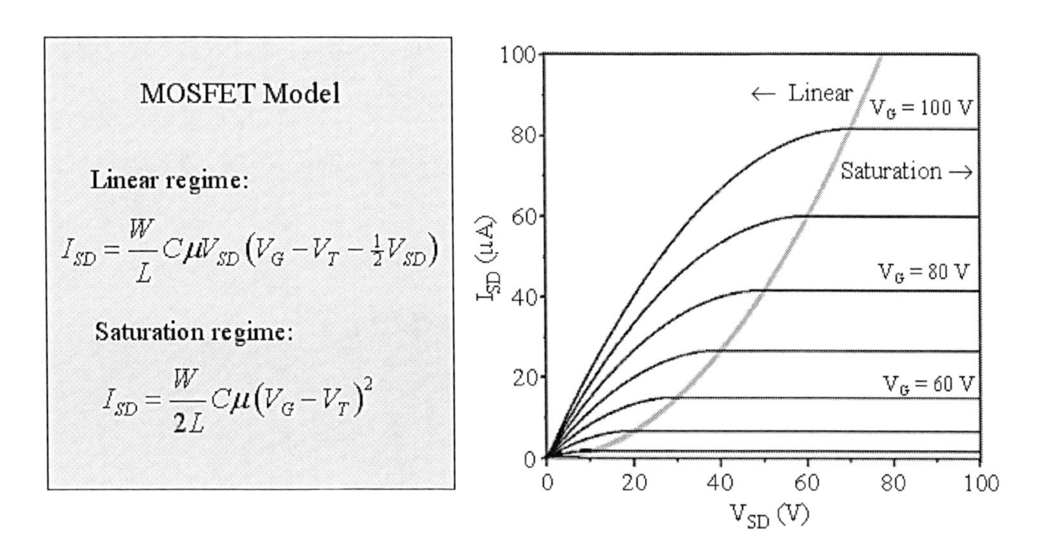

Figure 4. The MOSFET model. Here I_{SD} is the current from source to drain, V_{SD} is the voltage from source to drain, V_G is the gate voltage, V_T is the threshold voltage, L is the length of the transistor from source to drain in the direction that the current flows, W is the width of the transistor, C is the capacitance between the gate and the channel, and μ is the mobility. The MOSFET equations are plotted on the right for a transistor where $WC\mu/L = 3.3 \times 10^{-8}\ \Omega^{-1}$ and the threshold voltage is $V_T = 30$ V. The thick gray line separates the linear regime from the saturation regime.

Small conjugated molecules form regular crystalline structures where the conductivity is determined by the overlapping of the π–orbitals. The intermolecular bonds in molecular crystals are weaker than the interatomic bonds in inorganic semiconductors like silicon, so thermal fluctuations result in larger displacements in molecular crystals. These displacements disrupt the crystal order and result in lower mobilities than are observed in inorganic semiconductors. However, crystals of small molecules typically form more ordered films and have higher OFET mobilities compared to polymers.

Polymers are chains with good conductivity along the chains and weaker conductivity between the chains. Interchain electron transport takes place primarily due to π–orbital overlap between the chains. Sometimes the chains fold into ordered crystalline structures over short length scales but usually they have a microstructure that resembles cooked spaghetti and, thus, the conductivity is limited by disorder. While polymers are mostly soluble in organic solvents, devices based on small molecules are typically prepared by evaporation of the organic material due to their low solubility.

The performance of OFETs in the last 20 years has improved enormously.[1] Nowadays, mobilities of the same order as amorphous silicon (0.1-1 cm^2/Vs) are achieved in the best OFETs perfomance. Thiophene and, especially, acene derivatives are considered to be the benchmark in OFETs and most of the best mobilities have been found in these families of compounds. A mobility of 1.5 cm^2/Vs has been reported for pentacene,[2] and recently a mobility of 15 cm^2/Vs was reported for rubrene.[3] Some works have focused on directly functionalizing pentacene in order to tune its electronic properties[4] and enhance the intermolecular overlap.[5] Comparative studies of OFETs based on oligothiophenes regarding the deposition method[6] and their molecular structure have also been carried out.[7][8] All such studies are of great interest in order to gain a better understanding of the requirements that the organic semiconductors should accomplish to achieve high OFET performance. However, OFET performance depend on a large number of parameters and, therefore, it is essential to understand their role in the device operation and to have control over all of them in order to ensure reproducibility and to be able to establish comparative studies between the different devices. Below some of the key experimental variables for the OFETs fabrication are illustrated.

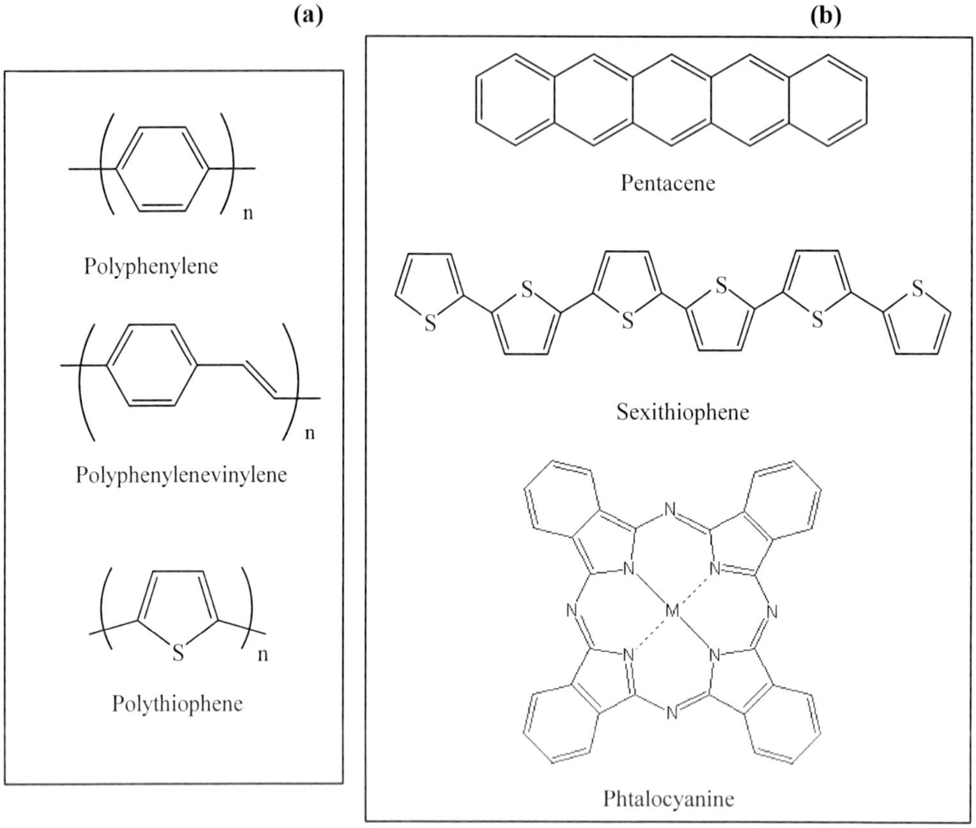

Figure 5. Molecular formula of organic semiconductors: a) conjugated polymers, b) low-molecular weight materials.

1. Device Configuration

The design of the device configuration will influence on the measured electronic properties. For instance, the choice of metal electrodes will affect the contact resistance. As mentioned before, it is important to choose the electrode metal according to the nature of the organic semiconductor to have efficient charge injection. That is, for an n-type material it will be more suitable to have a metal with a lower work function, whereas for p-type material a metal with high work function would be desired. The dielectric material will influence the electric field created along it, the current leakage through the gate insulator and the quality of the interface between the organic semiconductor and the dielectric. In addition, the characteristics of the dielectric can be modified with surface treatments. For example, growing a self-assembled monolayer of an organic molecule on silica will change the hydrophilic nature of the oxide to hydrophobic. It has been demonstrated that these treatments often have strong effects on the film structure and on the resulting electrical characteristics.[9]

Likewise, the source and drain electrodes can be evaporated on the top of the organic material (top contact configuration) or on the dielectric before depositing the organic semiconductor (bottom contact configuration). Usually, the electrical contact with the organic semiconductor is better in a top contact configuration. However, the evaporation of the metal might damage the organic material. This top-architecture does not permit the patterning of the electrodes through conventional lithography and, thus, the contacts have to be deposited through a shadow mask, which does not allow channel lengths shorter that tens of micrometers.[10] On the other hand, the protruding electrodes in the bottom-contact configurations might cause inhomogeneities during the film formation.

2. Integration of the Organic Material in the Device

The organic semiconductor can be deposited on the device employing solution based techniques like drop-casting or spin-coating. The use of such techniques is the most promising route to produce low manufacturing costs and to fabricate devices of large-area coverage. In addition, by combining them with stamping or printing techniques,[11] it is possible to pattern organic semiconductors eliminating the use of lithography.

However, since often organic semiconductors are not very soluble, an alternative deposition method is by sublimation of the organic material in a variety of vacuum depositions systems. Parameters such as pressure and substrate temperature determine the morphology and quality of the resulting films. Typically, the use of vapor phases allows for the obtaining of higher purity materials and, hence, the higher OFET mobilities have been achieved in this way.

3. Organic Semiconductor

Obviously, the choice of the organic semiconductor will mainly determine the device performance. Here, it will be important to take into account the intermolecular interactions that the molecules exhibit as the stronger the electronic coupling between neighboring molecules is, the higher mobilities will be achieved. In this way, it will be thus crucial to have materials with very high molecular order to avoid disruption of the electron conduction path.

Typically, solution processed polymers form complex microstructures, where microcrystalline domains are embedded in an amorphous matrix. The disordered matrix limits the charge transport resulting in low field-effect mobilities. For this reason, the most studied polymer for OFETs is poly(3-hexylthiophene) (P3HT), which has given the highest OFET mobility found for an organic polymer OFET, 0.1 cm^2/Vs.[12]2 This high mobility is related to structural order in the polymer film induced by the regioregular head-to-tail coupling of the hexyl side chains (Figure 6).[13]

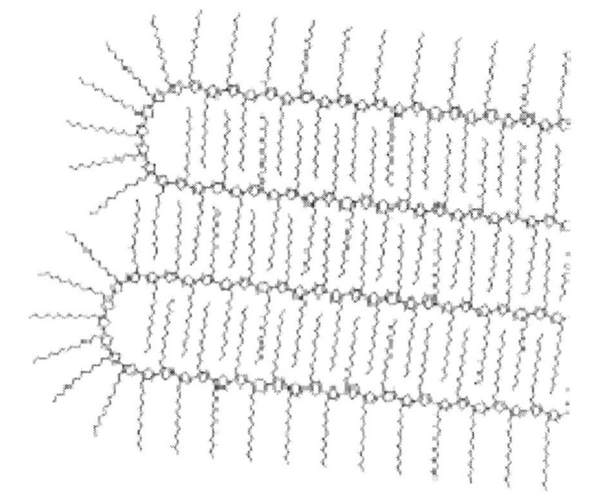

Figure 6. Schematic lamellar type ordering in P3HT films. Reprinted with permission from reference 13.

Additionally, when *p*-xylene or cyclohexanone are used as solvents, P3HT forms fibers in the solution, which can then be deposited on the substrate (Figure 7).[14] Hole mobilities in the range of 0.02- 0.06 cm^2/Vs have been reported very recently for single nanofibers and films of aligned fibers.[14]b-[15] The formation of similar fibers has been observed in other conjugated polymers and is believed to be governed by the π-stacking of the conjugated chains.[16]

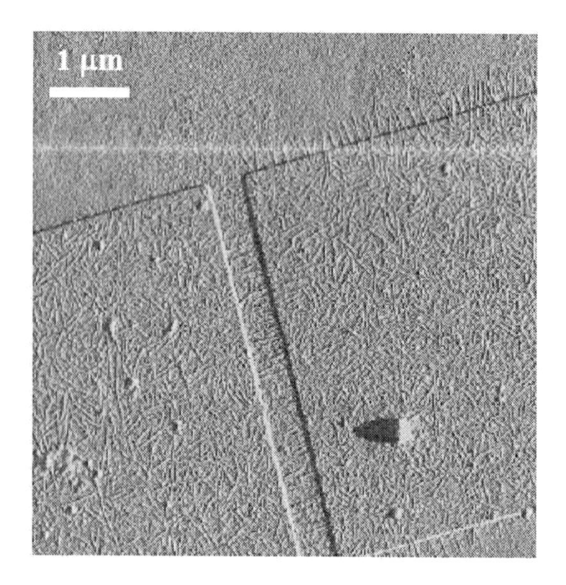

Figure 7. AFM image of a device prepared by drop casting a solution of P3HT in p-xylene whilst applying an ac voltage (2 V, 102 kHz) on the right electrode of the image. This resulted in the formation of aligned nanofibers. Reprinted with permission from reference [14b].

The importance of the molecular ordering for fabricating OFETs with high performance is not restricted to polymer based transistors. For instance, Dimitrakopoulos and Mascaro demonstrated that films of pentacene evaporated at different temperatures show a variety of crystallinity which correlates with the resulting OFET performance.[17]

The Langmuir-Blodgett technique, which consists in transferring on a substrate amphiphilic molecules ordered on the water surface-air interface, offers a unique approach of preparing well-ordered thin films. The application of this technique for the preparation of OFETs has already been demonstrated.[18][19]

The fundamental material characteristics of organic semiconductors are, however, most clearly measured in single-crystals owing to their higher molecular ordering and to the fact that they do not present grain boundaries that limit the mobility.[20] The highest OFET mobility reported to date has been observed in rubrene crystals (15 cm^2/Vs).[20] [a] All these organic crystals are typically grown from the vapor phase.

Additional parameters such as the purity or stability of the material and processability determine its potential for applications. For this reason, some attempts have focused on synthesizing soluble pentacene and oligothiophenes precursors which, after heating, are converted to their parents compounds,[21] since these latter show high OFET performances but very often have to be processed using vacuum-techniques due to their low solubility in common organic solvents.

Overall, to progress in this field towards device applications, it is very important to search for new molecules that are able to act as organic semiconductors, and also to develop easy methods to grow single crystals or films. Along these lines, we recently reported the preparation OFETs using single-crystals of tetrathiafulvalene (TTF) derivatives as the organic semiconductor.[22] These crystals were grown from solution by drop casting, a very simple method. Moreover, responding to the current needs of gaining a better understanding of the relationship between crystal structure and field-effect mobility, a systematic study of the dependence of the device performance on the crystal structure was carried out. This comparative study allowed for the focused investigation of the influence of the intermolecular interactions on the electronic transport properties.

OFETs Based on Tetrathiafulvalene (TTF) Derivatives

Since the discovery of the first organic metal tetrathiafulvalene-7,7',8,8'-tetracyano-quinodimethane (TTF-TCNQ) over thirty years ago,[23] tetrathiafulvalene (TTF) and its derivatives have been successfully used as building blocks for charge transfer salts giving rise to a multitude of organic conductors and superconductors. The crystallisation of TTF derivatives is governed by the π-π stacking, which permits, together with the S···S interactions, an intermolecular electronic transfer responsible for their transport properties. TTF derivatives are generally soluble in various solvents, are easily chemically modified, and are good electron donors. TTFs have already been extensively studied for the preparation of a wide range of molecular materials.[24] Considering thus all the above, TTF derivatives also promise to be good candidate molecules for the preparation of OFETs, due to the possibilities of synthesising tailored derivatives which can be easily processed, either in vacuum or from solution.[25]

With the aim of establishing a correlation between crystal structure and charge carrier mobility, we recently prepared OFETs based on neutral TTF crystals grown from solution.[22] We studied eight different TTF derivatives, which are shown in Figure 8, namely bis(ethylenedithio)-tetrathiafulvalene (BEDT-TTF), (ethylenethio)-(ethylenedithio)-tetrathiafulvalene (ETEDT-TTF), bis(ethylenethio)-tetrathiafulvalene (BET-TTF), (ethylenethio)-(thiodimethylene)-tetrathiafulvalene (ETTDM-TTF), dithiophene-tetrathiafulvalene (DT-TTF), (thiophene)(thiodimethylene)-tetrathiafulvalene (TTDM-TTF) and (ethylenethio)-(thiophene)-tetrathiafulvalene (ETT-TTF). Taking, thus, into account the crystal packing in which these molecules are arranged, this family of compounds can be classified into three groups. BEDT-TTF and ETEDT-TTF are from the first group (G1). Their supramolecular organisation consists of dimers sustained by hydrogen bonds, forming chains along the *a* axis due to lateral S···S interactions. The chains are arranged perpendicularly to each other avoiding therefore the formation of stacks. In the second crystal structure group (G2), BET-TTF and ETTDM-TTF crystallise forming chains of quasi planar molecules along the *a* axis interacting side-by-side. These chains stack into layers giving rise to a bidimenssional electronic structure. Finally, the molecules from group 3 (G3), DT-TTF, TTDM-TTF, and ETT-TTF, crystallise forming uniform stacks of almost planar molecules along the *b* axis. The interplanar distance between molecules of one stack is very short (3.56-3.66 Å).

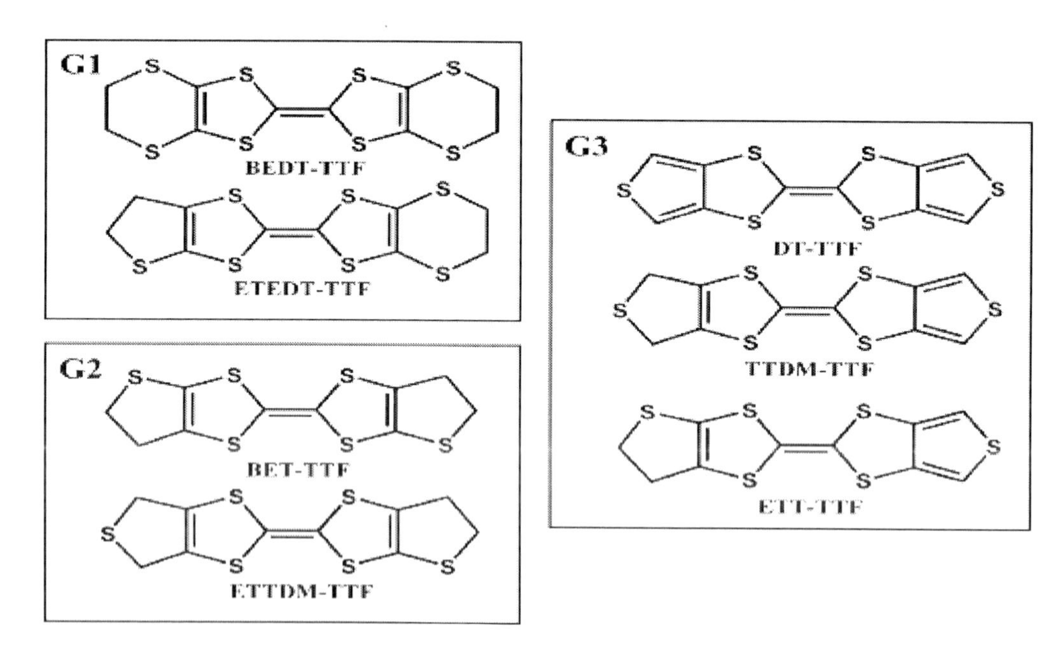

Figure 8. Molecular structures of the studied TTF derivatives for OFETs. The molecules are classified into three groups according to their crystal packing.

The OFET devices were prepared by drop casting a solution of the TTF derivative onto the gate insulator and the prefabricated source and drain gold electrodes. The solution was allowed to evaporate at room temperature which resulted in the formation of long regular crystals, some of which connected two of the microfabricated electrodes (Figure 9).

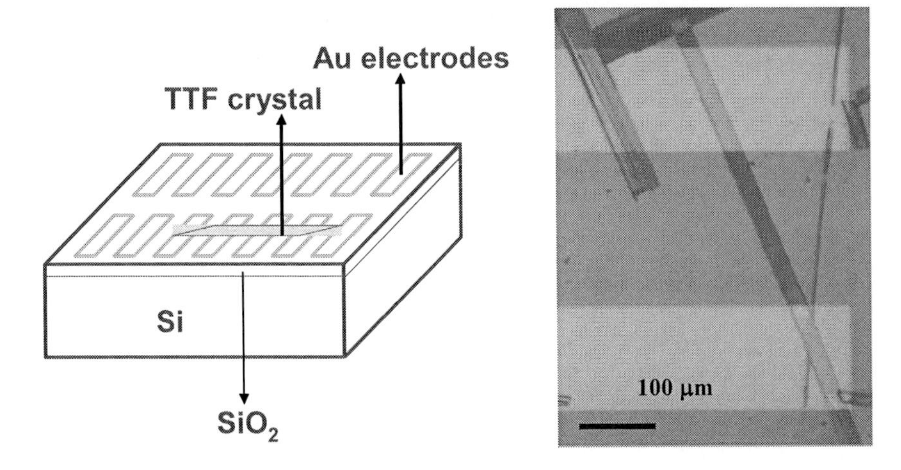

Figure 9. Left: Device configuration used. Right: Optical microscope image of a DT-TTF single crystal lying on the gold electrodes. Reprinted with permission from reference [22] a. Copyright 2004 American Chemical Society.

The electrical characteristics of the devices were studied by measuring the source-drain current versus the applied source-drain voltage for different gate voltages applied to the silicon substrate. Figure 10 shows the electrical measurements performed on a DT-TTF crystal. All the crystals exhibit a p-type behavior where the conductivity increases as a more negative gate is applied. We measured 67 different single-crystal OFETs using the TTF derivatives mentioned above. A clear correlation between crystal structure and device performance was found as a notable improvement of the charge carrier mobility was observed on going from group 1 to group 3. This trend suggests therefore that the crystal packing of group 3 is the most suitable for this family of materials for the preparation of OFETs. The highest mobility found was of 1.4 cm^2/Vs for a DT-TTF crystal. This is the highest mobility found for a solution processed material, which makes this material very interesting for potential applications.

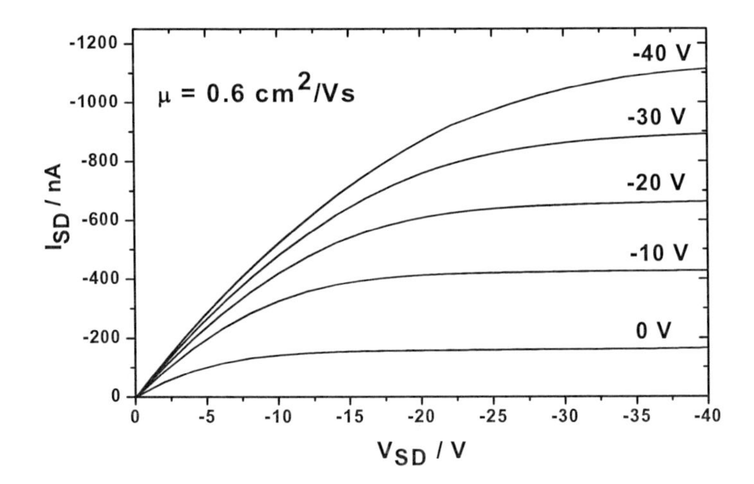

Figure 10. Source-drain current versus source-drain voltage at different gate voltages for a DT-TTF single crystal OFET. Reprinted with permission from reference [22b]. Copyright 2004 American Chemical Society.

At room temperature the charge mobility of semiconducting organic materials is often determined by a hopping transport process, which can be depicted as an electron or hole transfer reaction in which an electron or hole is transferred from one molecule to the neighboring one. Two major parameters determine self-exchange rates and, thus, the charge mobility:[26] (i) the electronic coupling between adjacent molecules, which needs to be maximized, and (ii) the reorganization energy (λ_{reorg}), which needs to be small for efficient charge transport. The reorganization energy of self-exchange in a hole-hopping material is defined as the sum of the geometrical relaxation energies of one molecule upon going from the neutral-state geometry to the charged-state geometry and the neighboring molecule upon going through the inverse process [$A(0) + A'(+) \rightarrow A(+) + A'(0)$]. These two portions of λ_{reorg} are typically nearly identical.[27] Previous theoretical works have attempted to explain the high mobility pentacene OFETs in terms of its low λ_{reorg}.[25a][28] We performed density functional (DF) calculations to calculate λ_{reorg} for all the studied TTF derivatives. All reported calculations were performed at a 6-31G(d,p)/B3LYP[29] level of theory using Gaussian 98.[30] In the neutral state, TTF derivatives are most energetically stable when in a distorted boat conformation, but adopt planar conformation for the +1 charged state.[31] First we calculated λ_{reorg} employing the stable boat conformation for each neutral isolated molecule and then, again, employing the closest local minimum energy structure for the neutral molecule which best approximated the conformation found in their crystal structure. For Group 2 and Group 3 crystal structures, the constituent TTF derivatives are found to be almost perfectly planar. Thus, in our calculations, to represent this packing-induce planarity, isolated neutral molecules were optimized to local energetic minimum planar conformations. For the Group 1 crystal packing, the TTF derivatives are not planar but adopt boat-like conformations. For BEDT-TTF we employed the lowest energy boat conformation, and, in the case of ETEDT-TTF, a minimum energy structure very close to the lowest energy boat conformation was found to be the best isolated molecular representation. For each calculation the lowest energy planar +1 molecular conformation was used. These results are summarized in Table 1. Comparing the resulting λ_{reorg} of all these molecules, we find that λ_{reorg} values calculated while always employing the lowest energy neutral molecule boat conformations do not follow any obvious pattern relating to the corresponding crystal structure. However, when employing the isolated neutral conformations, approximating the crystal packing-constrained molecular geometries, a trend of λ_{reorg} values emerges in line with that of the experimentally measured mobilities i.e. molecules with better mobilities tend to have lower λ_{reorg} values. The main reason for the difference in these two sets of calculations comes from the use of planar neutral molecules to represent the respective planar TTF derivates in the crystals of Group 2 and Group 3, which results in a notable drop of λ_{reorg}. This decrease in the λ_{reorg} values can be simply explained by the fact that the geometries of the neutral and charged planar molecules become closer, which is not the case for molecules of Group 1. The calculated differences in isolated molecular λ_{reorg} values, however, do not appear to fully explain the relatively large measured differences found in the mobilities, suggesting that intermolecular interactions should also be taken into account.

As mentioned above, the other parameter that determines mobilities, in addition to the reorganization energy, is the intermolecular electronic coupling, often estimated via the calculation of transfer integrals.[32] In Table 1 we thus further report transfer integrals for the TTF-derivative crystals displaying the highest mobility (and highest symmetry) in each crystal structure group. The single value, t_{max}, quoted for each Group is the maximum transfer integral for that crystal, based on a systematic search of all nearest neighbor dimeric possibilities for the strongest electronic coupling. Remarkably, the transfer integral values also follow the trend found for the experimentally measured mobilities i.e. the crystals with higher mobilites also have higher maximum transfer integrals. For DT-TTF, t_{max} is found along the packing crystal axis *b*, which also corresponds to the long crystal direction and, thus, the device channel direction. However, for BET-TTF, whereas the long crystal axis is known to correspond to the crystallographic *b*+*c* direction, t_{max} is found between the lateral S···S contacts present in the *ab* plane. Finally, for BEDT-TTF the interdimer t_{max} is also found to be between the lateral S·S along the *a* axis.

The extremely encouraging trends, in both λ_{reorg} and t_{max} values, are thus far based on calculations employing isolated molecules/dimers. Such studies can further benefit by taking into account a fuller representation of the true crystal environment e.g. the boat conformations in the crystals of Group 1 are stabilized by intermolecular hydrogen bonding. Previously we demonstrated that an additional drop in the λ_{reorg} value of DT-TTF is observed if one considers crystal-embedded molecules, rather than using solely their isolated planar molecular representations.[3] Following this line of reasoning, we hope in the future to further investigate the role of intermolecular interactions on TTF derivative crystal transport properties.

All these results are of great importance for the future design of new TTF derivatives with promising OFET performance. Following therefore this studied correlation, we prepared OFETs with the organic semiconductor dibenzo-TTF (DB-TTF), which crystallises similarly to the molecules in group 3 and has values of transfer integral (0.037 eV) and reorganisation energy (0.248 eV) close to the ones found for DT-TTF. Interestingly, the devices based on this material also displayed very high mobilities of the order of 0.1-1 cm^2/Vs.[34]

Table 1. Mean values of charge carrier mobilities (μ_m) and their standard deviations (in parenthesis), maximum transfer integrals of the symmetric derivatives (t_{max}), and isolated molecular reorganization energies of the TTF derivatives using the minimum energy neutral and +1 conformations (λ_{reorg}) and using the neutral local minimum energy conformation best approximating the crystal structure conformation (λ_{reorg} (c)).

TTF derivative		μ_m, cm^2/Vs	λ_{reorg}, eV	λ_{reorg} (c), eV	t_{max}, eV
G1	ETEDT-TTF	$1.4{\cdot}10^{-4}$ ($9.8{\cdot}10^{-5}$)	0.506	0.488	0.019
	BEDT-TTF	$1.2{\cdot}10^{-3}$ ($1.7{\cdot}10^{-3}$)	0.550	0.550	
G2	ETTDM-TTF	$1.8{\cdot}10^{-3}$ ($9.0{\cdot}10^{-4}$)	0.432	0.314	0.021
	BET-TTF	$1.5{\cdot}10^{-2}$ ($1.7{\cdot}10^{-2}$)	0.326	0.246	
G3	ETT-TTF	$6.2{\cdot}10^{-2}$ ($3.7{\cdot}10^{-2}$)	0.438	0.238	0.034
	TTDM-TTF	$1.5{\cdot}10^{-1}$ ($8.0{\cdot}10^{-2}$)	0.512	0.258	
	DT-TTF	$2.5{\cdot}10^{-1}$ ($4.0{\cdot}10^{-1}$)	0.574	0.238	

Other recent works have also prepared OFETs by evaporation of some TTF derivatives.[35] The best device performance employing vacuum techniques has been achieved with the dinaphto-tetrathiafulvelene analogue, which exhibited a maxim mobility of 0.42 cm^2/Vs, whereas the DB-TTF shows in this kind of device a maximum mobility of 0.06 cm^2/Vs,[34][d]which is one order of magnitude lower than the one found for the single crystal.

In summary, organic field-effect transistors are very promising for potential applications in large-area and low-cost electronics. However, further research is required to gain a better understanding and control over the parameters that influence the device performance. In addition, since most of the organic semiconductors are p conductors, searching for high performance and stable organic materials which are n conductors and which are ambipolar (conduction by holes and electrons) is also necessary to progress in the field.

We also described the influence of the crystal structure on the performance of single-crystal OFETs of TTF derivatives. A trend in mobilities through the different crystal structures was observed which was further strongly corroborated by calculations of both the molecular reorganization energies and the maximum intermolecular transfer integrals. The early obtained results with OFETs with TTF derivatives already point out the high potential of these materials, which can be easily processed either in vacuum or from solution.

REFERENCES

[1] C. D. Dimitrakopoulos, P. R. L. Malenfant, *Adv. Mater.* 2002, 14, 99. b) Y. Sun, Y. Liu, D. Zhu, *J. Mater. Chem.* 2005, 15, 53. c) Katz, H. E., *Chemistry of Materials* 2004, 16, 4748.

[2] S. F. Nelson, Y.-Y Lin, D. J Gundlach, T. N. Jackson, *Appl. Phys. Lett.* 1998, 72, 1854-1856.

[3] V. C. Sundar, J. Zaumseil, V.Podzorov, E. Menard, R. L. Willett, T. Someya, M. E. Gershenson, J. A. Rogers, *Science* 2004, 303, 1644-1646

[4] Meng, H. M.; Bendikov, M.; Mitchell, G.; Helgeson, R.; Wudl, F.; Bao, Z.; Siegrist, T.; Kloc, C.; Chen, C.-H. *Adv. Mater.* 2003, 15, 1090-1093.

[5] Anthony, J. E.; Brooks, J. S.; Eaton, D. L.; Parkin, S. R. *J. Am. Chem. Soc.* 2001, 123, 9482-9483. b) J. E. Anthony, D. L. Eaton, S. R. Parkin, *Org. Lett.* 2002, 4, 1, 15-18.

[6] Yanagi, H.; Araki, Y.; Ohara, T.; Hotta, S.; Ichikawa, M.; Taniguchi, Y. *Adv. Funct. Mater.* 2003, 13, 767-773.

[7] Halik, M.; Klauk, H.; Zschieschang, U.; Schmid, G.; Ponomarenko, S.; Kirchmeyer, S.; Weber, W. *Adv. Mater.* 2003, 15, 917-922.

[8] Videlot, C.; Ackermann, J.; Blanchard, P.; Raimundo, J.-M.; Frère, P.; Allain, M.; de Bettignies, R.; Levillain, E.; Roncali, J. *Adv. Mater.* 2003, 15, 306-310.

[9] D. J. Gundlach, J. A. Nichols, L. Zhou, T. N. Jackson, *Appl. Phys. Lett.* 2002, 80, 2925. b) I. Kymissis, C. D. Dimitrakopoulos, S. Purushothaman, *IEEE Trans. Electron Devices* 2001, 48, 1060.

[10] G. Horowitz, *J. Mater. Res.* 2004, 19, 1946.

[11] H. E. Katz, *Chem. Mater.* 2004, 16 (23), 4748. b) M. M. Ling, Z. Bao, Z. *Chem. Mater.* 2004,16 (23), 4824.

[12] Sirringhaus, P. J. Brown, R. H. Friend, M. M. Nielsen, K. Bechgaard, B. M. W. Langeveld-Voss, A. J. H. Spiering, R. A. J. Janssen, E. W. Meijer, P. Herwig, D. M. de Leeuw, *Nature* 1999, 401, 685.

[13] E. Mena-Osteritz, A. Meyer, B. M. W. Langeveld-Voss, R. A. J. Janssen, E. W. Meijer, P. Bäuerle, *Angew. Chem. Int. Ed.* 2000, 39, 2680

[14] K. J. Ihn, J. Moulton, P. Smith, *J. Polym. Sci. Part B: Polym. Phys.* 1993, 31, 735. b) M. Mas-Torrent, D. Den Boer, M. Durkut, P. Hadley, A. P. H. J. Schenning,. *Nanotechnology* 2004, 15, S265.

[15] J. A. Merlo, C. D. Frisbie, *J. Phys. Chem.* B 2004, 108, 19169; J. A. Merlo, C. D. Frisbie, *J. Polym. Sci. Part E: Polym. Phys.* 2003, 41, 2674.

[16] Ph. Leclère, E. Hennebicq, A. Calderone, P. Brocorens, A. C. Grimsdale, K. Müllen, J. L. Brédas, R. Lazzaroni, *Prog. Polym. Sci.* 2003, 28, 55.

[17] C. D. Dimitrakopoulos, D. J. Mascaro, *IBM J. Res. & Develop.* 2001, 45, 11.

[18] J. Paloheimo, P. Kulvalainen, H. Stubb, E. Vuorimaa, P. Yli-Lahti, *Appl. Phys. Lett.* 1990, 56, 1157. J. Matsui, S. Yoshida, T. Mikayama, A. Auki, T. Miyashita, *Langmuir* 2005, 21, 5343.

[19] Xu, G. Yu, W. Xu, Y. Xu, G. Cui, D. Zhang, Y. Liu, D. Zhu, *Langmuir* 2005, 21, 5391.

[20] V. Podzorov, S. E. Sysoev, E. Loginova, V. M. Pudalov, M. E. Gershenson, *Appl. Phys. Lett.* 2003, 83, 3504. b) V. C. Sundar, J. Zaumseil, V. Podzorov, E. Menard, R. L. Willett, T. Someya, M. E. Gershenson, J.A. Rogers, *Science* 2004, 303, 1644.R. W. I. De Boer, T. M. Klapwijk, A. F. Morpurgo, *Appl. Phys. Lett.* 2003, 83, 4345.M. Ichikawa, H. Yanagi, Y. Shimizu, S. Hotta, N. Suganuma, T. Koyama, Y. Taniguchi, *Adv. Mater.* 2002, 14, 1272.

[21] P. T. Herwig, K. Müllen, *Adv. Mater.* 1999, 11, 480. (b) G. H. Gelink, H. E. A. Huitema, E. van Veenendaal, E.Cantatore, L. Schrijnemakers, J. B. P. van der Putten, T. C. T. Geuns, M. Beenhakkers, J. B. Giesbers, B.-H Huisman, E. J. Meijer, E. Mena Benito, F. J. Touwslager, A. W. Marsman, B. J. E. van Rens, D. M. de Leeuw, Nature Materials 2004, 3, 106.A. Afzali, C. D. Dimitrakopoulos, T. L. Breen, *J. Am. Chem. Soc.* 2002, 124, 8812.

[22] M. Mas-Torrent, M. Durkut, P. Hadley, X. Ribas, C. Rovira *J. Am. Chem. Soc.* 2004, 126, 984. b) M. Mas-Torrent, P. Hadley, S. T. Bromley, X. Ribas, J. Tarrés, M. Mas, E. Molins, J. Veciana, C. Rovira, *J. Am. Chem. Soc.* 2004, 126, 8546.

[23] J. Ferraris, D. O. Cowan, V. V. Walatka and J. H. Perlstein,. *J. Am. Chem. Soc.*, 1973, 95, 948.

[24] See review papers in the issue *Chem. Rev.*, 2004, 104.

[25] M. Mas-Torrent, C. Rovira, *J. Mater. Chem.* in press.

[26] R. A. Marcus, *Rev. Mod. Phys.* 1993, 65, 599. b) N. E. Gruhn, D. A. da Silva Filho, T. G. Bill, M. Malagoli, V. Coropceanu, A. Kahn, J.-L. Brédas,. *J. Am. Chem. Soc.* 2002, 124, 7918.

[27] M. Malagoli, J.-L. Brédas, *Chem. Phys. Lett.* 2000, 327, 13.

[28] H. M. Meng, M. Bendikov, G. Mitchell, R. Helgeson, F. Wudl, Z. Bao, T. Siegrist, C. Kloc, C.-H Chen, *Adv. Mater.* 2003, 15, 1090. b) J. Cornil, J. Ph. Calbert, J.-L Brédas, *J. Am. Chem. Soc.* 2001, 123, 1250.

[29] D: Becke, *J. Phys. Chem.* 1993, *98*, 5648

[30] M. J. Frisch, et. al., *Gaussian* 98, Revision A.9, Gaussian, Inc.: Pittsburgh PA 1998.

[31] Demiralp, W. A. Goddard III, *J. Phys. Chem.* A 1997, 101, 8128.

[32] J. L. Brédas, J. P. Calbert, D. A. da Silvo Filho, J. Cornil, *J. Proc. Natl. Acad. Sci.* 2002, 99, 5804.

[33] S. T. Bromley, M. Mas-Torrent, P. Hadley, C. Rovira, *J. Am. Chem. Soc.* 2004, 126, 6544.

[34] M. Mas-Torrent, P. Hadley, S. T. Bromley, N. Crivillers, J. Veciana, C. Rovira, *Appl. Phys. Lett.* 2005, 86, 012110.

[35] M. Iizuka, Y. Shiratori, S. Kuniyoshi, K. Kudo and K. Tanaka, *Appl. Surf. Sci.* 1998, 130-132, 914. b) B. Noda, M. Katsuhara, I. Aoyagi, T. Mori and T. Taguchi, *Chem. Lett.*, 2005, 34, 392. c) M. Katsuhara, I. Aoyagi, H. Nakajima, T. Mori, T. Kambayashi, M. Ofuji, Y. Takanishi, K. Ishikawa, H. Takezoe and H. Hosono, *Synthetic Metals*, 2005, 149, 219. d) Naraso, J.-I. Nishida, S. Ando, J. Yamaguchi, K. Itaka, H. Koinuma, H. Tada, S. Tokito and Y. Yamashita, *J. Am. Chem. Soc.* 2005, 127,10142

In: Organic Semiconductors
Editors: Maria A. Velasquez

ISBN: 978-1-61209-391-8

Chapter 6

DEVELOPMENT OF NEW ORGANIC SEMICONDUCTORS FOR APPLICATION IN ORGANIC ELECTRONICS

***Chunyan Du*[1,2] *and Yunqi Liu*[1*]**

[1]Beijing National Laboratory for Molecular Sciences, Organic Solids Laboratory, Institute of Chemistry, Chinese Academy of Sciences, Beijing, China

[2]Graduate School of Chinese Academy of Sciences, Beijing, China

ABSTRACT

Electronic and optoelectronic devices using organic materials as active elements are attractive because they can take advantage of organic materials such as easy of functionality, light weight, low cost, and capability of thin-film, large-area, flexible device fabrication. Among them, organic field-effect transistors (OFETs), which consist of organic semiconductors, dielectric layers, and electrodes, are expected to be a promising technology for application in displays, sensors, and memories. Organic semiconductors play a key role in determining the device characteristics. Recent technological advances in OFETs have triggered intensive research into molecular design, synthesis, device fabrication, thin film morphology and transport of holes and electrons and so on. New organic semiconductors are currently being intensively studied, resulting in the development of a growing number of high-performance semiconducting materials with higher mobility than that of amorphous silicon (0.1 $cm^2V^{-1}s^{-1}$). Here, firstly, an introduction to OFET principles and history, as well as the charge transport mechanisms, film alignment and morphology and crystal growth processes are presented. Then we discussed the structural design/realization of recently developed high performance *p*- (hole-transporting) and *n*-channel (electron-transporting) semiconductors for OFETs. A survey of the reported molecules and correlations between their structure and transistor performance are presented. The device structures and dielectric gate insulator materials are also described. Besides, the influence of the device fabrication process, organic semiconductor/dielectric layer interface, and organic layer/electrode contact on the device performance was reviewed.

1. Introduction

Organic semiconductors have been widely investigated as active layers in optoelectronic devices recent years. Compared with their inorganic counterparts, organic semiconductors possess many unique advantages, such as low-cost, easy-processibility and low temperature device fabrication. [1] Since the first report in 1987, organic field-effect transistors (OFETs) [2–3] have received significant research interest, and remarkable progress has been made by a number of research groups. Recently, studies have demonstrated that numerous organic materials exhibit carrier mobility comparable to amorphous silicon (0.1 $cm^2V^{-1}s^{-1}$), [4] indicating OFETs have great promise for application in integrated circuits for large-area, flexible and ultralow-cost electronics, such as radio frequency identification (RFID) tags, smart cards, and organic active matrix displays. Organic semiconductors used as active layers in OFETs have experienced a dramatic improvement in performance, stability and also solution processibility over the past decade.

1.1. Principles and Operation Processes of Organic Field-Effect Transistors

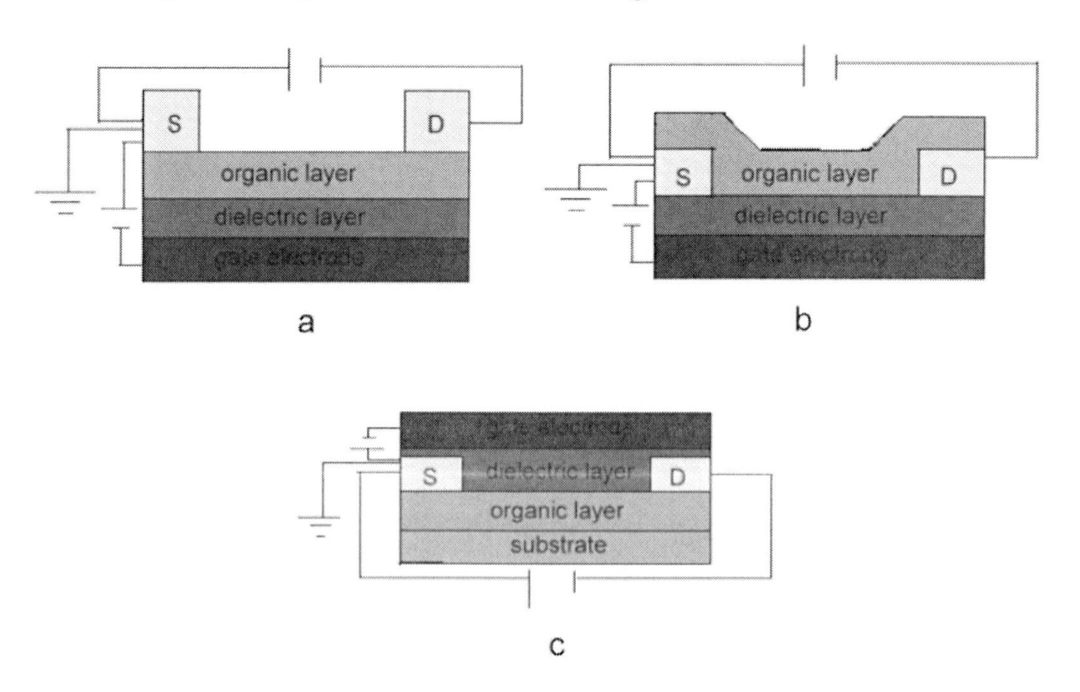

Figure 1. Schematic diagram of the device configuration: (a) top-contact configuration, (b) bottom-contact configuration, (c) top gate structure.

The most common OFET device configuration is that of a thin film transistor, in which the semiconducting layer is deposited on top of a dielectric with an underlying gate electrode (Figure 1). Charge-injecting source–drain (S–D) electrodes are defined either on the top of the organic layer through a shadow mask (top contact configuration, Figure 1a) or on the surface of the FET substrate before deposition of the semiconductor layer (bottom contact configuration, Figure 1b). For the above mentioned two device configurations, both the dielectric layer and the gate electrode are fabricated before the organic layer deposition. These configurations correspond to the bottom gate geometry. Also, OFET devices can be fabricated in the top gate configuration (Figure 1c) in which the gate electrode is the last device element deposited. The materials used as the gate dielectric, which also affect the device performance, are either inorganic dielectric materials, such as SiO_2, or organic dielectric materials, for example, insulating organic polymers. As for S–D electrodes, Au with excellent conductivity and high work function, which is nicely compatible with most *p*-type semiconductors, has been the most frequently used electrode materials. Recently, some conductive polymers and low-cost metals[5] such as Ag and Cu have been used as S–D electrodes with the aim to reduce the cost of OFET devices.

Small currents flow between the source and drain electrodes when there is no voltage application to the gate electrode; this state is referred to as the off-state of transistor. When negative voltage is applied to the gate electrode, hole carriers in the organic semiconductor layer become accumulated at the interface with the gate dielectric and hole transport takes place from the source to the drain electrode; this state corresponds to the on-state of transistor. This type of device is called a *p*-channel device. Correspondingly, for *n*-channel device, the positive source–drain voltages and gate voltages are biased; therefore, electrons are injected into the *n*-type active layer, accumulate in the conductive channel, and transport to the drain electrode. The current that flows from the source to the drain electrode under a given gate voltage (V_{SG}) increase almost linearly with the increasing S–D voltage (V_{SD}) and gradually becomes saturated. The basic relationships describing the OFET drain current are given in eqs 1 and 2:

$$(I_{SD})_{lin} = (W/L)\mu C_i(V_{SG} - V_T - V_{SD}/2)V_{SD} \quad (1)$$

$$(I_{SD})_{sat} = (W/2L)\mu C_i(V_{SG} - V_T)^2 \quad (2)$$

Where μ is the field-effect carrier mobility of the semiconductor, W the channel width, L the channel length, C_i the capacitance per unit area of the insulator layer, V_T the threshold voltage. The field-effect mobility μ, which is the key parameter to evaluate semiconductors, can be determined from the slope of the linear plots of $(I_{SD,sat})^{1/2}$ versus V_{SG}, according to eq 2. Besides, on/off ratio and threshold voltage are also parameters in characterizing an OFET. Specifically, the field-effect mobility quantifies the average charge carrier drift velocity per unit electric field, whereas the on/off ratio is defined as the drain source current ratio between the on and off states. The threshold voltage V_T is used to evaluate the amount of traps. Figure 2 shows the output and transfer characteristics of an OFET based on ABT (*p*-type). [6] Analysis of Figure 2 indicated a field effect mobility of 0.41 $cm^2V^{-1}s^{-1}$, an on/off ratio of 10^5. This device performance is comparable to that of the inorganic device based on the hydrogenated amorphous silicon.

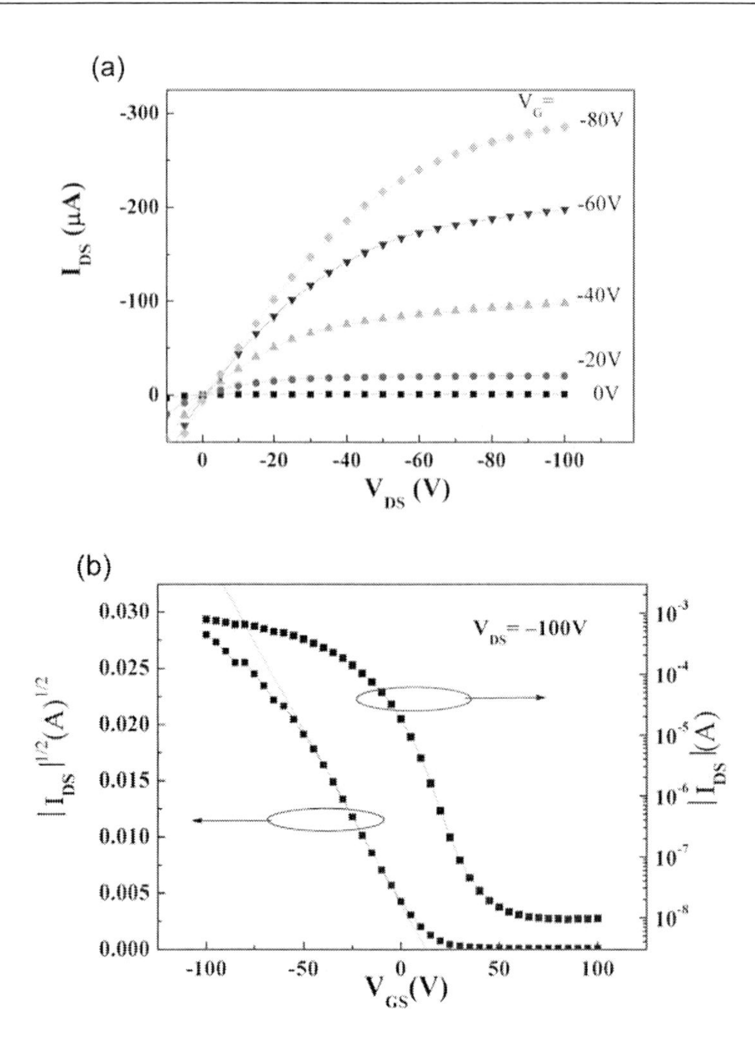

Figure 2. The output (a) and transfer (b) characteristics of an OFET based on an organic semiconductor ABT (*p*-channel) [6].

1.2. Deposition Technology of OFETs

For classical inorganic field-effect transistors, the α-Si layers were deposited mostly from the gas phase such as physical vapor deposition (PVD), chemical vapor deposition (CVD), sputtering depending on their low solubility and high vapor pressure. And high process temperatures above 350 °C are usually required for vapor deposition. However, in OFETs, the individual components (electrodes, semiconductors, insulators) can be processed by different techniques: on the one hand, they can be deposited from the gas phase similar to inorganic semiconductors; on the other hand, inexpensive solution techniques such as spin coating, inject printing, and screen printing are possible due to the flexibility and good solubility of organic materials in common solvents. Solution process of materials would allow large-area processing in roll-to-roll method and also bring cost advantages as compared to expensive vapor deposition methods. While the first OFETs adapted directly the construction of classical inorganic FET with only organic materials as semiconductor layer, in 1998 the scientists have succeeded in producing an integrated circuit that consisted entirely of organic materials. [7] This technology has opened up completely new areas for OFETs, where simple, low-cost OFETs could have myriad practical application.

1.3. Carrier Transport Mechanism

A number of methods have been developed to determine the charge carrier mobility, including time-of-flight (TOF) methods; [8] analysis of steady-state, trap-free, space-charge limited current (steady-state TF–SCLC method); [8,9] analysis of dark injection space-charge-limited transient current (DI–SCLC method); [8] analysis of the performance of OFETs (FET method);[8] measurement of transient electroluminescence (EL) by the application of step voltage (transient EL method); [10] and pulse radiolysis time-resolved microwave conductivity (PR–TRMC)[11] technique. Here, we mainly focus on charge transport in FET, which measure mobility parallel to the substrate and give more relevant estimations of charge mobility.

The mechanisms of charge transport in organic semiconductors are different from those for inorganic materials. While inorganic materials form energy band structures, that is, valence and conduction bands, organic semiconductors usually do not form energy bands as there are only weak intermolecular interactions such as van der Waals forces, hydrogen bonds, and π–π interactions in organic compounds. The exact nature of charge transport in organic materials is still open to debate. Generally, for organic disordered materials such as polymers and amorphous materials, hopping process has been widely accepted. That is, carrier transport is understood as a sequential redox process over molecules; electron are transferred from the anion radical of a molecule to the neutral molecule through lowest unoccupied molecular orbital (LUMO) for electron transport, and electrons are transferred from a neutral molecule to its cation radical through the highest occupied molecular orbital (HOMO) for hole transport. Charge transport in organic disordered materials indicate some characteristics such as temperature and electric-field dependence, which have been explained by Poole–Frenkel model,[12] small-polaron model, [13] and disorder formalism. [14] However, for molecular crystals, hopping process cannot satisfactorily explain their transport behavior. Charge transport in molecular crystals [15] characterized by high mobility value and small temperature dependence. The process of carriers transport in molecular crystal is generally thought to be between band process of inorganic materials and hopping process of organic semiconductors.

1.4. Solid-State Stacking in Organic Semiconductors and Thin Film Morphology

As discussed in section 1.3, charge transports in organic semiconductors prefer to occur along π-stacking axis through the overlapping π-orbital of adjacent molecules. Thus, the solid-state morphology of conjugated materials largely affects the efficiency of charge transport through the film. It is essential to understand the relationship between molecular packing and materials performance. For organic semiconductors used as semiconductor layer in OFETs, not only the single crystal structure but the thin film morphology plays an important role in the performance characteristics of devices. X-ray techniques have been widely used to characterize the solid-state structures of organic semiconductors.

Structural analysis of some popular OFET materials, e.g., pentacene, heteroacenes, and sexithiophene has shown that all these molecules crystallize into a herringbone structure, as illustrated in Figure 3. In this packing motif, molecules minimize π-orbital repulsion by adopting an edge-to-face arrangement forming two-dimensional layers, and high mobility have been achieved within this packing motif. [16]

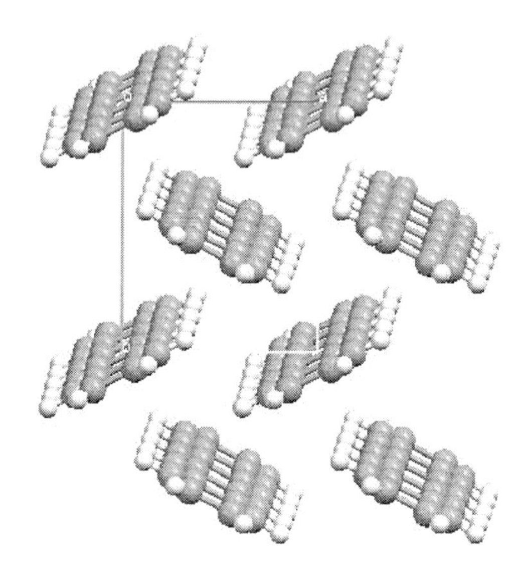

Figure 3. Illustration of herringbone packing structure of pentacene as viewed down the long axis of pentacene (the crystallographic file in CIF was obtained from CCDC via http://www.ccdc.cam.ac.uk/products/csd/request/).

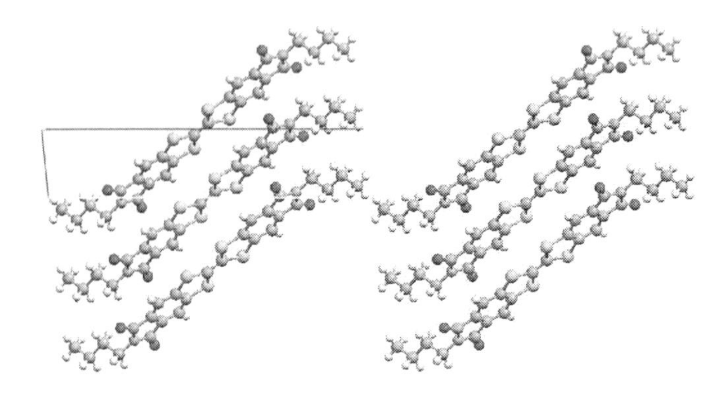

Figure 4. An illustration of face-to-face lamellar-like stacking structure of organic semiconductor dibenzotetrathiafulvalene bisimides. [19b].

However, there are also reports that the edge-to-face packing minimizes π–π overlap between adjacent molecules thus higher mobilities at room temperature might be achieved by designing conjugated molecules that stack face-to-face (π-stack) in the solid state. The face-to-face stacking not only increases the intermolecular interaction but also maintains the desirable 2-D characteristic of the herringbone structure. For example, when oligothiophenes are end-capped with linear alkyl chains, crystallization of the alkyl chains forces the conjugated units to adopt a more face-to-face arrangement, resulting in a lamellar-like structure where insulating alkyl chains separate the conjugated units. [17] An illustration of face-to-face stacking is shown in Figure 4. Recently, many oligomers such as fused-ring oligomers, silylethynylated pentacene and heteroacenes[18] and also TTF derivatives, [19] which have been reported to exhibit promising carrier mobility in OFETs, also adopt cofacial packing arrangement. To date, there are still no definite conclusion about the relationship between crystal packing and device performance, and further investigations are needed.

To study the thin film morphology of OFETs, which could tell us how the molecules will behave when processed into a thin film, many microscopy techniques are used including scanning electron microscopy (SEM), transmission electron microscopy (TEM), atomic force microscopy (AFM), and also scanning tunneling microscopy (STM). Generally, thin film deposition conditions and interactions between dielectric–semiconductor interface play a critical role in determining the final molecular orientation, film continuity, and crystalline grain size in the thin film. For most organic semiconductors, it is reported that the interconnectivity of the first several organic layers are more important to determining mobility than other obvious features such as grain size. [20] Besides, the orientation of the thin film was usually investigated by means of XRD analysis. The numbers of diffraction peaks indicate the degree of order and crystallinity of a thin film. Interlayer distances (*d*-spacings) can also be calculated from the first diffraction peak of 2θ, and from which the molecular orientation on the substrate could be determined. Such orientation that the organic molecular stand upright on substrate is reported to be favorable for achieving high mobility since the stacking direction is consistent with the direction of current flow. [6,21]

2. Newly Developed Organic Semiconductors

As mentioned previously, small molecules or oligomers as well as polymer are all suitable organic semiconductors for OFET applications. To date, many novel organic semiconductors have been developed and a few of review papers have summarized the developments of organic FET materials up to the year 2007. [22] Since then, new light was shed on air stability and flexible processibility of the organic semiconductors. In this chapter, we mainly focus on the newly developed organic semiconductors and correlations between their structure and transistor performance, classified by their chemical structures.

To achieve high field-effect performance, strong intermolecular interactions, proper energy levels, and highly ordered films are necessary. Acenes with extended π system and planar structure exhibit strong intermolecular overlap in the solid state thus effective carrier transport. Thiophenes-based materials indicate a variety of intra- and intermolecular interaction, such as van der Waals interactions, weak hydrogen bonds, π–π contacts, and sulfur–sulfur interactions. As a result, acenes and thiophenes derivatives are two promising classes of organic semiconductors for OFETs. [23].

2.1. Acenes and Heteroacenes

2.1.1. Anthracene and Heteroanthracene

Unsubstituted anthracene (compound 1) was not suitable for OFET due to its low molecular weight and high oxidation potential. However, functionalized anthracenes and heteroanthracene have been widely used as semiconductor layers in OFET (Chart 1). In 2003, Suzuki's group reported oligo(2,6-anthrylene) compounds 2–5.[24] Compound 5 afforded the maximum FET mobility of 0.18 $cm^2V^{-1}s^{-1}$ at the substrate temperature of 175 °C. And the author claimed that the addition of alkyl groups is more effective than the extension of π conjugation as the mobilities increased in the order 2<4 <3 <5. Since then, 2,6-sustituted anthracene derivatives 6–11 were synthesized and tested in OFET. [25] Among them, only compound 6 end-capped with 4-trifluormethylphenyl units[25a] exhibited n-type semi-conducting behavior with the electron mobility of 3.4×10^{-3} $cm^2V^{-1}s^{-1}$. The maximum hole mobility was obtained from compound 10 in the value of 1.3 $cm^2V^{-1}s^{-1}$. [25d]

From the theoretical point of view, it is pointed out that the heteroatom affects not only on the electronic structures but also on the solid-state structures which play a great role in determining charge carrier transport. Thus, many hetero atoms were introduced to acenes systems to construct heteroacenes derivatives. Early in 1997, Katz's group have reported benzodithiophene dimer (compound 12) [26] and achieved hole mobility as high as 0.04 $cm^2V^{-1}s^{-1}$. After that, Takimiya's group have synthesized a series of benzodichal-cogenophenes derivatives 13–17. [27,28] Compound 14 and 17 with selenophene atoms both showed higher mobilities than their sulfur counterparts 13 and 16, supporting the heavy chalcogen effect. Compared with the diphenylbenzodichalcogenophenes, structural isomers 18 and 19 showed low hole mobilities of 0.003 and 0.02 $cm^2V^{-1}s^{-1}$, respectively, indicating the importance of structure symmetry for molecular engineering design. [29] As a commercially available starting material, dibenzothiophene derivatives 20–22 were successfully synthesized by Hu's group [30] and comparable charge carrier mobility of 7.7×10^{-2} $cm^2V^{-1}s^{-1}$ was achieved.

As an analogy of anthracene, thienoacene with n = 3, that is named as dithieno[3,2-b:2,3-d]thiophene (DTT), attracted much attention due to the combination of the stability of the thiophene ring with the planarity of linear acenes in the thienoacene system. The dimer of DTT (compound 23) was firstly reported by Holmes's group and the crystal structure of 23 revealed a coplanar conformation with face-to-face π–π stacking which was thought to facilitate carrier transport and FET mobility of 0.05 $cm^2V^{-1}s^{-1}$ was obtained. [31] Since then, our group synthesized aryls or thiophene rings end-capped DTT derivatives 24–26. As compared with 23, we obtained much higher mobility as high as 0.42 $cm^2V^{-1}s^{-1}$ with phenyl substituted DTT derivatives 26. The other derivatives 24–25 with biphenyl or thiophene substituent also showed excellent FET performance with mobility above 0.1 $cm^2V^{-1}s^{-1}$. Remarkably, the devices fabricated with compounds 24–25 indicated high stability under ambient conditions. [32] Furthermore, we investigated the physicochemical properties and liquid crystalline behavior of DTT derivatives based on compound 26 with n-hexyloxy or n-dodecyloxy substituents (compound 27 and 28) [33], however, the FET devices based on the derivatives 27–28 exhibit lower mobility (0.02 $cm^2V^{-1}s^{-1}$) than the unsubstituted compound 26. Recently, Wang's group synthesized trans-1,2-(dithieno[2,3-b:3',2'-d]thiophene)ethene derivatives 30–33 with compound dithieno[2,3-b:3',2'-d]thiophene dimer 29 as comparison. Compound 32 with phenyl substituent demonstrates a very high carrier mobility up to 2.0 $cm^2V^{-1}s^{-1}$. [16c–d]

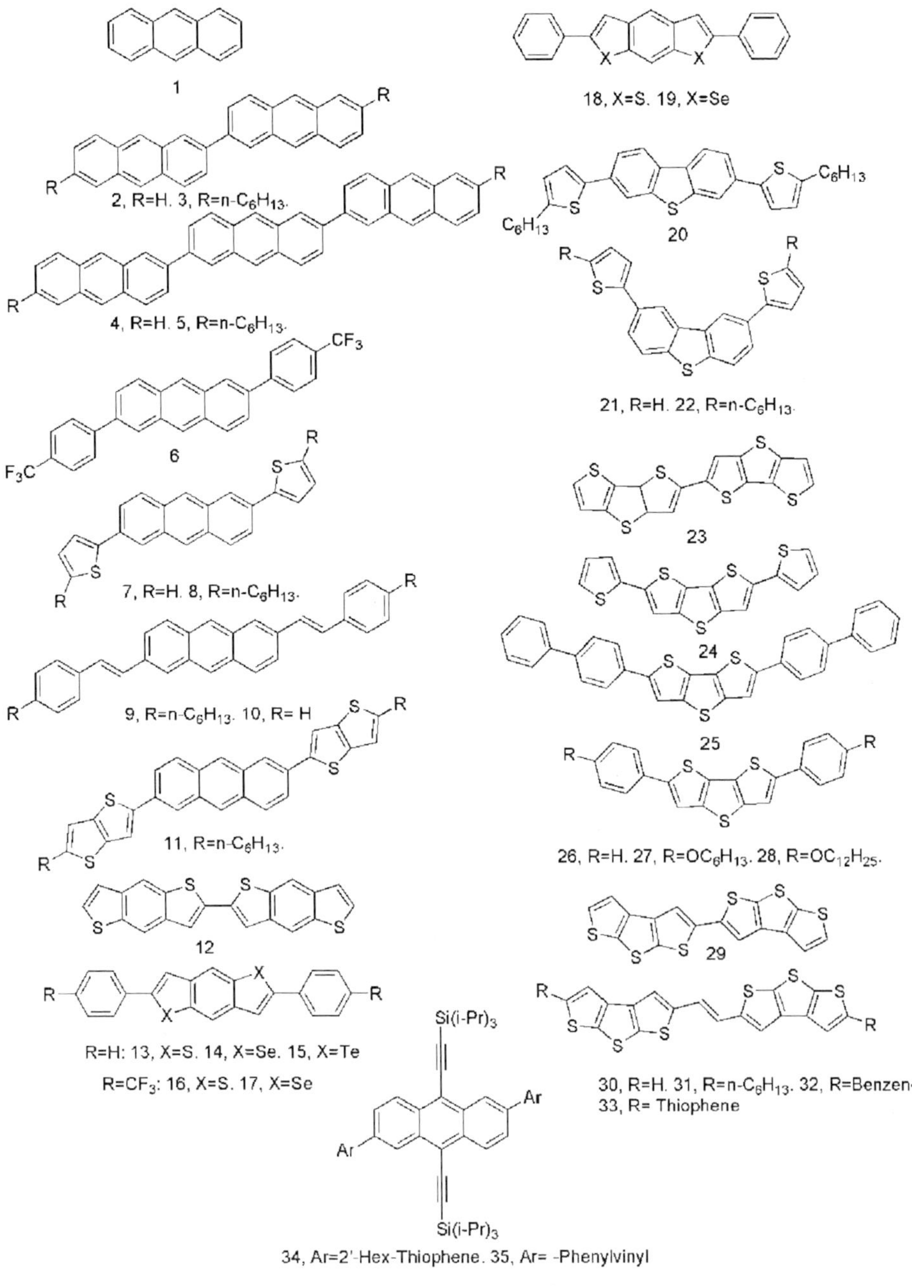

1

2, R=H. 3, R=n-C_6H_{13}.

4, R=H. 5, R=n-C_6H_{13}.

6

7, R=H. 8, R=n-C_6H_{13}.

9, R=n-C_6H_{13}. 10, R= H

11, R=n-C_6H_{13}.

12

R=H: 13, X=S. 14, X=Se. 15, X=Te

R=CF_3: 16, X=S. 17, X=Se

18, X=S. 19, X=Se

20

21, R=H. 22, R=n-C_6H_{13}.

23

24

25

26, R=H. 27, R=OC_6H_{13}. 28, R=$OC_{12}H_{25}$.

29

30, R=H. 31, R=n-C_6H_{13}. 32, R=Benzene.
33, R= Thiophene

34, Ar=2'-Hex-Thiophene. 35, Ar= -Phenylvinyl

Chart 1. Structures of antharene and heteroantharene referred in this chapter.

All these compounds mentioned above showed limited solubility in common solvents and FET devices were fabricated with vacuum evaporation. Compound 34 and 35 with bulky TIPS groups at the 9,10-position and aromatic groups at 2,6-position of anthracene core were the only reported solution processing anthracene derivatives. The mobility of compound 34 and 35-based devices were 4×10^{-3} and 3×10^{-4} $cm^2V^{-1}s^{-1}$, respectively. [34]

These results were an important demonstration of the utility of heteroacenes, and over the past few years the anthracene and heteroanthracene core has become common building blocks in organic electronics.

2.1.2. Tetracene and Heterotetracene (Chart 2)

Tetracene (36), like all unfunctionalized acenes, adopts an edge-to-face herringbone packing in the solid state. [35] Unlike anthracene, tetracene has been used in the construction of thin film FETs with hole mobility as high as 0.15 $cm^2V^{-1}s^{-1}$ have been reported.[36] However, with the ability to form large, high-quality single crystals by vapor growth or solution process, tetracene and its derivatives 37–41 has been much more studied in single-crystal FET devices. [37] Among them, rubrene (41), one of the most intensively studied tetracene derivatives, has become the benchmark of single crystal FETs with reported high mobility up to 30 $cm^2V^{-1}s^{-1}$. [38] Phenyl substitution of tetracene in this case leads to a strongly π-stacked arrangement in the solid state, and computational studies have shown that the particular intermolecular arrangement adopted by rubrene happens to be optimal for charge transport. Analysis of solid state stacking of compounds 39 and 40 which showed high mobility also indicated that slip π stack facilitate carrier transport. For tetracene derivatives with electron-deficient groups, only two examples compound 42a–d [39] and 43 were reported. For compounds 42a–d substituted with four fluorine atoms and a long-chain alkyl or alkoxy group, all of them showed strong π-stacking interactions in the solid state and derivative 42c indicated unique 2-D π-stacking. However, FET characteristics of these materials have not been reported yet. With the aim to develop *n*-type semiconductors for OFETs, Sakamoto and coworkers reported the synthesis and crystal structure of perfluorotetracene (43). [40] Compound 43 exhibit the common feature of close cofacial packing with minimum interplanar π–π distances of 3.27 Å and FET devices resulted in electron mobility of 2.4×10^{-4} $cm^2V^{-1}s^{-1}$. Although the electron mobility of perfluoro-tetracene is low, it was the first example of *n*-channel organic semiconductor based on tetracene.

Extending their work with benzodichalcogenophenes, Takimiya's group reported the synthesis of heterotetracene derivatives 44–45. [41] With extended conjugation length compared to benzodichalcogenophenes, compound 45a exhibited mobility as high as 2.0 $cm^2V^{-1}s^{-1}$. Furthermore, with the introduction of long alkyl chains, the group developed solution–processible organic semiconductors 45b and the maximum mobility up to 2.7 $cm^2V^{-1}s^{-1}$ was achieved for devices with *n*-$C_{13}H_{27}$ substituted derivatives. The authors attributed the high mobility of the compounds to 2D herringbone stacking and short S–S contacts.

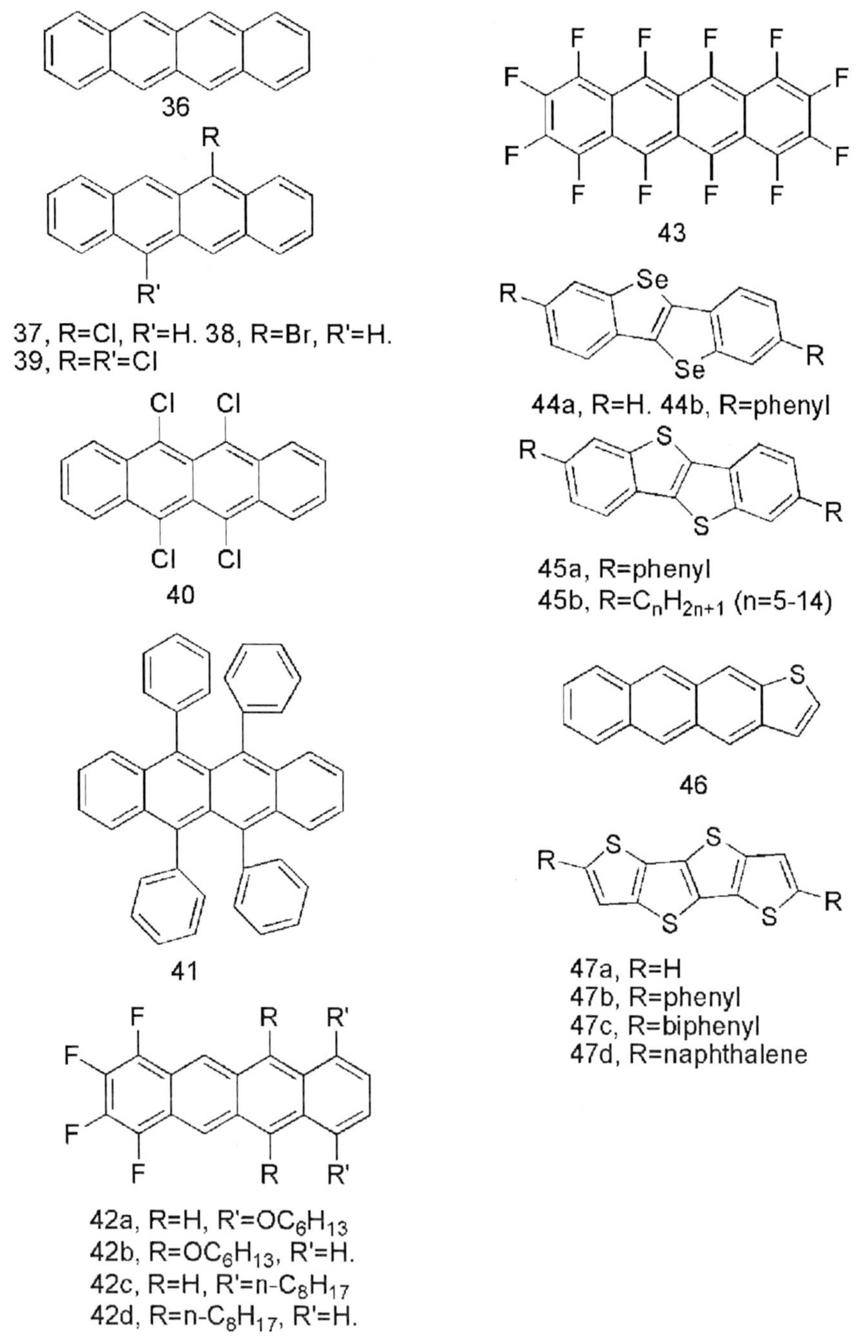

36

37, R=Cl, R'=H. 38, R=Br, R'=H.
39, R=R'=Cl

40

41

42a, R=H, R'=OC_6H_{13}
42b, R=OC_6H_{13}, R'=H.
42c, R=H, R'=n-C_8H_{17}
42d, R=n-C_8H_{17}, R'=H.

43

44a, R=H. 44b, R=phenyl

45a, R=phenyl
45b, R=C_nH_{2n+1} (n=5-14)

46

47a, R=H
47b, R=phenyl
47c, R=biphenyl
47d, R=naphthalene

Chart 2. Structures of tetracene and heterotetracenes referred in the chapter.

Asymmetrical tetracene-like compounds 46 with fused thiophene unit at one side was reported simultaneously by Bao's group [42] and Tao's group.[21] The compound 46 with a conjugation length between tetracene and anthracene was applied in thin film OFET devices and afforded mobility up to 0.15 $cm^2V^{-1}s^{-1}$, similar to tetracene. Tetrathienoacene (TTA, 47a) was an analogy of tetracene with four fused thiophene rings. However, electronic characterization of TTA was unsuccessful because of its low sublimation temperature which makes the thin film growth difficult. With our previous work on DTT derivatives, recently we synthesized tetrathienoacene (TTA) derivatives 47b–d. [43] Single crystal structure analysis of compound 47b indicated that π...π interactions and S...S contacts exist in the solid state structure. OFET devices were fabricated with these derivatives 47b–d and mobility up to 0.14 $cm^2V^{-1}s^{-1}$ was obtained for compound 47b. It should be noted that the devices structure were not optimized yet and better results should be obtained with optimized devices parameters.

2.1.3. Pentacene and Heteropentacene (Chart 3)

While rubrene sets the performance standard for single-crystal devices, pentacene (compound 48) is the benchmark for thin-film devices. Since the first report of OFET device based on pentacene, numerous research groups have studied and improved the performance of these devices.[44] Kelly and co-workers have obtained the highest hole mobility of pentacene with a poly(R-methylstyrene) gate dielectric, giving measured mobility >5.0 $cm^2V^{-1}s^{-1}$. [45] Efficient charge transport in pentacene has been attributed to high molecular order and the large grain sizes that can be obtained through optimization of the deposition conditions. Like tetracene, pentacene crystallizes in the classic acene herringbone motif (Figure 3), although it can potentially adopt any of a number of polymorphs. [46] However, pentacene suffers from the disadvantages of oxidative instability, extreme insolubility, and for display applications, a strong absorbance throughout the visible spectrum. [47] In recent years, many pentacene derivatives were developed to overcome the shortcomings of pentacene while remaining high performance. Meng, H and coworkers reported the synthesis and FET fabrication of 2,3,9,10-tetramethyl-pentacene (Me_4PENT, compound 49), which was the first reported alkyl substituted pentacene used in OFETs. Compared with pentacene, compound 49 indicated an apparent minor steric effect by the methyl substituent. It was noted that FETs fabricated with Me_4PENT showed higher mobility (0.31 $cm^2V^{-1}s^{-1}$) with the bottom-contact mode device than that with top-contact mode. [48]

Recently, pentacene derivatives 50–51 with terminal electron–withdrawing groups were reported. [49] FET devices were fabricated with these derivatives and the best performance was obtained for compound 51b, which exhibited a mobility up to 0.22 $cm^2V^{-1}s^{-1}$ and good stability at ambient conditions because of the introduction of substituent. Early in 2001, Anthony and coworkers reported the synthesis of bulky trialkylsilylethynyl moiety functionalized pentacene (TIPS-Pent, 52) derivatives, which showed improved solubility, stability and cofacial π-stacking in the solid state. [18a] FET devices fabricated with compound 52 by vacuum deposition showed a hole mobility of 0.4 $cm^2V^{-1}s^{-1}$. [50] Since then, many reports about TIPS-Pentacene and TIPS-heteroacenes have emerged. [51] Solution processed OFETs using compound 53b have exhibited mobility up to 1.0 $cm^2V^{-1}s^{-1}$. [18c] Compound 54 was reported to show ambipolar OFET characteristics and electron mobility of 0.37 $cm^2V^{-1}s^{-1}$ and hole mobility of 0.065 $cm^2V^{-1}s^{-1}$ were observed in inert atmosphere. [51f]

Another functionalized pentacene derivatives that exhibited novel electronic and intermolecular ordering properties is perfluoropentacene 55. Compound 55 was prepared in five steps from tetrafluorophthalic anhydride and hydroquinone and crystallized in a herringbone motif similar to pentacene but with a nearly 90° edge-to-face angle (compared with 52° for pentacene). FET devices fabricated with this compound exhibited electron mobility as high as 0.22 $cm^2V^{-1}s^{-1}$, demonstrating that perfluorination is a viable route to prepare *n*-type semiconductors from acenes. [40,52]

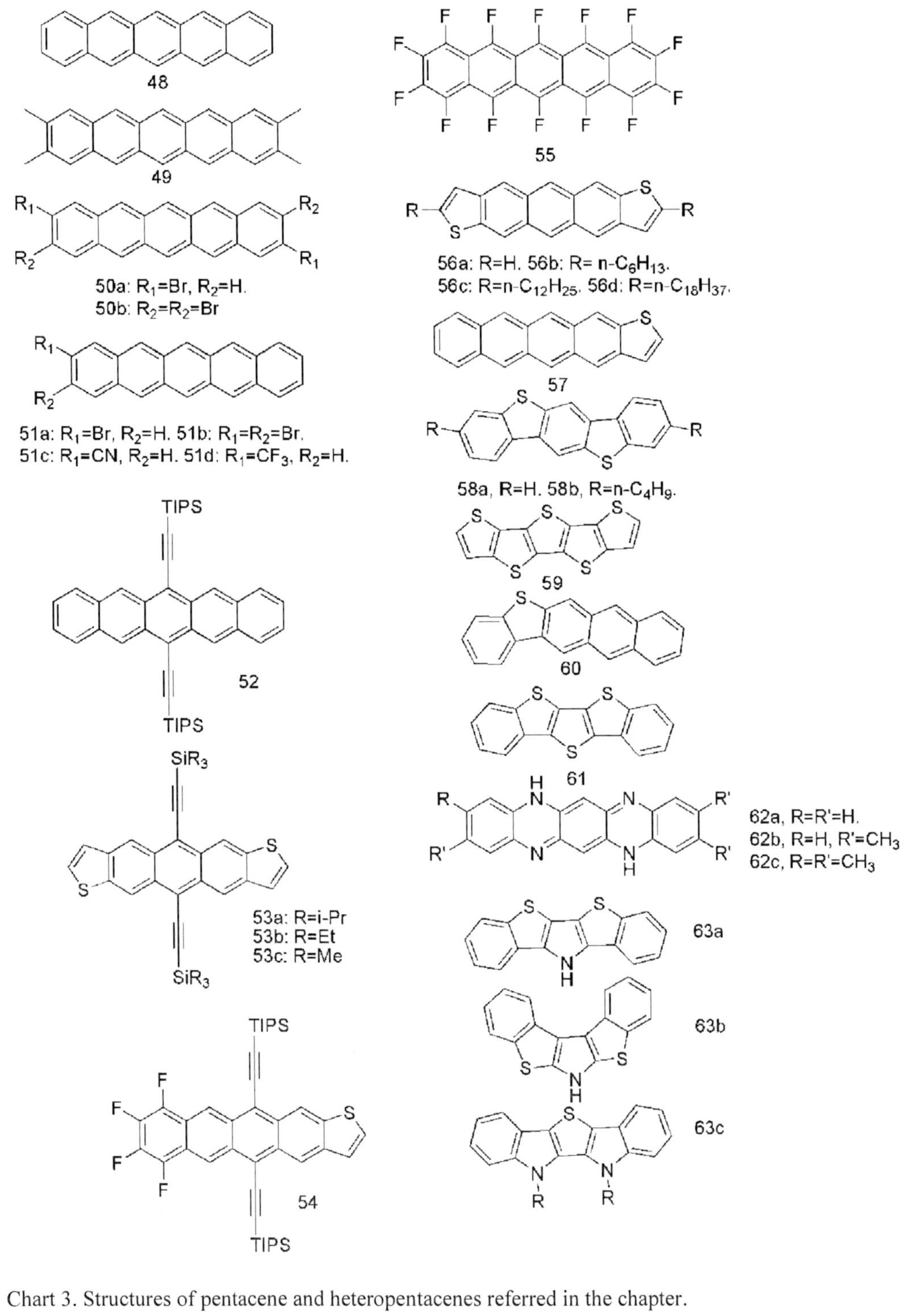
48
49
50a: R1=Br, R2=H.
50b: R2=R2=Br
51a: R1=Br, R2=H. 51b: R1=R2=Br.
51c: R1=CN, R2=H. 51d: R1=CF3, R2=H.
TIPS
52
SiR3
53a: R=i-Pr
53b: R=Et
53c: R=Me
54
55
56a: R=H. 56b: R= n-C6H13.
56c: R=n-C12H25. 56d: R=n-C18H37.
57
58a, R=H. 58b, R=n-C4H9.
59
60
61
62a, R=R'=H.
62b, R=H, R'=CH3
62c, R=R'=CH3
63a
63b
63c

Chart 3. Structures of pentacene and heteropentacenes referred in the chapter.

The other strategy to overcome the shortcomings of pentacene was introduction of hetero atoms such as sulfur, selenophene or nitrogen. In 1998, katz and coworkers reported the first heteropentacene, anthradithiophene (ADT, 56a), and dialkyl substituted derivatives, [53] which have been one of the most thoroughly studied heteropentacenes. The dihexyl and didodecyl ADT derivatives both yielded FET mobility of 0.15 $cm^2V^{-1}s^{-1}$ while the parent ADT only exhibit moderate mobility (0.09 $cm^2V^{-1}s^{-1}$) in a FET device. Furthermore, the alkyl–substituted materials were sufficiently soluble to allow film formation by solution casting, [54] affording mobility of 0.02 $cm^2V^{-1}s^{-1}$ in solution-processed devices. However, the synthesis of the ADT parent derivatives creates two isomers which are not easily separable. The isomers impurity may induce the disorder in the thin film. Thus, Bao's group reported the synthesis and FET characteristics of asymmetric heteropentacene containing fused thiophene units (compound 57), [42] with a conjugation length between tetracene and pentacene. FET devices fabricated with compound 57 showed high mobility up to 0.47 $cm^2V^{-1}s^{-1}$ at substrate temperature of 60 °C, and the high mobility of the devices were attributed to larger grains at substrate at higher temperature. Mullen's group synthesized heteropentacene derivatives, benzo[1,2-b:4,4-b']bis[b]benzothiophene (58a) and its alkyl substituted derivatives (58b). [55] FETs were fabricated via solution processing, giving a hole mobility of 0.01 and 0.001 $cm^2V^{-1}s^{-1}$ for 58a and 58b respectively. Interestingly, the carrier mobility of compound 58a obtained by solution processed devices was higher than that of vacuum deposited with the value of 2.4×10^{-3} $cm^2V^{-1}s^{-1}$. [56]

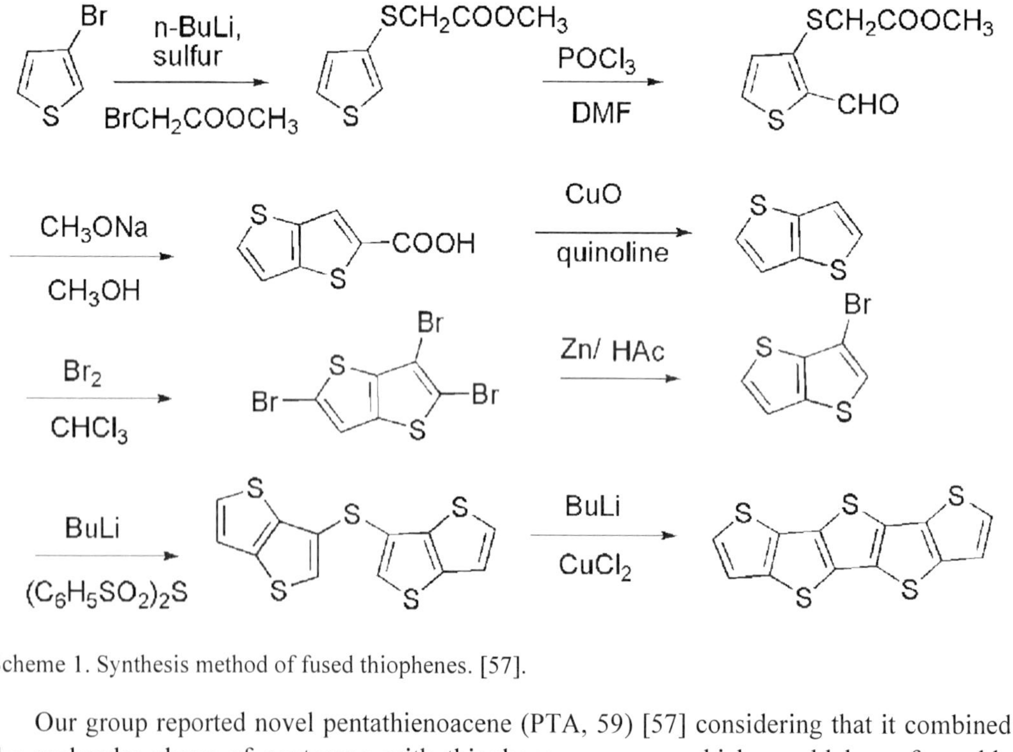

Scheme 1. Synthesis method of fused thiophenes. [57].

Our group reported novel pentathienoacene (PTA, 59) [57] considering that it combined the molecular shape of pentacene with thiophene monomer which would have favorable crystal packing geometry and also increased stability and nuclear aromaticity. PTA was synthesized as Scheme 1, which also indicated the synthesis of lower fused thiophene. The band gap of PTA was 3.29 eV, much higher than that of pentacene, indicating higher redox stability. FETs were fabricated with PTA and the best mobility achieved is 0.045 $cm^2V^{-1}s^{-1}$at a substrate temperature of 80 °C. The devices performances were believed to be further improved by purification of PTA. Furthermore, we synthesized a series of heteropentacenes 60–63 with the introduction of sulfur or/and nitrogen. Among them, compound 60[6] and 61 [58] with one or three fused thiophene units exhibited high hole mobility of 0.4–0.6 $cm^2V^{-1}s^{-1}$ at room temperature, comparable to pentacene devices under same conditions. A series of tetraazapentacene (62a–c) were synthesized with nitrogen replacing carbons in the phenyl rings of pentacene in order to avoid the facile oxidation of pentacene. [59] OFETs fabricated by using thermal evaporation of 62a and 62b gave mobilities both up to 0.02 $cm^2V^{-1}s^{-1}$, respectively. To investigate of the effect of both sulfur and nitrogen on heteropentacene, a series of fused heteroacenes 63a–63c have been synthesized. [60] FETs fabricated with 63a afforded mobility of 0.01 $cm^2V^{-1}s^{-1}$ while FET characteristics of the other derivatives 63b–c have not been reported yet.

2.1.4. Hexacene, Larger Acenes, and Larger Heteroacenes (Chart 4)

Although hexacene (compound 64) have been reported in the past, [61] it is still elusive in that the reported synthetic schemes are difficult to repeat. Lower molecular weight acenes (up to pentacene) are usually synthesized by reduction of the corresponding quinones. However, a similar approach used for higher acenes proved to be unsuccessful since the quinones were over reduced to hydrogenated acenes. Other methods of synthesis also failed for higher acenes, presumably because of their higher reactivity toward Diels–Alder additions which ultimately results in formation of either dimers or oxygen adducts (endoperoxides). [62] Recently, Strating–Zwanenberg photodecarbonylation have been used to achieve unequivocal synthesis of hexacene and heptacene, once again hexacene and heptacene were found to be extremely unstable in solution. However, irradiation of the dione precursors of hexacene and heptacene in the PMMA matrix using UV-LED array produced the acenes which were found to be stable up to 12 h and 4 h in the PMMA matrix for hexacene and heptacene, respectively. [63]

Expanding their work on soluble functionalized pentacene derivatives, Anthony's group reported the synthesis and characterization of silylethynylation hexacene and heptacene derivatives 65 and 66. [64] However, the TIPS group seems not to be sufficiently bulky to prevent Diels–Alder addition of the alkyne across reactive acenes, as neither of the TIPS-substituted hexacene and heptacene (65a and 66a) were stable in solution or in the solid state. Thus, the author chose the bulkier alkyne tri-tert-butylsilyl (TTBS) acetylene, leading to a moderate increase in stability of derivatives 65b, but solution of 66b still decomposed rapidly within a day even under oxygen-free conditions because of the larger reactivity of heptacene.

The higher oxidation potential and larger HOMO–LUMO gap per aromatic ring of the heteroacenes vs the carbocyclic acenes make the former class of compounds ideal candidates for exploring the chemistry of larger acenes. Pentaceno[2,3-b]thiophene (67), which is pentacene fused with a terminal thiophene ring, was synthesized as a hexacene-like molecule. [65] Compound 67 was stored at nitrogen gloves box as it was not stable under ambient conditions. FET devices were fabricated and measured entirely in nitrogen and the highest mobility of 0.574 $cm^2V^{-1}s^{-1}$ was obtained. Compounds dinaphtho-[2,3-b:2',3'-f]chal-cogenopheno[3,2-b]chalcogenophene (68a and 68b) were reported as another kinds of heterohexacene, and FET characteristics with mobility higher than 2.0 $cm^2V^{-1}s^{-1}$ were observed in 66a-based devices. [66] The two compounds also showed good stability with large energy gaps of 3.0 and 2.9 eV for 68a and 68b, respectively. However, these two compounds have lost the conjugation of hexacene, with largely bule-shift absorption wavelength.

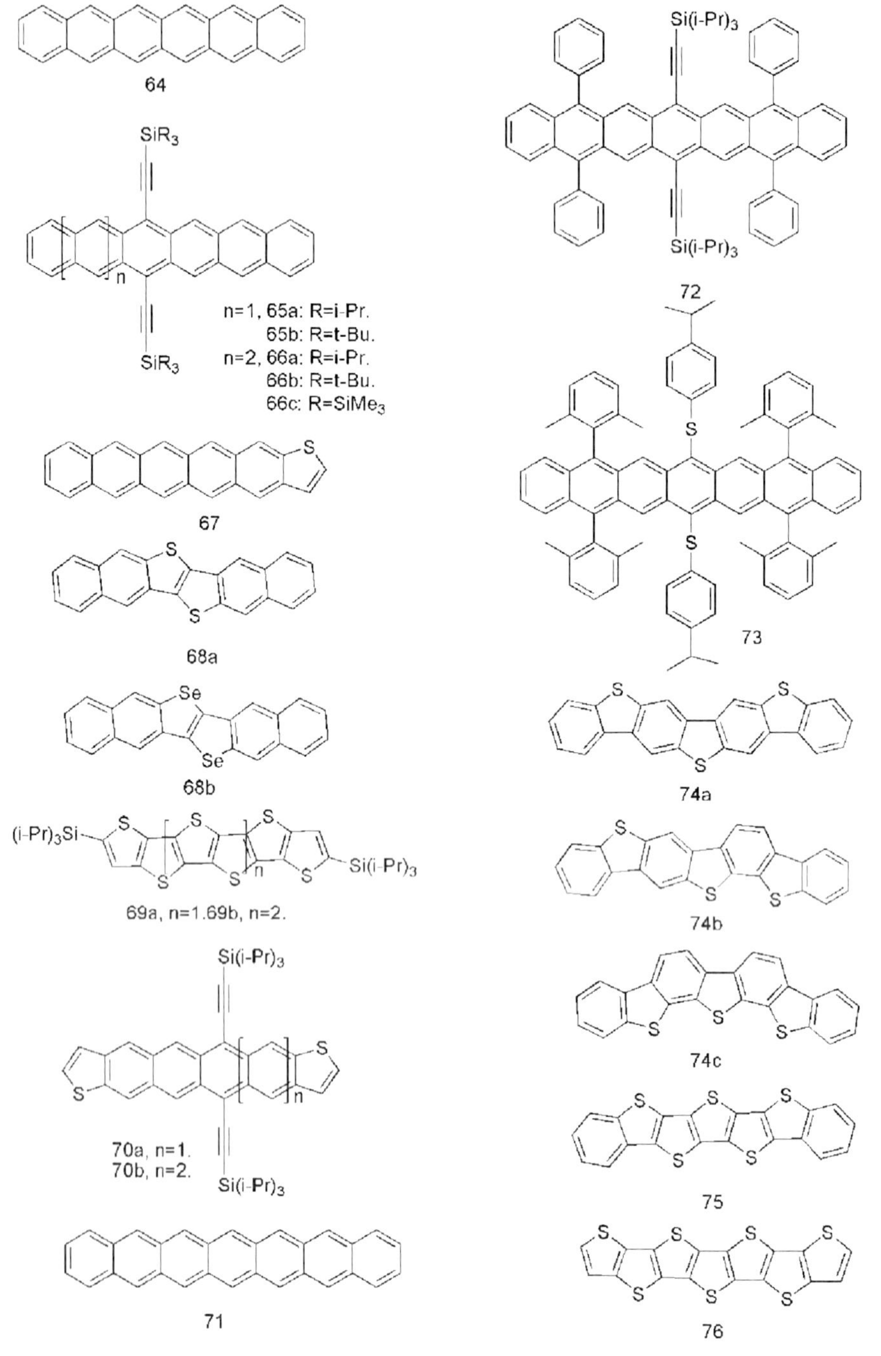
64
SiR3
n
SiR3
n=1, 65a: R=i-Pr.
65b: R=t-Bu.
n=2, 66a: R=i-Pr.
66b: R=t-Bu.
66c: R=SiMe3
S
67
S
S
68a
Se
Se
68b
(i-Pr)3Si
S
S
S
S
S
S
n
Si(i-Pr)3
69a, n=1.69b, n=2.
Si(i-Pr)3
S
S
n
70a, n=1.
70b, n=2.
Si(i-Pr)3
71
Si(i-Pr)3
Si(i-Pr)3
72
S
S
73
S
S
S
74a
S
S
S
74b
S
S
S
74c
S
S
S
S
S
75
S
S
S
S
S
S
S
76

Chart 4. Structures of higher acenes and heteroacenes referred in this chapter.

As a sulfur analogy of acenes, hexathienoacene and octathienoacene derivatives (69a and 69b) were synthesized with intramolecular double cyclization and dechalcogenation reaction. [67] These thienoacenes indicated blue-shift absorption and fluorescence spectra and better stability compared with their carbocyclic analogy.

Silylethynylated functionalized acenedithiophenes including tetradithiophene and pentadithiophene (70a–b) were prepared. [68] Compound 70a was stable in the solid state or in deoxygenated solution for weeks while 70b easily undergone dimerization within the third ring of one acene and the ethynyl group of a second acene, further indicating the TIPS group was not sufficiently to prevent Diels–Alder addition of the alkyne across reactive acenes in higher acenes and heteroacenes.

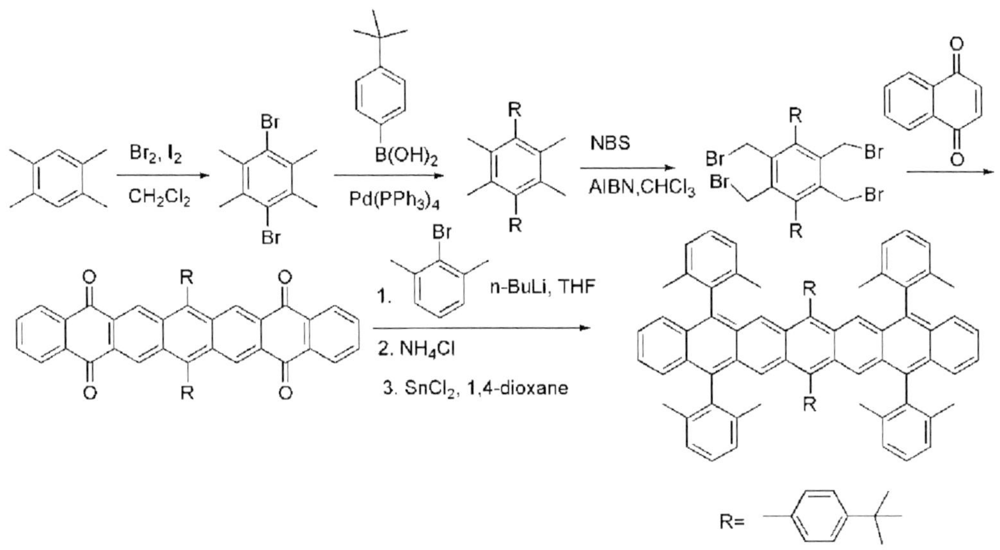

Br_2, I_2 / CH_2Cl_2; $B(OH)_2$ / $Pd(PPh_3)_4$; NBS / AIBN,$CHCl_3$; 1. n-BuLi, THF 2. NH_4Cl 3. $SnCl_2$, 1,4-dioxane; R=

Scheme 2. Synthesis of substituted heptacene derivative.[70]

To obtain stable heptacene derivatives, Wudl's group synthesized functionalized heptacene (72) with phenyl group and alkylsilylethynyl group substituents and found single crystal of compound 72 was sufficiently stable for over 21 days. Single crystal structure of 72 exhibited edge-to-face herringbone packing, however, there are no π–π interactions between the acene backones. [69] Recently, Miller's group reported another stable heptacene derivative 73 (Scheme 2) and demonstrated that not only alkylsilylethynyl groups but a combination of arylthio and *o*-dialkylphenyl substituents can also be utilized to produce a persistent heptacene derivative. [70] Both the derivatives possess small optical HOMO–LUMO gaps (1.35 and 1.37 eV for 72 and 73, respectively). Compound 74, prepared as a mixture of three isomers (74a–c) that could not be isolated by normal chemical means, was the first reported heteroheptacene with FET characteristics. [71] Devices made with compounds containing most isomer 74a from the mixture had hole mobility of 0.15 $cm^2V^{-1}s^{-1}$; without this careful control of sublimation conditions, the mixture of isomers is deposited, leading to poor film quality and hole mobility of less than 0.03 $cm^2V^{-1}s^{-1}$. The significant difference in geometry of the three isomers of 72 likely leads to different crystal packing arrangements for the molecules, thus deposition of the mixture of species would thus yield films with poor morphologies for device applications.

Dibenzopentathioacene (compound 75) has also been investigated as organic semiconductor. [72] Single crystal structure analysis indicated that compound 75 formed a face-to-face π-stacking structure along the c axis with the interplane distance of 3.50 Å. Single crystal FETs based on 75 afforded high hole mobility of 0.4 $cm^2V^{-1}s^{-1}$.

While heptacene 71 has remained an elusive compound, heptathienoacene 76 is very stable and soluble versions which can be easily isolated and characterized. [73] However, the FET characteristics of compound 76 have not been reported.

Until now, octacene, nonacene, and larger acene have never been isolated. They are predicted to have open-shell singlet ground states. [74] Increasing the proportion of carbocyclic rings in compounds such as 76 will allow an approximation of the properties of the fully carbocyclic higher acenes and will yield insight into functionality required to stabilize these reactive systems.

2.2. Oligothiophene and Other Heterocycles

2.2.1. Thiophene Oligomers and Co-Oligomers Linked by Single Bonds (Chart 5)

For many years oligothiophenes and their alkyl-substituents have been among the most intensely investigated organic semiconductors because of the synthetic versatility of the thiophene heterocycle. [76] The first printed organic transistor was fabricated by Garnier's group using sexithiophene as the active semiconductor layer. [77] Because of their ease of functionalization, oligothiophenes provide the opportunity to study systematic variations in molecular structure by controlling the number of repeat units in the conjugated backbone and/or varying the length and functionality of alkyl substituent.

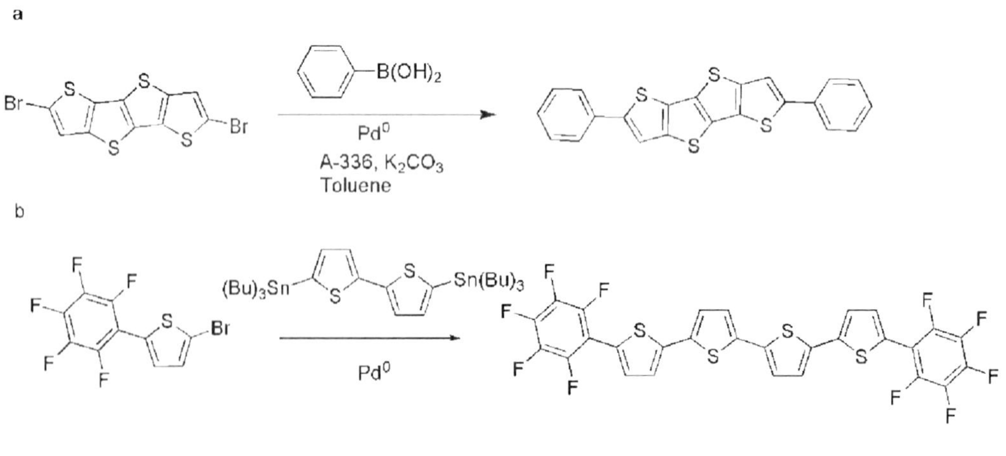

Scheme 3. Suzuki and Still couplings used to synthesize oligomers.

The most common methods used to construct thiophene oligomer include Suzuki, Yamamoto, Still, and Negishi couplings. [78] They involve a Pd- or Ni- mediated cross-coupling between an aromatic halide and an aromatic coupling partner substituted with a magnesium, boron, tin, halogen, or zinc group. By far, the most common reactions are the Still and Suzuki couplings; representative examples are shown in Scheme 3.

For unsubstituted oligothiophenes, oligomers with four (α-4T, compound 77a), five (α-5T, compound 78a), [79] six (α-6T, 79a), [80] and eight (α-8T, 80a) [81] thiophene rings have been examined. X-ray diffraction studies of these oligomers prove that they all display similar solid-state ordering in the bulk and in thin films. [82] They all have planar conformations and herringbone-type packing motifs and form polycrystalline films in which the molecules are orientated perpendicularly to the substrate. The charge mobility was found to be heavily dependent on orientation and morphology of the thin films. As one of the most widely studied oligothiophene, the hole mobility of α-6T has improved from 10^{-4} $cm^2V^{-1}s^{-1}$ to greater 0.1 $cm^2V^{-1}s^{-1}$, [83] while single crystal of the compound exhibit mobility up to 0.075 $cm^2V^{-1}s^{-1}$. [84] Charge mobility of 0.33 $cm^2V^{-1}s^{-1}$ have been reported for α-8T OFETs with films deposited at a substrate temperature of 120 °C. [85]

Many substituents have been introduced to functionalize the α-position of the thiophene ring, in order to increase the solubility or to influence the solid-state ordering of oligothiophenes. End substitution with alkyl chains has been found to be particularly useful, because it gives the molecules liquid-crystalline properties, which dramatically increases the ordering and enhances the charge mobility of the resulting evaporated films. Dihexyl-substituted oligomers with four (DHα-4T, 77b), [86] five (DHα-5T, 78b), [86b] six (DHα-6T, 79b), [87] and eight (DHα-8T, 80b)[88] thiophene rings have been synthesized and characterized.

The FET mobility of DHα-nT is typically reported to be between 0.02 and 1.0 $cm^2V^{-1}s^{-1}$ depending on the deposition conditions and dielectric layer used. [77,83a,89] Compared with the relative low mobilities of DHα-nT, Halik and coworkers reported a series of α, α'-didecyloligothiophenes (DDα-nT) containing four, five, and six thiophene rings. FET mobilities up to 0.5 $cm^2V^{-1}s^{-1}$ have been obtained for DDα-5T (78c) and DDα-5T derivatives and the authors claimed that the use of cross-linked poly(hydroxystryrene) as the dielectric layers is responsible for the high performance. [90] Furthermore, using poly-4-vinylphenol (PVP) as gate dielectric layer, the carrier mobility measured for DHα-6T (79b) increased to 1.0 $cm^2V^{-1}s^{-1}$ in top contact FETs, while that of DDα-5T (78c) and DDα-6T (79c) both increased to 0.5 $cm^2V^{-1}s^{-1}$ in bottom contact FETs. [83a]

The effect of end groups other than alkyl chains including aryls, acenes, heteroacenes, and electron-defficient groups has also been studied. A series of alkyl or cyclohexyl substituted fluorene end–capped oligothiophenes have been synthesized and characterized. [91] Derivatives 81b and 82 with bithiophene core both showed high mobility higher than 0.1 $cm^2V^{-1}s^{-1}$, higher than unsubstituted derivatives. The authors claimed that combining high and low band gap units (fluorene and thiophene) in a conjugated oligomer is a promising approach for realizing high charge transport mobility and good device performance.

Acenes and heteroacenes end groups with high oxidation potentials are also introduced into thiophene oligomers system in order to stabilize the oligomers. Derivatives 83–86 with bithiophene core were synthesized and investigated. [92–94] Compound 84 was synthesized using a soluble precursor in which solubility originated from the presence of trimethylsilyl groups. The measured mobility was 0.01 $cm^2V^{-1}s^{-1}$, higher than that of α-4T measured in the same conditions.

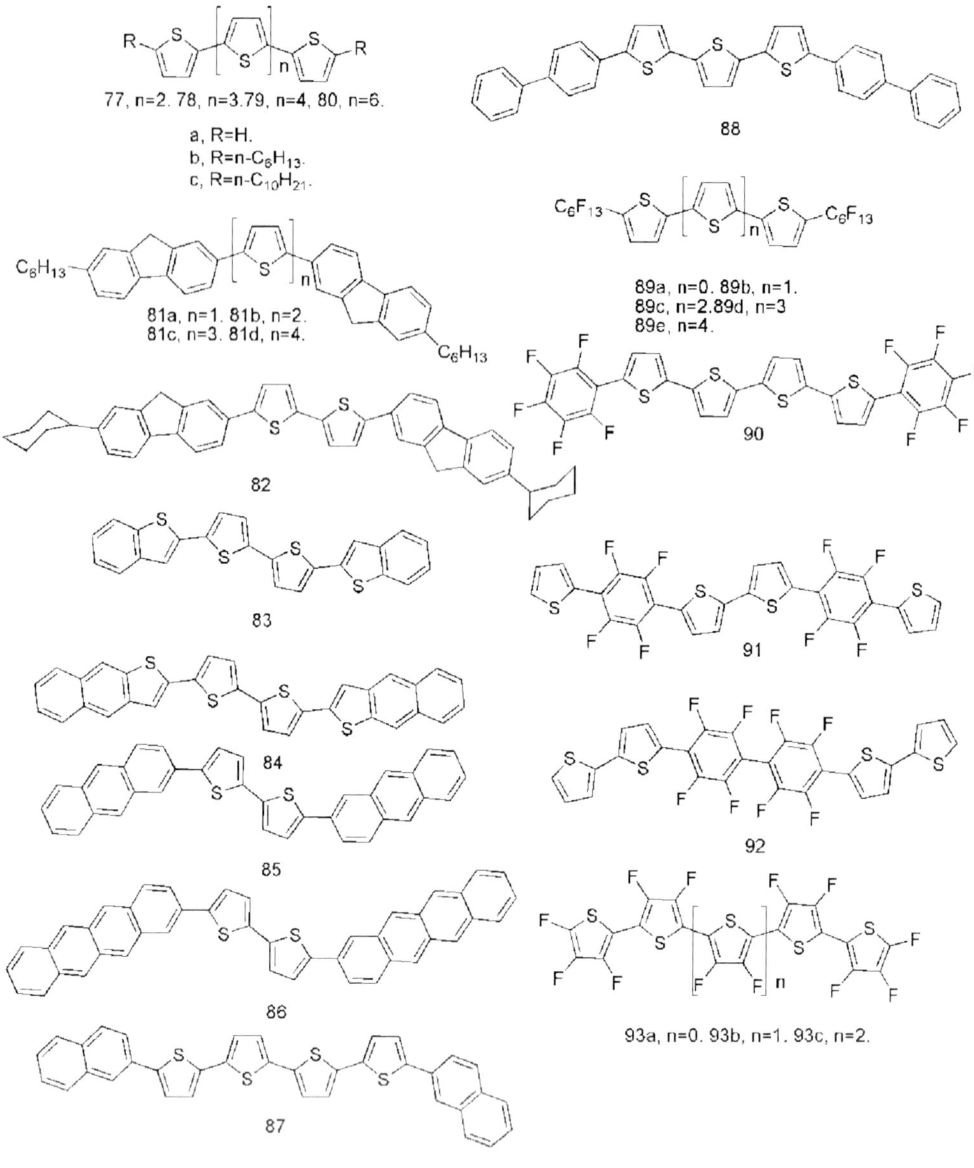

Chart 5. Structures of oligothiophenes linked by single bonds referred in this chapter.

Compound 85 with anthracene end-capped exhibited a FET mobility of 0.12 $cm^2V^{-1}s^{-1}$ and also a high degree of thermal stability. With more extended conjugation, tetracene analogue 86 showed a higher mobility (0.5 $cm^2V^{-1}s^{-1}$ with on/off ratio of 10^8). Recently, Geng's group synthesized naphthyl end-capped α-4T (87).[95] FETs fabricated with compound 87 afforded a mobility of 0.4 $cm^2V^{-1}s^{-1}$, similar to that of tetracene analogue 86. Very recently, biphenyl substituted thiophene oligomer 88 was investigated as active layers in organic single-crystal light-emitting transistors. [96] Typical ambipolar behavior was observed. The hole and electron mobilities up to 1.64 and 0.17 $cm^2V^{-1}s^{-1}$ have been obtained, respectively. Besides, the material exhibited high photoluminescence efficiencies of 80%.

Varieties of electron-deficient groups have been introduced to thiophene oligomers to obtain *n*-channel organic semiconductors. Marks and coworkers reported a family of α, ω-diperfluorohexyl-substituted thiophene oligomers (89a–e). [97] FET electron mobility of 0.02 $cm^2V^{-1}s^{-1}$ was obtained for DFH-6T (89e) with optimized device structure, which was the first *n*-type oligothiophene FET. The authors demonstrated that introduction of the end-fluoroalkyl chains onto the nT core substantially enhances thermal stability, volatility, and electron affinity vs the parent α, ω-dihexyl-substituted DH-nTs. Fluoroarene modified thiophene oligomers 90–92 were also synthesized to understand fluoro stabilization of the injected electrons. [98] However, only derivatives 90 behaved as *n*-channel semiconductor, giving electron mobility of 0.08 $cm^2V^{-1}s^{-1}$, while 91 and 92 still exhibited *p*-channel FET characteristics. The authors attributed the difference to the charge localization on the end rings, that is, electron-withdrawing end rings would preferentially accommodate a negative charge thus electron transport. They also demonstrated that reduction of the LUMO energy is necessary but not sufficient to promote majority electron transport. Perfluorinated oligothiophenes 93 were also synthesized by Sakamo and co-workers. [99] The derivatives have positive shifted redox potentials relative to the non-fluorinated derivatives and were found to adopt a face-to-face π-stacked structure, which has the potential to give rise to efficient charge mobility. However, transistor measurements have not been reported to date.

2.2.2. Thiophene Oligomers and Co-Oligomers Linked by Double and Triplet Bonds (Chart 6)

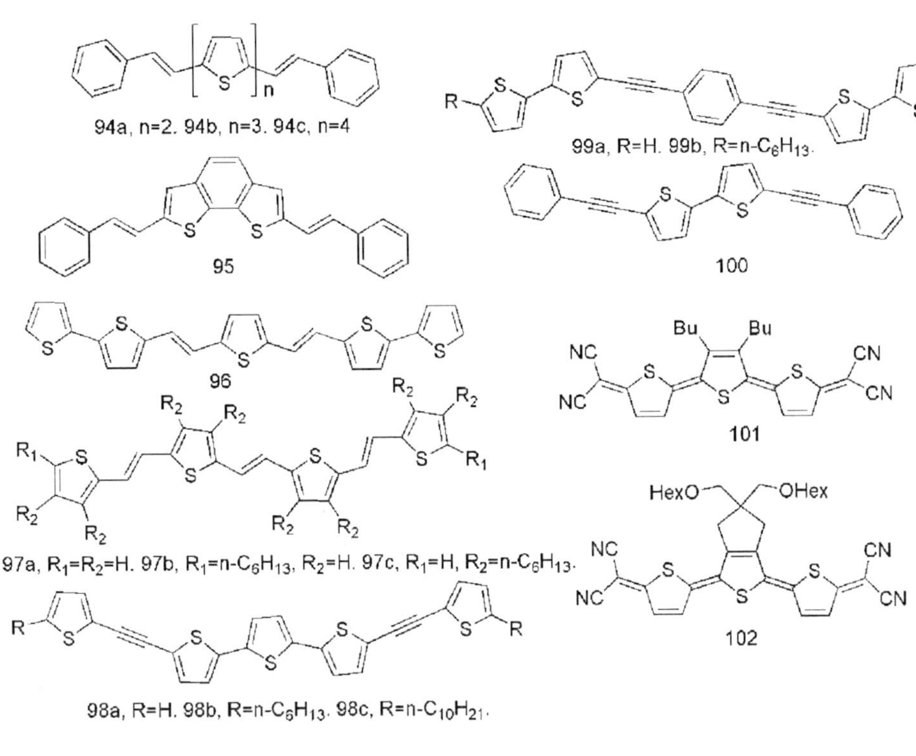

94a, n=2. 94b, n=3. 94c, n=4

99a, R=H. 99b, R=n-C_6H_{13}.

95

100

96

101

97a, R_1=R_2=H. 97b, R_1=n-C_6H_{13}, R_2=H. 97c, R_1=H, R_2=n-C_6H_{13}.

102

98a, R=H. 98b, R=n-C_6H_{13}. 98c, R=n-$C_{10}H_{21}$.

Chart 6. Structures of thiophene oligomers and co-oligomers linked by double and triplet bonds referred in this chapter.

α,ω-distyryl oligothiophenes 94 were synthesized and investigated as organic semiconducting layers in OFETs. [100] Compound 94c, with tetrathiophene core, not only showed high mobility up to 0.1 $cm^2V^{-1}s^{-1}$ and on/off ratio up to 10^5 but also were found to be exceptionally long-lived (more than 1 year of storage) and stable toward continuous operation, under ambient atmosphere. Notably, the compound with the highest HOMO energy levels was found to be more stable than shorter oligomers (94a and 94b) with lower HOMO levels, which was in contrast with previous observations. [91a, 101] Recently, bridged oligomer 95 was synthesized and compared with unbridged oligomer 94a. Interestingly, the compound 95 behaved as a 'kite' in bowl shaped system, different significantly from those of the unbridged analogue which is linear and coplanar. Hole mobilities as high as 0.1 $cm^2V^{-1}s^{-1}$ for OFET devices at T_{sub} = 80°C was achieved, higher by a factor of 5 than the unbridged coplanar oligomer 94a. The authors pointed that the reduced molecular conjugation length and the molecular shape are not limiting factors for the effective molecular overlap between adjacent molecules in the crystal structure. [102]

A series of unsubstituted (compound 96) [103] and substituted oligothienylenevinylenes (97a–c) [104] have been synthesized and characterized containing varying numbers of directly linked double bonds and thiophene units. Alternation of thiophene and a double bond was found to produce a faster decrease of the HOMO LUMO gap than incorporation of additional thiophene rings, and addition of two consecutive double bonds stabilizes the dicationic state but does not lower the HOMO LUMO gap further. In testing of these oligomers in OTFT devices, it was found that only the oligomers capable of reversible π-dimerization of radical cations had appreciable hole mobility. The hexyl-substituted derivative 97b was found to have a mobility of 0.055 $cm^2V^{-1}s^{-1}$, which is more than an order of magnitude higher than the unsubstituted derivative (97a).

Thiophene oligomers with triplet bonds link were also investigated. Quinquethiophenes 98 separated by acetylenic spacers have been investigated. For unsubstituted compound 98a, low mobility of 8×10^{-4} was obtained, [105] however, for 98b–c, the two compounds showed bipolar charge transport in TOF measurements while in FETs only hole mobility of 0.02 $cm^2V^{-1}s^{-1}$ was obtained. [106] Recently, Hu's group synthesized thiophene-phenylene oligomers containing triplet bonds (99–100) and derivative 99b exhibited the highest FET mobility of 0.084 $cm^2V^{-1}s^{-1}$. [107] Single crystal structures indicated that the compounds adopted a typical herringbone structure and introduction of carbon–carbon triple bonds did really decrease the steric repulsion between adjacent aromatic rings and provide the planar structure.

A terthiophene-based quinodimethanes stabilized by electron-withdrawing di-cyanomethylene groups at each end (101 and 102) were also synthesized to obtain *n*-channel organic semiconductors. [108] Derivative 101 was found to pack in a dimerized face-to-face π-stack, forming poly-crystalline films with the long axes of the molecules approximately perpendicular to the substrate. Optimized OFET measurements gave 101 field-effect electron mobilities as high as 0.2 $cm^2V^{-1}s^{-1}$ when measured under vacuum. When deposited at 135 °C, oligomer 101 exhibited ambipolar behavior giving hole mobilities of 10^{-4} $cm^2V^{-1}s^{-1}$ but a decreased electron mobility of $<10^{-4}$. With solubilizing bis(hexyloxymethyl)cyclopentane unit, derivative 102 showed good solubility in common solvents so solution processed OFETs were fabricated, affording high electron mobility up to 0.16 $cm^2V^{-1}s^{-1}$. [109] This was reported to be the best performance for solution processed *n*-channel thiophene-based material.

2.2.3. Other Heterocycle Oligomers and Co-Oligomers (Chart 7)

Because charge transfer is dependent on intermolecular orbital overlap, the addition of hetero atoms other than sulfur may have a positive impact on the transistor performance. The replacement of sulfur in thiophene ring has been attempted with other hetero atoms such as selenium and nitrogen, as well as the introduction of other heterocyclic oligomers. OFETs with quaterselenophene (103) as an active layer have been reported, demonstrating a mobility of 3.6×10^{-3} $cm^2V^{-1}s^{-1}$. [110]

Thiazole rings in thiophene oligomers are known to reduce the steric interaction due to the absence of one hydrogen atom. [111] In addition, thiazole is an electron withdrawing heterocycle and hence, imparts stability to oxygen. [112] Thiophene/thiazole or thiazolothiazole co-oligomers (104, 105) have been synthesized and characterized. Compound 104 with lower electron-donating ability than the parent sexithiophene showed a maximum mobility of 0.011 $cm^2V^{-1}s^{-1}$, lower than that of corresponding DHα6T. [113] The authors attributed the lower mobility of compound 104 to less efficient charge injection, caused by the decreased electron–donating ability of the heterocycle oligomers, and a possibly mismatch between the work function of the electrodes and the HOMO levels of the compound. Compound 105 with thiazolothiazole core exhibited similar mobility as compound 104. [114] By introducing trifluoromethylphenyl group, thiophene/thiazole or thiazolothiazole co-oligomers have behaved as high performance *n*-type organic semiconductors. FET devices fabricated with derivatives 106 showed electron mobility up to 0.30 $cm^2V^{-1}s^{-1}$. [115] Interestingly, with bithiazole as core, compound 107 afforded the highest electron mobility of 1.83 $cm^2V^{-1}s^{-1}$ while compound 108 did not show any FET characteristics. [116] Single crystal structures analysis indicated that the two-dimensional structure of compound 107 play an important role in the high mobility.

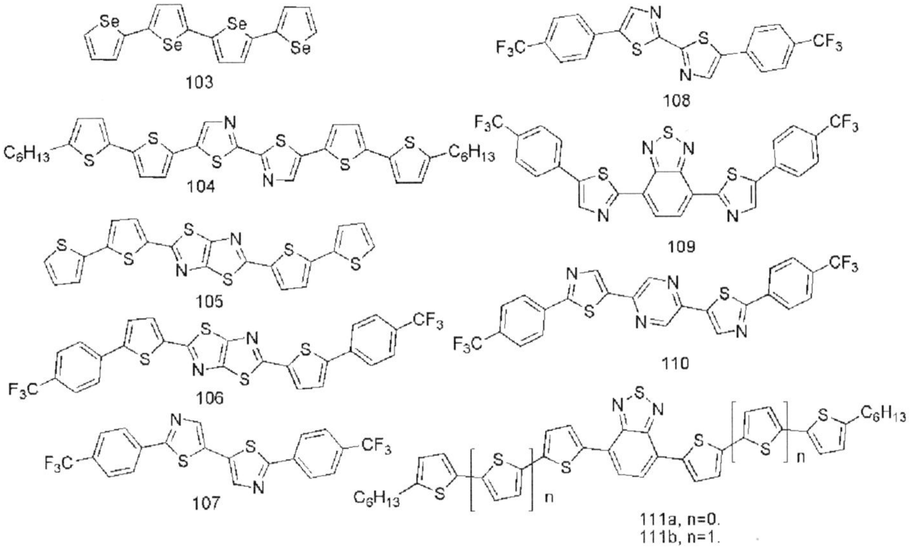

Chart 7. Structures of heterocycle oligomers referred in this chapter.

Other heterocycle oligomers with trifluoromethylphenyl end groups have been reported as n-type semiconductors. FET devices using benzothiadiazole core derivative 109 [117] afforded electron mobility of 0.068 $cm^2V^{-1}s^{-1}$ while the pyrazine core derivative 110 [118] indicated mobility of 0.04 $cm^2V^{-1}s^{-1}$. Liquid crystalline semiconductor oligomers with benzothiadiazole have also been investigated. [119] However, hole mobility of only 5×10^{-4} and 2×10^{-4} $cm^2V^{-1}s^{-1}$ were obtained for derivatives 111a and 111b.

2.2.4. Tetrathiafulvalene (TTF) Derivatives (Chart 8)

TTF and its derivatives have been intensively investigated as building blocks for charge-transfer salts, producing a multitude of organic conductors and superconductors, as well as for the preparation of a wide range of molecular materials. [75,120] It was recently shown that TTF derivatives can also be used in OFETs. [121] Single crystal FETs of dithiophene-tetrathiafulvalene (DT-TTF, 112) have been prepared by drop casting from solution and mobility as high as 3.6 $cm^2V^{-1}s^{-1}$ was obtained. [122] Calculations of the transfer integral and the reorganization energy of these materials taking into account their crystal packing were in agreement with the experimental results; that is, the materials showing higher mobility, have lower reorganization energy and larger transfer integral. [123] TTF derivatives with fused aromatic rings (benzo-113, naphtha-114, and quinoxalino-115) have also been synthesized, with mobilities of 0.1 $cm^2V^{-1}s^{-1}$ and 0.2 $cm^2V^{-1}s^{-1}$ obtained for single crystals of 113 and 115. [124]

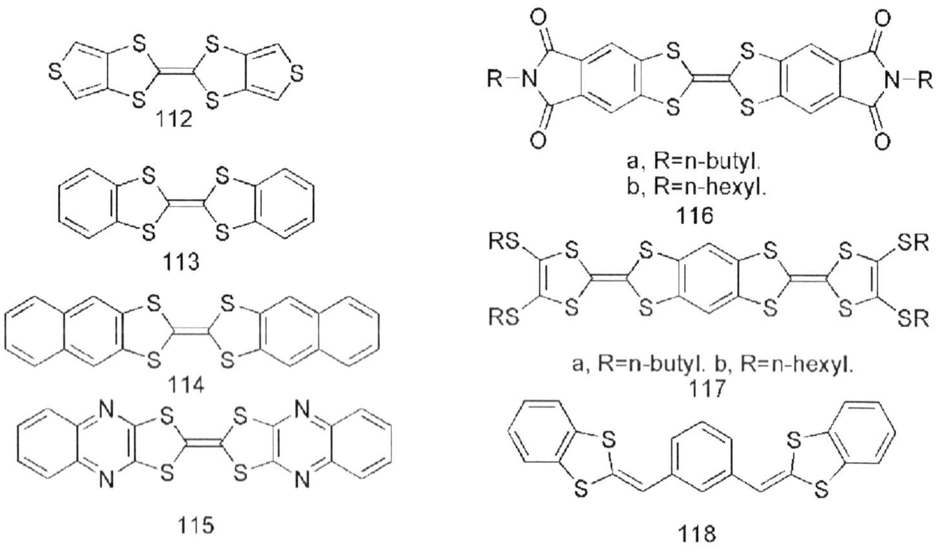

Chart 8. Structures of TTF derivatives for OFETs referred in the chapter.

Dibenzotetrathiafulvalene bisimides derivatives (116) were recently reported, with the aim to lower the electron-donating property of TTF derivatives thus decrease the conductivity of the device in the off state by introducing of electron deficient bisimides groups. Thin film FETs have been prepared and mobility up to 0.40 $cm^2V^{-1}s^{-1}$ and even high on/off ratios of 10^6–10^8 have been achieved. [19b] Benzene–fused bis(tetrathiafulvalene) compounds [125] and TTF derivatives with insertion of an aromatic unit [126] (compounds 117–118) have also been synthesized. Solution processed FET devices fabricated with 117b showed mobility up to 0.02 $cm^2V^{-1}s^{-1}$ while vacuum evaporated films of 118 exhibited mobility of 0.081 $cm^2V^{-1}s^{-1}$.

2.3. PDI and NDI Small Molecules (Chart 9).

Perylene tetracarboxylic dianhydride (PTCDA) and naphthalene tetracarboxylic dianhydride (NTCDA) and their family of imide derivatives have also been studied in detail by many groups. When measured in vacuum, the unsubstituted PTCDA (119) [127] and NTCDA (124) [128] showed *n*-channel mobility of 10^{-4} $cm^2V^{-1}s^{-1}$ and 3×10^{-3} $cm^2V^{-1}s^{-1}$, respectively. The substituted perylene and naphthalene tetracarboxylic diimides (PTCDI, 120–123, and NTCDI, 125–128) can be easily synthesized from commercially available starting materials. The energy levels of alkyl substituted PTCDI (120a–120f) [129] are similar to unsubstituted versions (3.4 eV and 5.4 eV for LUMO and HOMO referenced to vacuum level.) [130] The highest reported electron mobility for alkyl substituted PTCDI derivative 120b reached a value of 1.3 $cm^2V^{-1}s^{-1}$ under vacuum, when corrected for contact resistance. [129a] A record mobility of 2.1 $cm^2V^{-1}s^{-1}$ was reported for 120d after annealing at 140°C. [131] The stability and performance for 120b and 120d FETs was improved greatly using optimized thin film growth rates and sulfur-modified electrodes. [132]

Core-substituted perylene diimides have also been functionalized at the four bay positions with different groups (121–123). Compounds 121a–121c with electron withdrawing cyano groups showed good solubility in organic solvents and have a lower LUMO than 120b, suggesting improved *n*-carrier stability. Crystal structure analysis of compound 121b revealed a face-to-face, slipped π-stacked structure with a minimum interplanar spacing of 3.4 Å. Electronic measurements were performed on compound 121 that were a mixture of isomers cyanated at the 1,7- and 1,6- positions of the perylene core. Mobilities as high as 0.1, 0.64, and 0.14 $cm^2V^{-1}s^{-1}$ were found for 121a, 121b [133] and 121c [134], respectively. A series of bay-perfluoroalkylated perylene bisimides were also synthesized via copper coupling reactions. Single crystal structure analysis of compound 122 indicated that the perylene cores are highly twisted owing to the steric encumbrance effect of perfluoroalkyl substituents. FET devices based on 122 afforded electron mobility of 0.053 $cm^2V^{-1}s^{-1}$ in vacuum and 0.052 $cm^2V^{-1}s^{-1}$ in ambient air, indicating good air-stability owing to electron-withdrawing fluorinated substituents. [135]

Four chloro or fluoro-substituted PDI derivatives (123a and 123c) were also synthesized to elucidate the effect of successive bay substitution. [136] Derivative 123a without strong electron–withdrawing groups showed mobility up to 0.18 $cm^2V^{-1}s^{-1}$ and also good air stability [136a] while derivative 123c with four electron-withdrawing fluorine atoms only indicated mobility of 0.03 $cm^2V^{-1}s^{-1}$. As contrast, two fluorine atoms substituted compound 123b afforded high mobility up to 0.349 $cm^2V^{-1}s^{-1}$. [137] The authors attributed the poor performance of 123c to morphological issues or poorer intermolecular interaction within the grains due to the twisting of the PDI skeleton.

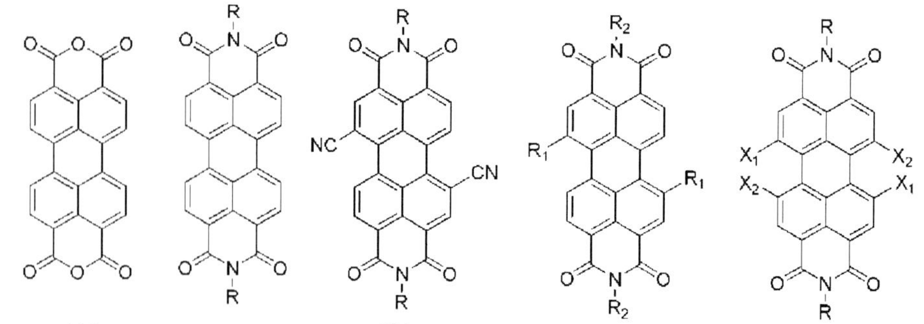

119

120
R=n-C_5H_{11}. (a)
=n-C_8H_{17}. (b)
= n-$C_{12}H_{25}$ (c)
= n-$C_{13}H_{27}$.(d)
= n-C_6H_5. (e)
= (f)

121
R= (a)
= n-$CH_2C_3F_7$. (b)
=n-C_8H_{17}. (c)

122. R_1=n-C_4F_9.
R_2=n-C_4H_9.

123a, X_1=X_2=Cl. R=H.
123b, X_1=F, X_2=H.R=n-$CH_2C_3F_7$.
123c, X_1=X_2=F, R=n-$CH_2C_3F_7$.

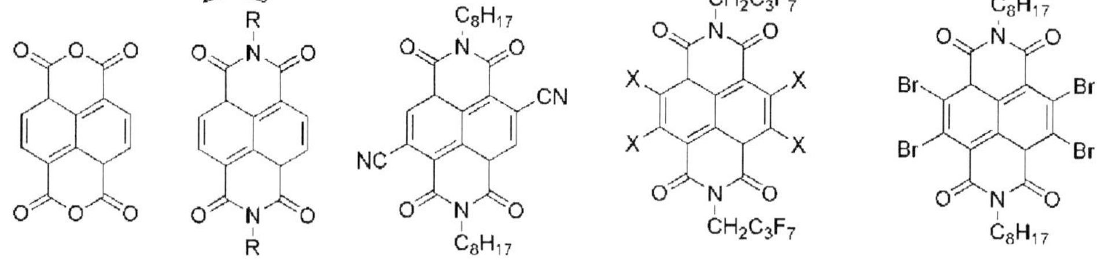

124

125a, R=n-$CH_2C_7F_{15}$.
125b, R=n-$CH_2C_3F_7$.
125c, R=n-C_8H_{17}.
125d, R=cyclohexyl

126

X=

127a, n=1. 127b, n=2. 127c, n=3.

128

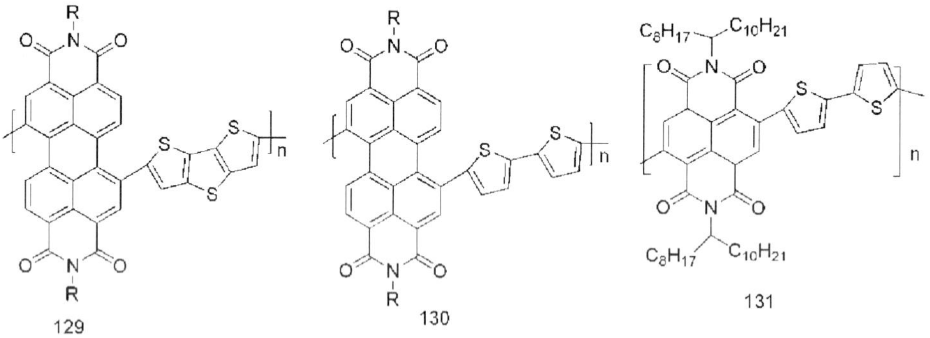

129

130

131

Chart 9. Structures of small molecules and polymers including PDI or NDI unit referred in this chapter.

Alkyl substituted versions of NTCDI showed mobility of 0.16 $cm^2V^{-1}s^{-1}$ for derivative 125a when measured under vacuum, but no observable electron mobility when measured in ambient. Similarly, the incorporation of fluoroalkyl groups on the side chains also stabilizes NTCDI thin films and allow for *n*-channel operation under ambient conditions. Mobilities as high as 0.06 $cm^2V^{-1}s^{-1}$ were obtained for 125b and 0.03 $cm^2V^{-1}s^{-1}$ for 125c with on/off ratio on the order of 10^5. [138] Recently, cyclohexyl end groups in NDI was found to make a dramatic improvement in field effect mobility and mobility near 6 $cm^2V^{-1}s^{-1}$ was obtained for cyclohexyl derivative 125d when measured in Ar atmosphere while only 0.16 $cm^2V^{-1}s^{-1}$ was obtained for the linear *n*-hexyl substituted analogue under similar device fabrication conditions. [139]

Core-cyanated NDI semiconductors (126) were also synthesized in similar procedure with PDI derivatives. Electron mobilities up to 0.15 $cm^2V^{-1}s^{-1}$ and 0.11 $cm^2V^{-1}s^{-1}$ were obtained for compound 126 when measured in vacuum and ambient conditions, respectively, suggesting that ambient stability can be obtained via polycyanation to achieve reduction potentials about 0 V vs S.C.E. [140]

Four substituted NDI derivatives with varied length of thiophene substituents (127a–c) have been reported. [141] These hybrid compounds indicated low-lying LUMO levels (about –4.5 eV) which were not affected by thiophene side chains. However, the FET mobilities observed were very low for these compounds, remaining about 10^{-6} $cm^2V^{-1}s^{-1}$. 2,3,6,7-tetrabromo NDI (127) has also been reported by two groups and the derivative showed lower lying LUMO level relative to that of unsubstituted analogue which was thought to benefit the air-stable *n*-type OFETs, however, OFET characterizations have not been reported. [142]

Polymer semiconductors containing PDI and NDI units have been reported recently. These polymers combined the high electron affinities, high electron mobilities of imides and the planarity, high hole mobilities of fused thiophenes units. D–A polymer 129 based on alternating dithienothiophene and perylene diimide have been synthesized and applied in OFET and all polymer solar cells. [143] Electron mobility up to 0.013 $cm^2V^{-1}s^{-1}$ was obtained for compound 129. Very recently, copolymers based on bithiophene and PDI or NDI (130, 131) were reported and characterized in detail. [144] Very large and stable electron mobility up to 0.85 $cm^2V^{-1}s^{-1}$ for 131-based FETs have been obtained in ambient with Au electrode and polymeric dielectrics. All-printed polymeric complementary inverters have also been demonstrated with 131 and P3HT. While for polymer 130, electron mobility of 0.002 $cm^2V^{-1}s^{-1}$ was reported.

2.4. Thiophene Polymers And Copolymers (Chart 10).

Polymers are attractive materials for solution processable organic semiconductors because of their high solubility and good film forming properties. The choice of synthetic route for the preparation of the polymer is a critical factor in determining its ultimate performance, since molecular weight, polydispersity, defects in the polymer backbone, and impurity levels are all influenced by the choice of synthetic methods, and these factors can have a significant influence on electrical performance. There are mainly four synthetic routes developed for the preparation of conjugated polymers: 1. Chemical oxidation, in the presence of certain oxidizing agents such as ferric chloride, thiophene reacts with itself to form a polymer. [145] 2. Electrochemical oxidation routes, which involve the in situ polymerization of monomer solution on a working electrode in the presence of a suitable electrolyte, are most useful for the study of insoluble polymers. [146] 3. Dehalogenative polymerization, firstly reported by Yamamoto, means that the polymerization of halogenated thiophenes by heating with an excess of zero-valent nickel catalyst in the presence of triphenylphosphine. The high yield, functional-group compatibility, and simplicity of the reaction have made it widely utilized in the synthesis of many conjugated polymers. [147] 4. Transition-Metal Catalyzed cross-coupling methodologies. Various organometallic groups such as organatin (Still reaction), organoboron (Suzuki reaction), organozinc (Negishi reaction), and organomagnesium (Grignard reaction) have been explored. The latter two are reactive intermediates that cannot be readily isolated and stored, so they tend to be formed in situ during the polymerization process. Nevertheless, both methods have been particularly successful for the synthesis of poly(3-hexylthiophene) (P3HT) with regioregularity of over 98%. [148]

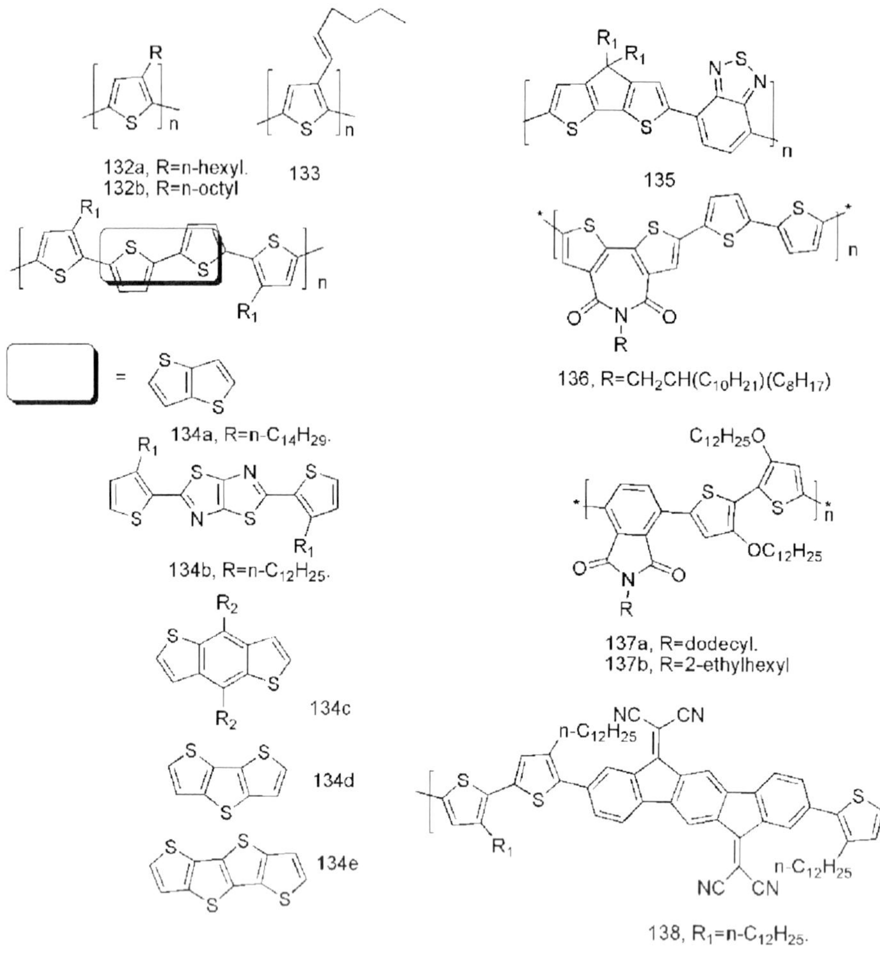

132a, R=n-hexyl.
132b, R=n-octyl
133
135
136, R=$CH_2CH(C_{10}H_{21})(C_8H_{17})$
134a, R=n-$C_{14}H_{29}$.
134b, R=n-$C_{12}H_{25}$.
134c
134d
134e
137a, R=dodecyl.
137b, R=2-ethylhexyl
138, R_1=n-$C_{12}H_{25}$.

Chart 10. Structure of polymers including thiophene unit referred in this chapter.

One of the first solution–processed organic semiconductors used for FETs was poly(thiophene) (PT, 132). Since then various substituents have been incorporated on the polymer backbone to impart functionality, increase solubility, or induce self-assembly. Among them, poly(3-hexylthiophene) (P3HT, 132a) is the most widely studied. [16a,149] P3HT with a high regioregularity (>95% HT linkages) adopt lamellae arrangement with an edge-on orientation, giving mobilities of 0.05–0.2 $cm^2V^{-1}s^{-1}$ and on/off ratios close to 10^6. [149,150] Effects of solvents polarity, alkyl chain length, and modification of the side chains on the field effect mobilities have been investigated. [150a] Poly(3–octylthiophene) (132b) was reported to exhibit mobility up to 0.22 $cm^2V^{-1}s^{-1}$. [151] Poly(3–alkenylthiophene) (133), with the change of a single bond to double bond on the substituent alkyl group of P3HT, has been synthesized to construct conjugated-side-chain substituted polythiophenes. Mobility of 1.44×10^{-3} $cm^2V^{-1}s^{-1}$ was obtained, comparable to the P3HT with the similar stereoregularity and molecular weight. [152] Polythiophenes involving thienylene vinylene as bridges have also been investigated and these polymers showed improved mobility in FETs and high power conversion efficiency in polymer solar cells. The vinylene-bridges were found to be beneficial for the hole "hopping" from one PT backbone to another. [153]

However, PTs showed poor air stability due to oxygen dopant and it was reported that the stability can be improved by increasing its ionization potential (IP). [154] The IP is partially dependent on effective π-conjugation length of the polymer backbone, which can be controlled by reducing π overlap between adjacent thiophene rings or introducing less conjugated unit in the backbone. Recent years, a series of thiophene copolymers have been synthesized and stable FETs were fabricated under ambient conditions. Liquid-crystal polythiophene 134a with thieno[3,2-b]thiophene core was designed to assemble into large crystalline domains on crystallization. Good transistor stability under static storage and operation in a low–humidity air environment was demonstrated, with charge-carrier field effect mobility of 0.2–0.6 $cm^2V^{-1}s^{-1}$ achieved under nitrogen. [155] A significant reduction in contact resistance by employing Pt electrode places the field-effect mobility of compound 134a at 1 $cm^2V^{-1}s^{-1}$, indicating that systematic optimization on device level is needed to reach the ultimate mobilities in these materials. [156] Polythiophene 135b bearing thiazolothiazole units in the backbone was also synthesized with the consideration that this donor–acceptor system could enhances the intramolecular charge transfer which can enhance charge carrier mobility. [114,115,157] Devices with channel length of 10 μm afforded the highest mobility of 0.14 $cm^2V^{-1}s^{-1}$ for 134b, and greater stability against oxidative doping as compared with P3HT was also demonstrated. [158] S. Ong's group reported high performance polymer semiconductor 134c with mobility up to 0.15–0.25 $cm^2V^{-1}s^{-1}$ even without postdeposition thermal annealing. [159] Furthermore, they reported polymer 134d incorporating dithienothiophene moiety, which not only offers high FET mobility up to 0.3 $cm^2V^{-1}s^{-1}$ but also other important processing and performance attributes including room temperature processability in environmentally friendlier nonchlorinated solvent. [160] All the results indicated that the fused unit could cause a reduction in electron delocalization over the polymer backbone, widening the band gap with respect to regioregular P3HT. Recently, He and coworkers further demonstrated that high performance polymer semiconductors can be achieved by increasing the rigidity of the thiophene monomer through the use of alkyl substituted TTA core that consists of four fused thiophene rings. Field-effect mobility of 0.33 $cm^2V^{-1}s^{-1}$ and excellent environmental stability was obtained for polymer 134e. [161]

Polycyclopentadithiophenes, a class of solution processable, thiophene based analogues of the polyfluorenes, have also been investigated as charge transport materials in OFETs. [162] Hole mobility up to 0.17 $cm^2V^{-1}s^{-1}$ has been achieved for polymer 135 when controlling film fabrication conditions. [163]

Besides *n*-type PDI and NDI-based thiophene copolymers mentioned above, N-alkyl-2,2'-bithiophene-3,3'-dicarboximide unit (BTI) and phthalimide unit were also introduced to construct D–A polymer 136–137. The polymers still exhibited *p*-channel FET performance and the hole mobilities of 8×10^{-3} $cm^2V^{-1}s^{-1}$ was obtained for polymer 136. [164] For polymer 137a, mobility up to 0.2 $cm^2V^{-1}s^{-1}$ have been achieved, which was attributed to extensive domain interconnectivity of the films. [165]

A series of thiophene copolymers incorporating electron-withdrawing moiety indenofluorenebis(dicyanovinylene) have also been reported. Interestingly, polymer 138 exhibited ambipolar carriers transport with electron and hole mobilities of ca. 2×10^{-4} $cm^2V^{-1}s^{-1}$. [166] Although ambipolar behaviors have been observed for several polymers by using various gate dielectric surface treatments, these measurements were carried out under nitrogen, [167] thus polymer 138 was reported to be the first example of an air-stable, highly soluble ambipolar semiconducting polymer.

2.5. Polymers Not Based Thiophene Rings (Chart 11)

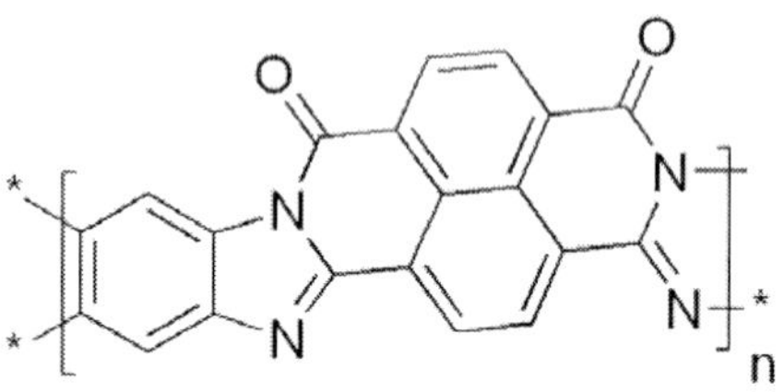

BBL, 139

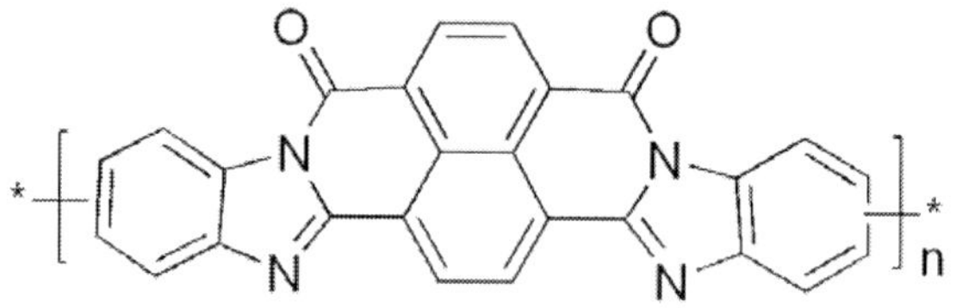

BBB, 140

Chart 11. Structures of polymers not including thiophenes referred in the chapter.

A few of polymers without thiophene unit have been investigated as active layers in OFETs, although most of these materials exhibited much lower mobilities. However, polymer 139 (BBL), the first reported *n*-type polymer semiconductor, exhibited electron mobility up to 0.1 $cm^2V^{-1}s^{-1}$ in the linear regime, while the non-ladder derivative 140 (BBB) showed only mobility of 10^{-6} $cm^2V^{-1}s^{-1}$ when cast under the same conditions. The large differences in mobility were attributed to morphological differences in thin films between the two polymers. The BBL has more efficient π-stacking and greater intermolecular order resulting in high degree of crystallinity, while the BBB films were shown to be completely amorphous. [168]

2.6. Two, Three Dimensional and Macrocyclics Semiconductors (Chart 12)

Triarylamine-based organic semiconductors have been widely investigated for optoelectronic applications. [169] As a matter of fact, these materials possess many attributes that meet the requirements for OFET applications. Unfortunately, these materials usually exhibit low field effect mobilities due to their amorphous film states. Amorphous polytriarylamines (PTAA) is a widely studied polymer, and a maximum mobility close to

0.01 $cm^2V^{-1}s^{-1}$ was obtained. [170] Recently, a cyclic triphenylamine dimer (141) was synthesized, both the mobility (0.015 $cm^2V^{-1}s^{-1}$) and the on/off ratio (10^7) of derivative 141 are an order of magnitude larger than those of the linear triphenylamine derivatives (142). The property of good polycrystalline film formation should be responsible for the efficient carrier transport. [171].

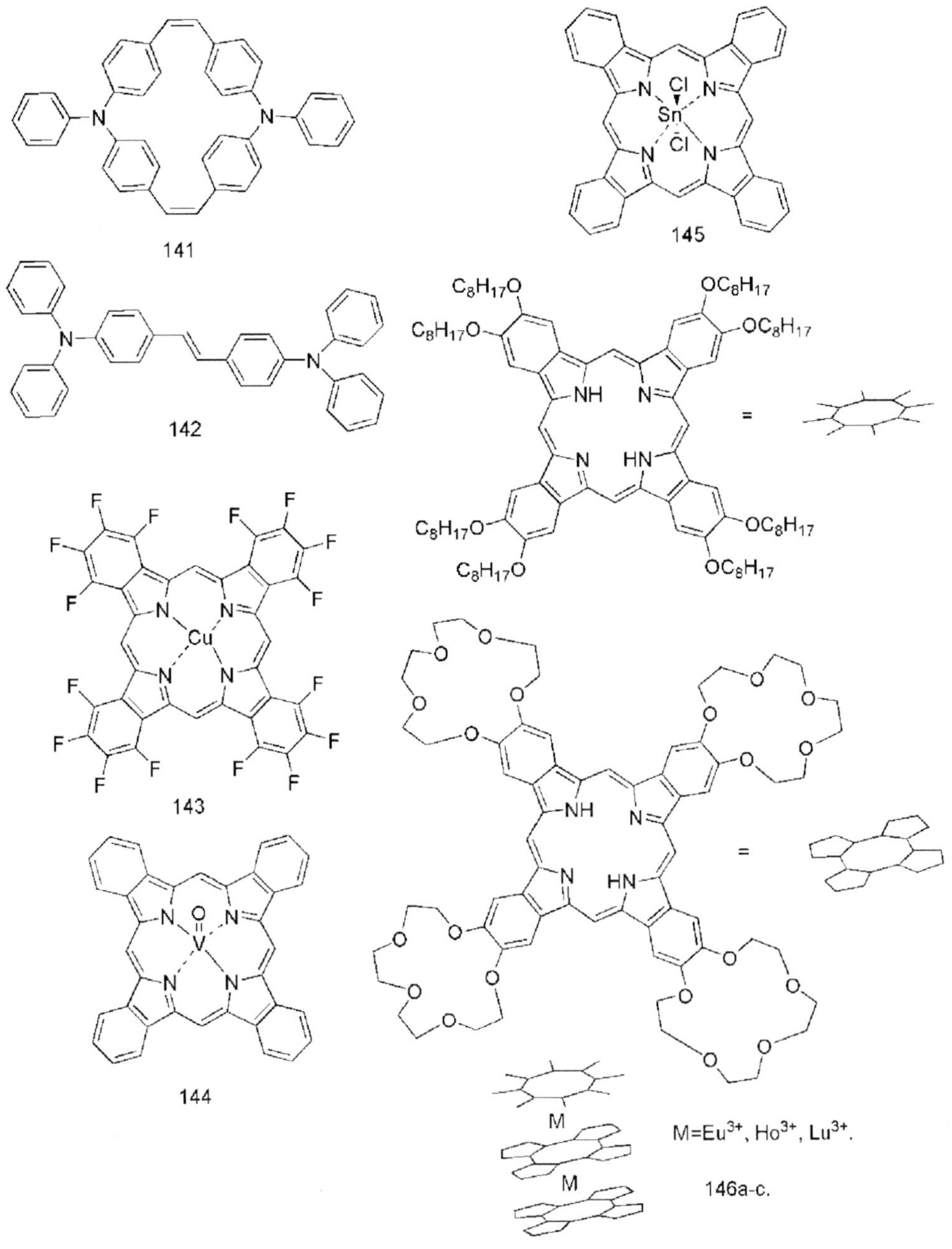

141

142

143

144

145

$M=Eu^{3+}, Ho^{3+}, Lu^{3+}$.

146a-c.

Chart 12. Structures of macrocycle derivatives referred in this chapter.

Phthalocyanine (Pc) is a planar and aromatic macrocycle, which consists of four isoindole units presenting 18 π electrons. A remarkable feature is their versatility. Many different metal elements can replace the hydrogen atoms of the central cavity. Usually, strong π–π interactions exist in the solid state of Pc and the interaction could contribute to carrier transport. Interestingly, Pc derivatives could exhibit different carrier transport (hole or electron) with varied metals in central cavity. The field-effect properties of phthalocyanine derivatives such as CuPc, [172] FePc, [173] hexadecafluorinated copper phthalocyanine ($F_{16}CuPc$, 143), [174] vanadyl phthalocyanine (VOPc, 144) and tin dichloride phthalocyanine ($PcSnCl_2$, 145) have been reported. In particular, hole mobility up to 1.5 $cm^2V^{-1}s^{-1}$ has been realized with VOPc, while electron mobility of 0.11 and 0.30 $cm^2V^{-1}s^{-1}$ have been obtained from $F_{16}CuPc$ and $PcSnCl_2$, respectively, via weak epitaxial growth of large-area highly ordered organic films. [175]

Devices based on tris(phthalocyaninato)rare earth tripledecker complexes (146a–c) have been fabricated by using LB technique and high mobilities of 0.2–0.6 $cm^2V^{-1}s^{-1}$ have been obtained. Strong intermolecular π–π stacking and intermolecular *J* aggregation may be responsible for the unexpectedly positive results. [176].

3. Physical Chemistry of OFETS

Since the molecular arrangement and the thin film morphology of an organic semiconductor are largely determined by the properties of the interface between the organic film and the insulator or electrode, a great deal of research has focused on interface engineering. [177] Here we mainly focus on common features about organic semiconductor/dielectric layer interface contact and organic semiconductors/electrode interface contacts.

3.1. Organic Semiconductor/Dielectric Layer Interface Contacts

The dielectric layer is an important element of OFETs. Its chemical composition, surface morphology (roughness), and dielectric properties have dramatic influence on the condition of the organic thin films and the device performance. [170] Silicon dioxide (SiO_2) is widely used as the dielectric layer because of its excellent insulating property and stability as well as its low surface roughness. However, small grains are usually obtained when organic semiconductors are deposited on the bare SiO_2 surface. It limits the improvement of the device performance. Furthermore, SiO_2-based devices usually need higher operation voltage, which will result in excessive power consumption. Devices performances would be improved by modification of the SiO_2 surface using a self-assembly monolayer such as octadecyltrichorosilane (OTS) [178] or hexamethyldisilazane (HMDS) [179] or introduction of a buffer layer between the organic semiconductor and the dielectric layer. [180] Recently, fabrication of OFET using a phenyltrimethoxysilane (PhTMS) modified SiO_2 insulator greatly improves the device electrical properties over those with plain or OTS modified SiO_2, particularly improves the carrier mobility, the subthreshold slope, and channel resistance resulted from reduced density of charge trapping states at the semiconductor/insulator interface. The pentacene OFETs with modification from PhTMS achieves carrier mobility of 1.03 $cm^2V^{-1}s^{-1}$. [181] Sol-gel silica gate-dielectric was also reported to have positive effect on devices performance enhancement not only for *p*-channel small molecule/polymer but also for *n*-channel fullerene derivative. The low surface energy coupled with the smooth surface of sol-gel silica promotes in-plane π-stacking order along with improved crystallinity for organic semiconductors, thus leading to devices performance enhancement. [182]

Low operating voltage would be achieved when a dielectric layer with higher dielectric capacitance was used, such as barium zirconate titanate (BAT), barium strontium titanate (BST), Ta_2O_5 and Al_2O_3. [183].

Polymer dielectric layer, which can be processed by solution approaches, such as spin-coating, casting, or printing, at room temperature and under ambient conditions, is attractive for use in flexible devices. Recently, a number of polymer dielectric materials have been developed. [144b,184] Using the polymer dielectric layer usually leads to higher performance, lower operating voltage, or both. Varied organic film morphology can be formed on different polymer dielectric layers. The grain sizes of the pentacene deposited on HMDS-modified SiO_2/Si or polystyrene (PS) substrates are much larger than those observed on the PVA substrates. [185] Figure 5 indicates the different morphology of organic semiconductor DHCO-4T films grown on different gate dielectrics.

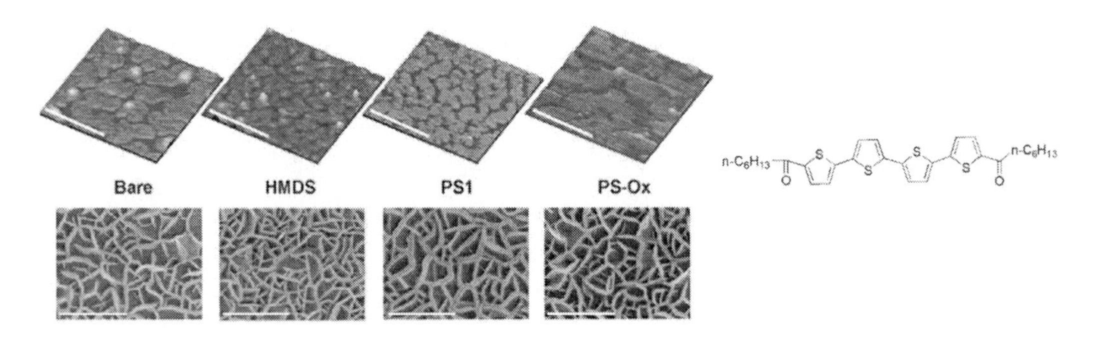

Figure 5. AFM and SEM images of DHCO-4T films grown on the indicated gate dielectrics, Scale bars denote 1 μm and the structure of DHCO-4T, with HMDS = hexamethyldisilazane, PS1 = 24 nm polystyrene thin film, PS-Ox = polystyrene film treated with an O_2 plasma. (from Marks, T. J. et al., J Am Chem Soc, 2006, 128, 12851–12869. Copyright 2006, American Chemical Society).

One of the most revealing studies on the importance of the dielectric/semiconductor interface was presented by Chua and coworkers. [167] They found that removing interface traps can unveil electron mobility in organic semiconductors previously thought to show only hole transport. This result suggests that the semiconducting properties of an organic material are highly dependent on the nature of the dielectric and the dielectric interface.

3.2. Organic Layer/Electrode Contacts

Carrier injection from the S–D electrode into the organic layer is very important for OFET devices. The carrier injection barrier mainly depends on the barrier between the work function of the metal electrode and the HOMO or LUMO energy level of the organic semiconductors. Comparing the work function of the injecting metal with the HOMO/LUMO levels of a semiconductor can help to determine whether charge injection is likely and whether high or low contact resistance is to be expected. *p*-Type organic semiconductors typically have HOMO levels between −4.9 and −5.5 eV, resulting in ohmic contact with high work-function metals such as gold (5.1 eV) and platinum (5.6 eV). [186] *n*-Type materials usually have LUMO levels between −3 and −4 eV and should have better contact with low work-function metals such as calcium and lithium. For example, Burgi et al. demonstrated that the work function of gold (φ = 5.1 eV) is well aligned with the HOMO level of P3HT (4.8 eV), which leads to a very low contact resistance, while for copper (φ = 4.6 eV) the contact resistance was several orders of magnitude higher and almost no charge injection from aluminum (φ = 4.0 eV) was observed. [187] Selecting the proper S D electrodes is an effective approach to reduce the carrier injection barrier. Up to now, gold has been the most widely used S/D electrode in most FET devices because compatibility with most organic semiconductors. Also, the gold surface is relatively clean and stable without native oxide, which may help to form a good contact with an organic layer. Gold, however, may not be preferred in the real application because, as a noble metal, it is expensive and difficult to etch for patterning. Besides, the work function of Au is energetically incompatible with the LUMO energy level of the organic semiconductors, thus leading to large contact resistance. Although highly incompatible of LUMO levels, high electron mobility have been reported with Au as S–D electrodes, [144b] indicating the complication of contact resistance and carrier injection. As a contact metal, silver, copper, calcium, palladium and indium tin oxide (ITO) were studied.[188] High performance pentacene FETs have also been fabricated with NiOx as a S/D electrode. The normalized contact resistance of NiOx is comparable with other metals even though the material itself is an insulator. [189].

However, contact resistance is not simply determined by the metal and the semiconductor but also by the device structure, that is, whether a top contact or bottom contact geometry is employed. The position of the injecting electrodes with respect to the gate electrode also plays a role in determining the contact resistance. Devices with top-contact geometry usually exhibit good electrode–semiconductor contact and efficient carrier injection. Most of the reported high performance organic semiconductors adopted top-contact device configuration. However, the top contact geometry exhibits a significant obstacle to large-area application, that is, incompatibility with the photolithographic process. The bottom contact configuration is a feasible geometry in practical application. Unfortunately, the devices with bottom contact suffer from lower performance due to bad metal/organic contact. The bad contact can be properly resolved by self-assembly monolayer (SAM) containing -SH. [190] The work function of a metal can change with a SAM on top of the metal surface and also the morphology of the semiconducting layers grown on a SAM surface can be different from the layer grown directly on a metal surface.

The introduction of a buffer layer between the electrode and the active layer can also improve the device performance. When 7,7,8,8-tetracyanoquin-odimethane (TCNQ) was inserted between the metal electrode and the organic layer, the device performance was obviously improved. [79] The introduced layer reduced the injection barrier and impeded electrode metal penetration. By modifying the electron/organic semiconductors interface with TCNQ in the form of charge-transfer complexes (Ag-TCNQ or Cu-TCNQ), we have resolved the energy mismatch of Ag or Cu S–D electrodes with organic semiconductors and the worse interface contact in the bottom-contact configuration. [5b] The pentacene transistors with Ag-TCNQ modified Ag bottom-contact electrodes exhibit outstanding properties, comparable to those of the Au top-contact device.

CONCLUSION AND OUTLOOK

In the chapter, we first introduced the principles, operation processes and basic device structure of OFETs, then we reviewed the chemical design, structure, properties, and device performance of newly developed organic semiconductors for OFETs. Specific examples of materials underscoring the importance of chemical synthetic techniques, rational design of functional molecules, molecular/polymer solid-state packing and conformation, and charge transport on the corresponding OFET device performance have been presented. Besides, the interface contacts including the organic semiconductor/dielectric layer interface contacts and organic layer/electrode contacts of OFETs were discussed.

From the discussion we can conclude that organic electronics are newly emerging fields of science and technology that cover chemistry, physics, and materials science. Electronic and optoelectronic devices using organic materials are attractive because of the materials characteristics of light weight, potentially low cost, and capability of large-area, flexible device fabrication. Over the past few years, painstaking progress has been made in organic semiconductors and their devices, whose mobilities are comparable to those of amorphous silicon FETs. The progress exhibited great potential of organic materials for realizing low-cost, flexible electronic devices.

However, there are key requirements where existing materials fall short. First, new organic semiconductors with high carrier mobility, excellent stability, and low cost (particularly for n–type semiconductors) are still desirable. Because we still do not have a complete understanding of the fundamental limits of charge mobility through organic materials and the true effect that oxygen, moisture, and light has on device performance. Organic semiconductors with high environmental stability could withstand the influence induced by environmental conditions such as light irradiation and O_2 doping, etc. The operational stability is related to the operation lifetime of devices in the operational state. When the OFETs are in the operational state, the high carrier density in the active channel can prevent a decrease in the device performance. Both the high environmental and operational stabilities are necessary to enable the applications of OFETs. Solution processing of high performance organic semiconductors is always a subject with great opportunity to achieve low-cost, flexible, organic electronics. A standardized method for evaluating new organic semiconductors and predicting chemical structures that would give superior properties have not been developed.

For the devices fabrication, the relationship between the device fabrication process and device performance requires more exact examination because the organic layers, the dielectric layer, and the electrodes all influence the device performance. Great attention has been paid to the dielectric layers and conductive electrodes necessary to complete the all-organic electronic devices. For further industry application, the devices must have good reproducibility and low deviation in device performance. However, this information has been ignored in most literatures and only the best mobility was reported. It is misleading because one working device may have a different morphology from the bulk of the film. Reporting the reproducibility and the standard deviation of devices performance should become routine in future research of OFETs.

In short, both opportunities and challenges exist in further development of OFET toward applications. Development of *n*-channel and ambipolar FETs opens up new opportunities for a better understanding of the device physics of organic semiconductors. It is still a challenge to realize flexible OFETs with both stability, high performance and low cost which were the true benefits of organic electronics.

5. Table of Device Characteristics

Table 5.1. FET mobilities, on/off ratios and device structure of each organic semiconductor described in the chapter.

compound	Mobility[a]/on/off/voltage	device configuration	reference
1	Not observed		
2	0.013 (h) /10^4/	Top contact	24
3	0.13 (h) /10^4	Top contact	24
4	0.072 (h) /10^5	Top contact	24
5	0.18 (h) /10^4	Top contact	24
6	3.4×10^{-3} (e) /10^4/+75	Top contact	25a
7	4.5×10^{-3} (h) /10^3/–40 0.063 (h) /10^5/–6.3	Top contact Top contact	25a 25b
8	0.50 (h) /10^7/–10.4	Top contact	25b
9	1.28 (h) /10^7/–15.4	Top contact	25c
10	1.3 (h) / 10^7/–16	Not reported	25d
11	0.14 (h)/10^6/–14	Top contact	25e
12	0.04 (h) /10^5	Not reported	26
13	0.081 (h)/10^3	Top contact	27
14	0.17 (h)/10^5	Top contact	27
15	0.0073 (h)/10^3	Top contact	27
16	0.044 (e)/10^6/31	Top contact	28
17	0.1 (e)/10^5/46	Top contact	28
18	3×10^{-3} (h)	Top contact	29
19	0.02 (h)	Top contact	29
20	7.7×10^{-2} (h) /10^7/–53.3	Top contact	30
21	6.3×10^{-4} (h) /6×10^4/–49.9	Top contact	30
22	Not observed		30
23	0.05 (h)/10^8	Top and bottom contact	31
24	0.14 (h)/10^4/3.6	Top contact	32
25	0.12 (h)/5×10^5/–20.2	Top contact	32
26	0.42 (h)/5×10^6/–23.4	Top contact	32
27	0.017 (h) /8×10^3/–4	Top contact	33
28	0.018 (h) /10^4/–4	Top contact	33
29	0.005 (h)/ 10^5/	Top contact	16d
30	0.89 (h)/ 10^7/	Top contact	16d
31	0.038 (h)/ 10^4/–12	Top contact	16c

Table 5.1. (continued)

compound	Mobility[a]/on/off/voltage	device configuration	reference
32	2.0 (h)/ 10^{8}/–31	Top contact	16c
33	0.002 (h)/ 10^{4}/–13.8	Top contact	16c
34	4×10^{-3}(h)(solution process)/10^{6}	Top contact	34
35	3×10^{-4} (h) /10^{5}	Top contact, solution process	34
36	1.3 (h) (single crystal)	Not reported	36
37	1.4×10^{-4} (h) (single crystal)/10^{3}	Not reported	37a
38	0.3(h) (single crystal)/10^{2}	Not reported	37a
39	1.6(h) (single crystal)/10^{5}	Not reported	37a
40	1.7(h) (single crystal)/10^{4}/–12	Not reported	37b
41	30 (h) (single crystal)	Not reported	38
42a–42d	Not reported		39
43	2.4×10^{-4} (e)	Top contact	40
44a 44b	Not observed 0.31 (h)/10^{6}/–20	 Top contact	41 41
45a 45b	2.0 (h)/10^{8}/ 2.75 (h)/10^{7}/–27	Top contact Solution process	41 41
46	0.15 (h)/10^{6}/–14	Top contact	21, 42
47a 47b 47c 47d	Not reported 0.14 (h)/10^{6}/–19 0. 1 (h)/10^{6}/–15 0.09 (h)/10^{6}/–28	 Top contact Top contact Top contact	43
48	>5.0 (h)/10^{8}	Not reported	45
49	0.30 (h)/6.3×10^{3}/12.5 0.26 (h) /5×10^{5}/7.0	Bottom contact at 85 °C Top contact at 105 °C	48
50a 50b	0.069 (h) /3.7×10^{5}/–16 Not obtained	Top contact/80 °C	49
51a 51b 51c 51d	0.13 (h) /2.5×10^{4}/–11 0.23 (h) /1.9×10^{4}/–9 0.032 (h) /5.5×10^{3}/–13 0.03 (h) /1.6×10^{5}/–27	Top contact/80 °C Top contact/80 °C Top contact/80 °C Top contact/80 °C	49
52	0.4 (h) /10^{6}	Not reported	50
53a 53b 53c	<10^{-4} (h) 1.0 (h) /10^{7} Not obtained	Solution process Solution process Solution process	18c
54	0.37 (e) 0.065 (h)	Not reported	51f
55	0.22(e)/10^{5}	Top contact, in vacuum	40, 52
56a 56b 56c	0.09 (h) 0.15 (h) 0.15 (h)	Top and bottom contact	53
57	0.47 (h) /1.3×10^{6}/7	Top contact	42
58a 58b	0.01 (h) /10^{5} 0.001 (h) /10^{5}	Bottom contact Solution processes	55, 56 55
59	0.045 (h) /10^{3}	Top contact	57
60	0.42 (h) /10^{5}	Top contact	6
61	0.51 (h) /4.5×10^{6}/–67	Top contact	58

Table 5.1. (continued)

compound	Mobility[a]/on/off/voltage	device configuration	reference
compound	Mobility[a]/on/off/voltage	device configuration	reference
62a 62b 62c	0.02 (h) /10^2/7.5 0.01 (h) /10^5/–9.4 2×10^{-5} (h) /10^2/–8.5	Top contact Top contact Top contact	59
63a 63b–c	0.012 (h) /10^5/–20.8 Not reported	Top contact	60
64–66	Not reported		61–64
67	0.574 (h) /3×10^3/20	Top contact, in nitrogen	65
68a 68b	2.9 (h) /10^7/–11 1.9 (h) /5×10^6/–7.5	Top contact Top contact	66
69–73	Not reported		67–70
74	0.15 (h)	Top contact	71
75	0.4 (h) (single crystal FET)	Top contact	72
76	Not reported		73
77a 77b 77c	2.5×10^{-3} (h) /–15 0.03–0.23(h)/10^5[b] 0.2 (h) /10^5	Top contact Not reported Bottom contact	79 86 83a,90
78a 78b 78c	1.5×10^{-3} (h) /5 0.02–0.10(h)/10^4 0.5 (h) /10^5	Top contact Top contact Bottom contact	79 86b 83a, 90
79a 79b 79c	0.07–0.1 (h) /10^3 0.03–1.0 (h) /10^4 0.5 (h) /10^5	Bottom contact top contact Bottom contact	80, 83, 84 77, 83a, 87, 89 83a,90
80a 80b	0.33 (h) 0.03 (h)	Top contact Top contact	85 88
81a 81b 81c 81d	0.012 (h) /10^3/ 0.14 (h) /2×10^4 0.025 (h) /10^2 0.023 (h) /10^2	Top contact Top contact Top contact Top contact	91a
82	0.17 (h) /8×10^5	Top contact	91b
83	0.011 (h) /10^2/–5	Top contact	92
84	0.01 (h) /29/–12	Top contact	93
85	0.12 (h) /10^8	Top contact	94
86	0.5 (h) /10^8	Top contact	94
87	0.4 (h) /10^5/–5.5	Top contact	95
88	1.64 (h) 0.17 (e)	Au, Ca two electrodes Top contact	96
89a 89b 89c 89d 89e	Not measured Not measured 0.048 (e) /10^5/25 0.026 (e) /6×10^4/35 0.02 (e) /10^5/35	 Top contact Top contact Top contact	97b 97b 97b 97b 97a
90	0.08 (e) /10^5	Top contact	98
91	0.01 (h) /10^4	Top contact	98
92	4×10^{-5} (h) /10^2	Top contact	98
93	Not reported		99

Table 5.1. (continued)

compound	Mobility[a]/on/off/voltage	device configuration	reference
94a 94b 94c	0.02 (h) /10^5 0.02 (h) /10^4 0.1 (h) /10^3	Top contact Top contact Top contact	100
95	0.1 (h) /10^6/–12	Top contact	102
96	0.01 (h)	Top contact	103
97a 97b 97c	1.4×10^{-3} (h) 0.055 (h) 1.1×10^{-6} (h)	Top contact Top contact Top contact	104
98a 98b 98c	8×10^{-4} (h) /10 0.02 (h) 0.02 (h)	Bottom contact Top gate Top gate	105 106 106
99a 99b	0.056 (h) /1.3×10^6/–6.34 0.084 (h) /10^5/–11.8	Top contact Top contact	107
100	0.028 (h) /3.9×10^5/1.72	Top contact	107
101	0.2 (e)	Top contact	108
102	0.16 (e)/10^3/–1.2	Top contact, solution process	109
103	3.6×10^{-3} (h)	Top contact	110
104	0.011 (h)	Top and bottom contact	113
105	0.02 (h) /10^4/–7	Top contact	114
106	0.3 (e) /10^6/60	Top contact	115
107	1.83 (e) /10^4/78	Top contact	116
108	Not observed		
109	0.068 (e) /3×10^4/15	Top contact	117
110	0.04 (e) /5.1×10^3/36 3×10^{-3} (e) /2.7×10^7/41	Top contact Botom contact	118
111a 111a 111b	2×10^{-4} (h) /10^4 5×10^{-4} (h) /10^6 2×10^{-4} (h) /10^5	Cast films (glassy, 140 °C) Cast films (glassy, 180 °C) Vacuum films	119
112	3.6 (h)	Single crystal FET	122
113	0.1 (h) /10^6	Bottom contact	124a
114	0.38 (h) /160/10	Bottom contact	124b
115	0.2 (h) /10^6/36	Top contact	124b
116a 116b	0.094 (h) /10^6 0.4 (h) /10^7	Top contact	19b
117b	0.02 (h)	Top contact, Solution process	125
118	0.081 (h) /5×10^4	Top contact	126
119	10^{-4} (e)	Not reported	127
120a 120b 120c 120d 120e 120f	0.055 (e) 0.6 (e) /10^5 1.3 (e) 0.52 (e) 0.6 (e) 1.5×10^{-5} (e) 1.9×10^{-4} (e) /10^2	Top contact Bottom contact Top contact Top contact Not reported Bottom contact Top and bottom contact	129a 129b 129a 129a 131 129d 91b
121a 121b 121c	0.1 (e) /10^5 0.64 (e) /10^4 0.14 (e)	Top contact Top contact Not reported	133 133 134

Table 5.1. (continued)

compound	Mobility[a]/on/off/voltage	device configuration	reference
122	0.053 (e) /3.4×10^{7}/27	Top contact	135
123a 123b 123c	0.18 (e) /10^{6} 0.349 (e) /3.5×10^{7}/14.3 0.032 (e) /8.1×10^{5}/1.53	Top contact Top contact Top contact	136 137 137
124	3×10^{-3} (e)	Not reported	128
125a 125b 125c 125d	0.16 (e) /10^{5} 0.06 (e) 0.03 (e) 6.2 (e) /6×10^{8}/58	Top contact Top contact Top contact Top contact	138 138 138 139
126	0.15 (e)	Top contact	140
127a 127b 127c	8.6×10^{-6} (e) /10^{4} 4.4×10^{-6} (e) /4×10^{2} 3.1×10^{-6} (e) /80	Bottom contact Bottom contact Bottom contact	141
128	Not reported		142
129	0.013 (e) /10^{4}	Top contact solution processed	143
130	0.002 (e)	Top contact, solution processed	144a
131	0.85 (e) /10^{7}/10	Bottom contact, solution processed	144b
132a 132b	0.05–0.2 (h) 0.22 (h)	Not reported	16a, 149, 150 151
133	1.44×10^{-3} (h)	Top contact	152
134a 134b 134c 134d 134e	0.6 (h) 0.14 (h) /2×10^{6} 0.25 (h) 0.3 (h) /10^{7}/−14.2 0.33 (h) /10^{5}/−10	Not reported Bottom contact Top contact Bottom contact Top contact	155 158 159 160 161
135	0.17 (h) /10^{5}	Bottom contact	163
136	8×10^{-3} (h) /2×10^{7}/−12	Top contact	164
137a 137b	0.28 (h) /10^{6}/24 0.036 (h) /10^{6}/31	Top contact Top contact	165
138	(h) 2×10^{-4}/10^{4}/−10 (e) 2×10^{-4}/10^{4}/5	Top contact Top contact	166
139	0.1 (e)	Not reported	168
140	10^{-6} (e)	Not reported	168
141	0.015 (h) /10^{7}	Top contact	171
142	2×10^{-4} (h) /10^{5}	Top contact	171
143	0.11 (e)	Not reported	174
144	1.5 (h)	Top contact	175a
145	0.3 (e)/10^{6}	Top contact	175c
146a 146b 146c	0.60 (h) 0.40 (h) 0.24 (h)	Top contact Top contact Top contact	176

[a] "h" means hole mobility, while "e" means electron mobility. [b] Reported mobility range under different conditions in different references.

REFERENCES

[1] (a) Inokuchi, H. *Org Elect*, 2006, 7, 62–76. (b) He, G. S.; Tan, L.-S.; Zheng, Q.; Prasad, P. N. *Chem Rev,* 2008, 108, 1245–1330. (c) Hwang, S.-K.; Moorefield, C. N.; Newkome, G. R. Chem Soc Rev, 2008, 37, 2543–2557. (d) Shirota, Y.; Kageyama, H. *Chem Rev,* 2007, 107, 953–1010.

[2] Bao, Z; Locklin, *J. Organic Field Effect Transistors*, CRC Press, Taylor & Francis Group, Boca Raton, London, New York. 2007.

[3] (a) Di, C.; Yu, G.; Liu, Y.; Zhu, D. *J Phys Chem B*. 2007, 111, 14083–14096. (b) Sun, Y.; Liu, Y.; Zhu, D. *J Mater Chem*. 2005, 15, 53–65.

[4] (a) Frechet, J. M. J.; Murphy, A. R. Chem Rev. 2007, 107, 1066–1096. (b) Takimiya, K.; Kunugi, Y.; Ostubo, T. *Chem Lett*. 2007, 36, 578–583.

[5] (a) Wu, Y.; Li, Y.; Ong, B. S. *J Am Chem Soc*. 2006, 128, 4202–4203. (b) Di, C.; Yu, G.; Liu, Y.; Xu, X.; Wei, D.; Song, Y.; Sun, Y.; Wang, Y.; Zhu, D.; Liu, J.; Liu, Y.; Wu, D. *J Am Chem Soc* 2006, 128, 16418–16419.

[6] Du, C.; Guo, Y.; Liu, Y.; Qiu, W.; Zhang, H.; Gao, X.; Liu, Y.; Qi, T.; Lu, K.; Yu, G. *Chem Mater*. 2008, 20, 4188–4190.

[7] Drury, C. J.; Mutsaers, M. J.; Hart, C. M.; Matters, M.; deLeeuw, D. M. *Appl Phys Lett*. 1998, 73, 108.

[8] (a) Lampert, M. A.; Mark, P. Current Injection in Solids. Academic Press. New York, 1970. (b) Mort, J., Pai, D. M. *Photoconductivity and Related Phenomena*. Elsevier, New York, 1976.

[9] An, Z.; Yu, J.; Jones, S. C.; Barlow, S.; Yoo, S.; Domercq, B.; Prins, P.; Siebbeles, L. D. A.; Kippelen, B.; Marder, S. R. *Adv Mater*. 2005, 17, 2580–2583.

[10] (a) Hosokawa, C.; Tokailin, H.; Higashi, H.; Kusumoto, T. *Appl Phys Lett*. 1992, 60, 1220–1222. (b) Hosokawa, C.; Tokailin, H.; Higashi, H.; Kusumoto, T. *Appl Phys Lett*. 1993, 63, 1322–1324. (c) Redecker, M.; Bassler, H.; Horhold, H. H. *J Phys Chem B* 1997, 101, 7398–7403. (d) Wong, T. C.; Kovac, J.; Lee, C. S.; Hung, L. S.; Lee, S. T. *Chem Phys Lett*. 2001, 334, 61–64.

[11] Schouten, P. G.; Warman, J. M.; de Haas, M. P. *J Phys Chem*. 1993, 97, 9863–9870.

[12] (a) Gill, W. D. *J Appl Phys*. 1972, 43, 5033–5040. (b), Frenkel, *J. Phys Rev*. 1938, 54, 647–648.

[13] Schein, L. B.; Mack, J. X. *Chem Phys Lett*. 1988, 149, 109–112.

[14] (a) Bassler, H. Phys Status Solidi B 1981, 107, 9–54. (b) Bassler, H. *Phys Status Solidi B* 1993, 175, 15–56.

[15] Schein, L. B. *Phys Rev B*. 1977, 15, 1024–1034.

[16] (a) Sirringhaus, H.; Brown, P. J.; Friend, R. H.; Nielsen, M. M.; Bechgaard, K.; Langeveld-Voss, B. M. W.; Spiering, A. J. H.; Janssen, R. A. J.; Meijer, E. W.; Herwig, P.; de Leeuw, D. M. *Nature*, 1999, 401, 685–688. (b) Osterbacka, R.; An, C. P. Jiang, X. M. Vardeny, Z.V. *Science*, 2000, 287, 839–842. (c) Zhang, L.; Tan, L.; Wang, Z.; Hu, W.; Zhu, D. *Chem Mater*, 2009, 21, 1993–1999. (d) Tan, L.; Zhang, L.; Jiang, X.; Yang, X.; Wang, L.; Wang, Z.; Li, L.; Hu, W.; Shuai, Z.; Li, L.; Zhu, D. *Adv Funct*

Mater. 2009, 19, 272–276. (e) Lee, K.; Heeger, A. *J. Synth Met.* 2002, 128, 279–282. (f) Izawa, T.; Miyazaki, E.; Takimiya, K. *Adv Mater.* 2008, 20, 3388–3392.

[17] Lovinger, A. J.; Kata, H. E.; Dodabalapur, A. *Chem Mater.* 1998, 10, 3275–3277.

[18] (a) Anthony, J. E.; Brooks, J. S.; Eaton, D. L.; Parkin, S. R. *J Am Chem Soc.* 2001, 123, 9482–9483. (b) Payne, M. M.; Odom, S. A.; Parkin, S. R.; Anthony, J. E. *Org Lett.* 2004, 6, 3325–3328. (c) Payne, M. M.; Parkin, S. R.; Anthony, J. E.; Kuo, C.–C.; Jackson, T. N. *J Am Chem Soc.* 2005, 127, 4986–4987.

[19] (a) Gao, X.; Wu, W.; Liu, Y.; Qiu, W.; Sun, X.; Yu, G.; Zhu, D. *Chem Comm.* 2006, 2750–2752. (b) Gao, X.; Wang, Y.; Yang, X.; Liu, Y.; Qiu, W.; Wu, W.; Zhang, H.; Qi, T.; Liu, Y.; Lu, K.; Du, C.; Shuai, Z.; Yu, G.; Zhu, D. *Adv Mater.* 2007, 19, 3037–3042.

[20] Kelly, T. W.; Baude, P. F.; Gerlach, C.; Ender, D. E.; Muyres, D.; Haase, M. A.; Vogel, D. E.; Theiss, S. D. *Chem Mater.* 2004, 16, 4413–4422.

[21] Valiyev, F.; Hu, W.-S.; Chen, H.-Y.; Kuo, M.-Y.; Chao, I.; Tao, Y.-T. *Chem Mater.* 2007, 19, 3018–3026.

[22] see early review papers about FET materials: (a) Katz, H. E.; Bao, Z.; Gilat, S. *Acc Chem Res.* 2001, 34, 359–369. (b) Murphy, A. R.; Frechet, J. M. *Chem Rev.* 2007, 107, 1066–1096. (c) Zaumseil, J.; Sirringhaus, H. *Chem Rev.* 2007, 107, 1296–1323.

[23] (a) Marseglia, E. A.; Grepioni, F.; Tedesco, E.; Braga, D. *Mol Cryst Liq Cryst.* 2000, 348, 137–151. (b) Barbarella, G.; Zambianchi, M.; Bongini, A.; Antolini, L. *Adv Mater.* 1993, 5, 834–838.

[24] Ito, K.; Suzuki, T.; Sakamoto, Y.; Kuboto, D.; Inoue, Y.; Sato, F.; Tokito, S. *Angew Chem Int Ed.* 2003, 42, 1159–1162.

[25] (a) Ando, S.; Nishida, J.; Fujiwara, E.; Tada, H.; Inoue, Y.; Tokito, S.; Yamashita, Y. *Chem Mater.* 2005, 17, 1261–1264. (b) Meng, H.; Sun, F.; Goldfinger, M. B.; Jaycox, G. D.; Li, Z.; Marshall, W. J.; Blackman, G. S. *J Am Chem Soc.* 2005, 127, 2406–2407. (c) Meng, H.; Sun, F.; Goldfinger, M. B.; Gao, F.; Londono, D. J.; Marshal, W. J.; Blackman, G. S.; Dobbs, K. D.; Keys, D. E. *J Am Chem Soc.* 2006, 128, 9304–9305. (d) Klauk, H.; Zschieschang, U.; Weitz, R. T.; Meng, H.; Sun, F.; Nunes, G.; Keys, D. E.; Fincher, C. R.; Xiang, Z. *Adv Mater.* 2007, 19, 3882–3887. (e) Kim, H. S.; Kim, Y.-H.; Kim, T.-H.; No, Y.-Y.; Pyo, S.; Yi, M. H.; Kim, D. Y.; Kwon, S.-K. *Chem Mater.* 2007, 19, 3561–3567.

[26] Laquindanum, J. G.; Katz, H. E.; Lovinger, A. J.; Dodabalapur, A. *Adv Mater.* 1997, 9, 36–39.

[27] Takimiya, K. Kunugi, Y.; Konda, Y.; Niihara, N.; Otsubo, T. *J Am Chem Soc.* 2004, 126, 5084–5085.

[28] Takimiya, K.; Kunugi, Y.; Ebata, H.; Otsubo, T. *Chem Lett.* 2006, 35, 1200–1201.

[29] Takimiya, K.; Konda, Y.; Ebata, H.; Otsubo, T.; Kunugi, Y. *Mol Cryst Liq Cryst.* 2006, 455, 361–365.

[30] Gao, J.; Li, L.; Meng, Q.; Li, R.; Jiang, H.; Li, H.; Hu, W. *J Mater Chem.* 2007, 17, 1421–1426.

[31] Li, X-C.; Sirringhaus, H.; Garnier, F.; Holmes, A. B.; Moratti, S. C.; Feeder, N.; Clegg, W.; Teat, S. J.; Friend, R. H. *J Am Chem Soc.* 1998, 120, 2206–2207.

[32] Sun, Y.; Ma, Y.; Liu, Y.; Lin, Y.; Wang, Z.; Wang, Y.; Di, C.; Xiao, K.; Chen, X.; Qiu, W.; Zhang, B.; Yu, G.; Hu, W.; Zhu, D. *Adv Funct Mater*. 2006, 16, 426–432.

[33] Zhang, S.; Guo, Y.; Xi, H.; Di, C.; Yu, J.; Zheng, K.; Liu, R.; Zhan, X.; Liu, Y. *Thin Solid Films*. 2009, 517, 2968–2973.

[34] Park, J.-H.; Chung, D. S.; Park, J.-W.; Ahn, T.; Kong, H.; Jung, Y. K.; Lee, J.; Yi, M. H.; Park, C. E.; Kwon, S.-K.; Shim, H.-K. *Org Lett*. 2007, 9, 2573–2576.

[35] Robertson, J. M.; Sinclair, V. C.; Trotter, *J. Acta Cryst*. 1961, 14, 697.52.

[36] Cicoira, F.; Santato, C.; Dinelli, F.; Murgia, M.; Loi, M.; Biscarini, F.; Zamboni, R.; Heremans, P.; Muccini, M. *Adv Func Mater*, 2005, 15, 375–380.

[37] (a) Moon, H.; Zeis, R.; Brokent, E.-J.; Besnard, C.; Lovinger, A. J.; Siegrist, T.; Kloc, C.; Bao, Z. *J Am Chem Soc*. 2004, 126, 15322–15323. (b) Chi, X.; Li, D.; Zhang, H.; Chen, Y.; Garcia, V.; Garcia, C.; Siegrist, T. *Org Elect*. 2008, 9, 234–240.

[38] Uno, M.; Tominari, Y.; Takeya, *J. Org Elect*. 2008, 9, 753–756.

[39] Chen, Z.; Muller, P.; Swager, T. M. *Org Lett*. 2006, 8, 273–276.

[40] Sakamoto, Y.; Suzuki, T.; Kobayashi, M.; Gao, Y.; Inoue, Y.; Tokito, S. *Mol Cryst Liq Cryst*. 2006, 444, 225–232.

[41] (a) Takimiya, K.; Kunugi, Y.; Konda, Y.; Ebata, H.; Toyoshima, Y.; Otsubo, T. *J Am Chem Soc*. 2006, 128, 3044–3050. [b] Ebata, H.; Izawa, T.; Miyazayi, E.; Takimiya, K.; Ikeda, M.; Kuwabara, H.; Yui, T. *J Am Chem Soc*. 2007, 129, 15732–15733. (c) Izawa, T.; Miyazaki, E.; Takimiya, K. *Adv Mater*. 2008, 20, 3388–3392.

[42] Tang, M. L.; Okamoto, T.; Bao, Z. *J Am Chem Soc*. 2006, 128, 16002–16003.

[43] Liu, Y.; Wang, Y.; Wu, W.; Liu, Y.; Xi, H.; Wang, L.; Qiu, W.; Lu, K.; Du, C.; Yu, G. *Adv Funct Mater*. 2009, 19, 772–778.

[44] See recent reports about pentacene: (a) Nelson, S. F.; Lin, Y. Y.; Gundlach, D. J.; Jackson, T. N. *Appl Phys Lett*. 1998, 72, 1854–1856. (b) Gundlach, D. J.; Lin, Y. Y.; Jackson, T. N.; Nelson, S. F.; Schlom, D. G. *IEEE Electron Device Lett*. 1997, 18, 87–89. (c) Klauk, H.; Halik, M.; Zschieschang, U.; Schmid, G.; Radlik, W.; Weber, W. *J Appl Phys*. 2002, 92, 5259–5263. (d) Klauk, H.; Halik, M.; Zschieschang, U.; Eder, F.; Schmid, G.; Dehm, C. *Appl Phys Lett*. 2003, 82, 4175–4177. (e) Volkel, A. R.; Street, R. A.; Knipp, D. *Phys Rev B* 2002, 66, 195336–195343.

[45] Kelley, T. W.; Muyres, D. V.; Baude, P. F.; Smith, T. P.; Jones, T. D. *Mater Res Soc Symp Proc*. 2003, 771, L6.5.

[46] (a) Campbell, R. B.; Monteath Robertson, J.; Trotter, *J. Acta Cryst*. 1961, 14, 705–711. (b) Mattheus, C. C.; Dros, A. B.; Baas, J.; Meetsma, A.; de Boer, J. L.; Palstra, T. T. M. *Acta Cryst*. 2001, C57, 939–941.

[47] Yamada, M.; Ikemoto, I.; Kuroda, H. *Bull Chem Soc Jpn*. 1988, 61, 1057–1062.

[48] Meng, H.; Bendikov, M.; Mitchell, G.; Helgeson, R.; Wudl, F.; Bao, Z.; Siegrist, T.; Kloc, C.; Chen, C.-H. *Adv Mater*, 2003, 15, 1090–1093.

[49] Okamoto, T.; Senatore, M. L.; Ling, M.-M.; Mallik, A. B.; Tang, M. L.; Bao, Z. *Adv Mater*. 2007, 19, 3381–3384.

[50] Sheraw, C. D.; Jackson, T. N.; Eaton, D. L.; Anthony, J. E. *Adv Mater*. 2003, 15, 2009–2011.

[51] See recent papers: (a) Anthony, J. E. *Chem Rev*. 2006, 106, 5028–5048. (b) Subramanian, S.; Park, S. K.; Parkin, S. R.; Podzorov, V.; Jackson, T. N.; Anthony, J. E. *J Am Chem Soc*. 2008, 130, 2706–2707. (c) Allard, S.; Forster, M.; Souharce, B.;

Thiem, H.; Scherf, U. *Angew Chem Int Ed*. 2008, 47, 4070–4098. (d) Anthony, J. E. *Angew Chem Int Ed*. 2008, 47, 452–483. (e) Lim, J. A.; Lee, H. S.; Lee, W. H.; Cho, K. *Adv Funct Mater*. 2009, 19, 1515–1525. (f) Tang, M. L.; Reichardt, A. D.; Miyaki, N.; Stoltenberg, R. M.; Bao, Z. *J Am Chem Soc*. 2008, 130, 6064–6065.

[52] Sakamoto, Y.; Suzuki, T.; Kobayashi, M.; Gao, Y.; Fukai, Y.; Inoue, Y.; Sato, F.; Tokito, S. *J Am Chem Soc*. 2004, 126, 8138–8140.

[53] Laquindanum, J. G.; Katz, H. E.; Lovinger, A. J. *J Am Chem Soc*. 1998, 120, 664–672.

[54] Katz, H. E.; Li, W.; Lovinger, A. J.; Laquindanum, *J. Synth Met*. 1999, 102, 897–899.

[55] Gao, P.; Beckmann, D.; Tsao, H. N.; Feng, X.; Enkelmann, V.; Pisula, W.; Mullen, K. *Chem Comm*. 2008, 1548–1550.

[56] Ebata, H.; Miyazaki, E.; Yamamoto, T.; Takimiya, K. *Org Lett*. 2007, 9, 4499–4502.

[57] Xiao, K.; Liu, Y.; Qi, T.; Zhang, W.; Wang, F.; Gao, J.; Qiu, W.; Ma, Y.; Cui, G.; Chen, S.; Zhan, X.; Yu, G.; Qin, J.; Hu, W.; Zhu, D. *J Am Chem Soc*. 2005, 127, 13281–13286.

[58] Gao, J.; Li, R.; Li, L.; Meng, Q.; Jiang, H.; Li, H.; Hu, W. *Adv Mater*. 2007, 19, 3008–3011.

[59] Ma, Y.; Sun, Y.; Liu, Y.; Gao, J.; Chen, S.; Sun, X.; Qiu, W.; Yu, G.; Cui, G.; Hu, W.; Zhu, D. *J Mater Chem*. 2005, 15, 4894–4898.

[60] (a) Qi, T.; Qiu, W.; Liu, Y.; Zhang, H.; Gao, X.; Liu, Y.; Lu, K.; Du, C.; Yu, G.; Zhu, D. *J Org Chem*. 2008, 73, 4638–4643. (b) Qi, T.; Guo, Y.; Liu, Y.; Xi, H.; Zhang, H.; Gao, X.; Liu, Y.; Lu, K.; Du, C.; Yu, G.; Zhu, D. *Chem Comm*. 2008, 6227–6229.

[61] (a) Bailey, W. J.; Liao, C.-W. *J Am Chem Soc*. 1955, 77, 992–993. (b) Satchell, M. P.; Stacey, B. E. *J Chem Soc C*. 1971, 3, 468. (c) Angliker, H.; Rommel, E.; Wirz, *J. Chem Phys Lett*. 1982, 87, 208–212.

[62] Fang, T. *Heptacene, Octacene, Nonacene, Supercene and Related Polymers*. Ph D, Thesis, University of California, Los Angeles, CA, 1986.

[63] (a) Mondal, R.; Adhikari, R. M.; Shah, B. K.; Neckers, D. C. *Org Lett*. 2007, 9, 2505–2508. (b) Mondal, R.; Shah, B. K.; Neckers, D. C. *J Am Chem Soc*. 2006, 128, 9612–9613.

[64] Payne, M. M.; Parkin, S. R.; Anthony, J. E. *J Am Chem Soc*. 2005, 127, 8028–8029.

[65] Tang, M. L.; Mannsfeld, S. C. B.; Sun, Y.-S.; Becerril, H. A.; Bao, Z. *J Am Chem Soc*. 2009, 131, 882–883.

[66] Yamamoto, T.; Takimiya, K. *J Am Chem Soc*. 129, 2224–2225.

[67] Okamoto, T.; Kudoh, K.; Wakamiya, A.; Yamaguchi, S. *Chem Eur J*. 2007, 13, 548 – 556.

[68] Payne, M. M.; Parkin, S. R.; Anthony, J. E. *J Am Chem Soc*. 2005, 127, 8028–8029.

[69] Chun, D.; Cheng, Y.; Wudl, F. *Angew Chem Int Ed*. 2008, 47, 8380–8385.

[70] Kaur, I.; Stein, N. N.; Kopreski, R. P.; Miller, G. P. *J Am Chem Soc*. 2009, 131, 3424–3425.

[71] Sirringhaus, H.; Friend, R. H.; Wang, C.; Leuninger, J.; Mullen, K. *J Mater Chem*. 1999, 9, 2095–2101.

[72] (a) Okamoto, T.; Kudoh, K.; Wakamiya, A.; Yamaguchi, S. *Org Lett*. 2005, 7, 5301–5304. (b) Yamada, K.; Okamoto, T.; Kudoh, K.; Wakamiya, A.; Yamaguchi, S.; Takeya, *J. Appy Phys Lett*. 2007, 90, 072102–072104. (c) Yoshida, H.; Watazu, Y.; Sato, N.; Okamoto, T.; Yamaguchi, S. *Appl Phys A*. 2009, 95, 185–191.

[73] Zhang, X.; Cote, A. P.; Matzger, A. J. *J Am Chem Soc*. 2005, 127, 10502–10503.
[74] Bendikov, M.; Houk, K. N.; Duong, H. M.; Starkey, K.; Carter, E. A.; Wudl, F. *J Am Chem Soc*. 2004, 126, 7416–7417.
[75] For an excellent recent review of acenes, along with a detailed description of the issue surrounding higher acenes, see: Bendikov, M.; Wudl, F.; Perepichka, D. F. *Chem Rev*. 2004, 104, 4891–4945.
[76] Brbarella, G.; Melucci, M.; Sotgiu, G. *Adv Mater*. 2005, 17, 1581–1593.
[77] Garnier, F. Hajlaoui, R.; Yassar, A.; Srivastava, P. *Science*, 1994, 265, 1684–1686.
[78] Anthony, J. E.; Heeney, M.; Ong, B. S. *MRS Bulletin*, 2008, 33, 698–705.
[79] Hajlaoui, R.; Horowitz, G.; Garnier, F.; Arce-Brouchet, A.; Laigre, L.; El, Kassmi, A.; Demanze, F.; Kouki, F. *Adv Mater*. 1997, 9, 389–391.
[80] Dodabalapur, A.; Torsi, L.; Katz, H. E. *Science* 1995, 268, 270–271.
[81] Hajlaoui, R.; Fichou, D.; Horowitz, G.; Nessakh, B.; Constant, M.; Garnier, F. *Adv Mater*. 1997, 9, 557–561.
[82] (a)Lovinger, A. J.; Davis, D. D.; Dodabalapur, A.; Katz, H. E.; Torsi, L. Macromolecules 1996, 29, 4952–4957. (b) Lovinger, A. J.; Davis, D. D.; Dodabalapur, A.; Katz, H. E. *Chem Mater*. 1996, 8, 2836–2838.
[83] (a)Halik, M.; Klauk, H.; Zschieschang, U.; Schmid, G.; Ponomarenko, S.; Kirchmeyer, S.; Weber, W. *Adv Mater*. 2003, 15, 917–922. (b) Garnier, F. Horowitz, G.; Peng, X. Z.; Fichou, D. *Synth Met*. 1991, 45, 163–171.
[84] Horowitz, G. Garnier, F.; Yassar, A.; Hajlaoui, R.; Kouki, F. *Adv Mater*, 1996, 8, 52–54.
[85] Hajlaoui, M. E.; Garnier, F.; Hassine, L.; Kouki, F.; Bouchriha, H. *Synth Met*. 2002, 129, 215–220.
[86] (a)Katz, H. E.; Lovinger, A. J.; Laquindanum, J. G. *Chem Mater*. 1998, 10, 457–459. (b) Li,W.; Katz, H. E.; Lovinger, A. J.; Laquindanum, J. G. *Chem Mater*. 1999, 11, 458–465.
[87] Garnier, F.; Yassar, A.; Hajlaoui, R.; Horowitz, G.; Deloffre, F.; Servet, B.; Ries, S.; Alnot, P. *J Am Chem Soc*. 1993, 115, 8716–8721.
[88] Hajlaoui, R.; Fichou, D.; Horowitz, G.; Nessakh, B.; Constant, M.; Garnier, F. *Adv Mater*. 1997, 9, 557–561.
[89] Katz, H. E.; Dodabalapur, A.; Torsi, L.; Elder, D. *Chem Mater*. 1995, 7, 2238–2240.
[90] (a)Ponomarenko, S.; Kirchmeyer, S. *J Mater Chem*. 2003, 13, 197–202. (b) Halik, M.; Klauk, H.; Zschieschang, U.; Schmid, G.; Radlik, W.; Ponomarenko, S.; Kirchmeyer, S.; Weber, W. *J Appl Phys*. 2003, 93, 2977–2981. (c) Ponomarenko, S. A.; Kirchmeyer, S.; Halik, M.; Klauk, H.; Zschieschang, U.; Schmid, G.; Karbach, A.; Drechsler, D.; Alpatova, N. M. *Synth Met*. 2005, 149, 231–235.
[91] (a)Meng, H.; Zheng, J.; Lovinger, A. J.; Wang, B.-C. Patten, P. G. V.; Bao, Z. *Chem Mater*. 2003, 15, 1778–1787. (b) Locklin, J.; Li, D.; Mannsfeld, S. C. B.; Borkent, E.-J.; Meng, H.; Advincula, R.; Bao, *Z. Chem Mater*. 2005, 17, 3366–3374.
[92] Deman, A.-L.; Tardy, J.; Nicolas, Y.; Blanchard, P.; Roncali, *J. Synth Met*. 2004, 146, 365–371.
[93] Nicolas, Y.; Blanchard, P.; Roncali, J.; Allain, M.; Mercier, N.; Deman, A.-L.; Tardy, *J. Org Lett*. 2005, 7, 3513–3516.

[94] Merlo, J. A.; Newman, C. R.; Gerlach, C. P.; Kelly, T. W.; Muyres, D. V.; Fritz, S. E.; Toney, M. F.; Frisbie, C. D. *J Am Chem Soc*. 2005, 127, 3997–4009.

[95] Tian, H.; Shi, J.; Yan, D.; Wang, L.; Geng, Y.; Wang, F. *Adv Mater*. 2006, 18, 2149–2152.

[96] Bisri, S. Z.; Takenobu, T.; Yomogida, Y.; Shimotani, H.; Yamao, T.; Hotta, S.; Iwasa, Y. *Adv Funct Mater*. 2009, 19, **1728–1735.**

[97] **(a)**Facchetti, A.; Deng, Y.; Wang, A.; Koide, Y.; Sirringhaus, H.; Marks, T. J.; Friend, R. H. *Angew Chem Int Ed.* 2000, 39, 4547**–4551**. (b) Facchetti, A.; Mushrush, M.; Katz, H. E.; Marks, T. *J. Adv Mater*, 2003, 15, 33–38.

[98] Facchetti, A.; Yoon, M.-H.; Stern, C. L.; Katz, H. E.; Marks, T. J. *Angew Chem Int Ed.* 2003, 42, 3900–3903.

[99] (a)Sakamoto, Y.; Komatsu, S.; Suzuki, T. *J Am Chem Soc.* 2001, 123, 4643–4644. (b) Osuna, R. M.; Ortiz, R. P.; Delgado, M. C. R.; Sakamoto, Y.; Suzuki, T.; Hernandez, V.; Navarrete, J. T. L. *J Phys Chem B* 2005, 109, 20737–20745.

[100] Videlot-Ackermann, C.; Ackermann, J.; Brisser, H.; Kawamura, K.; Yoshimoto, N.; Raynal, P.; Kassmi, A. E.; Fages, F. *J Am Chem Soc.* 2005, 127, 16346–16347.

[101] Meng, H.; Bao, Z.; Lovinger, A. J.; Wang, B.-C.; Mujsce, A. M. *J Am Chem Soc*. 2001, 123, 9214–9215.

[102] Didane, Y.; Mehl, G. H.; Kumagai, A.; Yoshimoto, N.; Videlot-Ackermann, C.; Brisset, H. *J Am Chem Soc.* 2008, 130, 17681–17683.

[103] Dimitrakopoulos, C. D.; Afzali-Ardakani, A.; Furman, B.; Kymissis, J.; Purushothaman, S. *Synth Met*. 1997, 89, 193–197.

[104] (a)Frere, P.; Raimundo, J. M.; Blanchard, P.; Delaunay, J.; Richomme, P.; Sauvajol, J. L.; Orduna, J.; Garin, J.; Roncali, J. *J Org Chem*. 2003, 68, 7254–7265. (b) Videlot, C.; Ackermann, J.; Blanchard, P.; Raimundo, J.-M.; Frere, P.; Allain, M.; Bettignies, R.; Levillain, E.; Roncali, *J. Adv Mater*. 2003, 15, 306–310.

[105] Ostoja, P.; Maccagnani, P.; Gazzano, M.; Cavallini, M.; Kengne, J. C.; Kshirsagar, R.; Biscarini, F.; Melucci, M.; Zambianchi, M.; Barbarella, G. *Synth Met.* 2004, 146, 243–250.

[106] Breemen, A. J. J. M.; Herwig, P. T.; Chlon, C. H. T.; Sweelssen, J.; Schoo, H. F. M.; Setayesh, S.; Hardeman, W. M.; Martin, C. A.; Leeuw, D. M.; Valeton, J. J. P.; Bastiaansen, C. W. M.; Broer, D. J.; Popa-Merticaru, A. R.; Meskers, S. C. *J Am Chem Soc.* 2006, 128, 2336–2345.

[107] Meng, Q.; Gao, J.; Li, R.; Jiang, L.; Wang, C.; Zhao, H.; Liu, C.; Li, H.; Hu, W. *J Mater Chem.* 2009, 19, 1477–1482.

[108] (a)Pappenfus, T. M.; Chesterfield, R. J.; Frisbie, C. D.; Mann, K. R.; Casado, J.; Raff, J. D.; Miller, L. L. *J Am Chem Soc*. 2002, 124, 4184–4185. (b) Chesterfield, R. J.; Newman, C. R.; Pappenfus, T. M.; Ewbank, P. C.; Haukaas, M. H.; Mann, K. R.; Miller, L. L.; Frisbie, C. D. *Adv Mater*. 2003, 15, 1278–1282.

[109] Handa, S.; Miyazaki, E.; Takimiya, K.; Kunugi, Y. *J Am Chem Soc*. 2007, 129, 11684–11685.

[110] Kunugi, Y.; Takimiya, K.; Yamane, K.; Yamashita, K.; Aso, Y.; Otsubo, T. *Chem Mater.* 2003, 15, 6–7.

[111] Mitschke, U.; Debaerdemaeker, T.; Bauerle, P. *Eur J Org Chem*. 2000, 3, 425–437.

[112] Hong, X. M.; Katz, H. E.; Lovinger, A. J.; Wang, B.-C. Raghavachari, K. *Chem Mater.* 2001, 13, 4686–4691.
[113] Li, W.; Katz, H. E.; Lovinger, A. J.; Laquindanum, J. G. *Chem Mater*. 1999, 11, 458–465.
[114] Ando, S.; Nishida, J.; Inoue, Y.; Tokito, S.; Yamashita, Y. *J Mater Chem*. 2004, 14, 1787–1790.
[115] Ando, S.; Nishida, J.; Tada, H.; Inoue, Y.; Tokito, S.; Yamashita, Y. *J Am Chem Soc.* 2005, 127, 5336–5337.
[116] Ando, S.; Murakami, R.; Nishida, J.; Tada, H.; Inoue, Y.; Tokito, S.; Yamashita, Y. *J Am Chem Soc.* 2005, 127, 14996–14997.
[117] Akhtaruzzaman, M.; Kamata, N.; Nishida, J.; Ando, S.; Tada, H.; Tomura, M.; Yamashita, Y. *Chem Comm.* 2005, 3183–3185.
[118] Kojima, T.; Nishida, J.; Tokito, S.; Tada, H.; Yamashita, Y. *Chem Comm*. 2007, 1430–1432.
[119] Melucci, M.; Favaretto, L.; Bettini, C.; Gazzano, M.; Camaioni, N.; Maccagnani, P.; Ostoja, P.; Monari, M.; Barbarella, G. *Chem Eur J.* 2007, 13, 10046–10054.
[120] (a)Segura, J. L.; Martin, N. *Angew Chem Int Ed.* 2001, 40, 1372–1409. (b) Bryce, M. R. *J Mater Chem*, 2000, 10, 589–598. (c) Williams, J. M.; Ferraro, J. R.; Thorn, R. J.; Carlson, K. D.; Geiser, U.; Wang, H. H.; Kini A. M.; Whangbo, M.-H. *Organic Superconductors (Including Fullerenes): Synthesis, Structure, Properties, and Theory,* Prentice Hall, Englewood Cliffs, New Jersey, 1992.
[121] Mas-Torrent, M.; Rovira, C. *J Mater Chem*. 2006, 16, 433–436.
[122] Leufgen, M.; Rost, O.; Gould, C.; Schmidt, G.; Geurts, J.; Molenkamp, L. W.; Oxtoby, N. S.; Mas-Torrent, M.; Crivillers, N.; Veciana, J.; Rovira, C. *Org Elect*. 2008, 9, 1101–1106.
[123] Bromley, S. T.; Hadley, P.; Rovira, C. *J Am Chem Soc*. 2004, 126, 6544–6545. (b) Mas-Torrent, M.; Hadley, P.; Bromley, S. T.; Ribas, X.; Tarres, J.; Mas, M.; Molins, E.; Veciana, J.; Rovira, C. *J Am Chem Soc*. 2004, 126, 8546–8553.
[124] (a)Mas-Torrent, M.; Hadley, P. *Appl Phys Lett*. 2005, 86, 012110/1–012110/3. (b*) J Am Chem Soc*. 2005, 127, 10142–10143.
[125] Gao, X.; Wu, W.; Liu, Y.; Qiu, W.; Sun, X.; Yu, G.; Zhu, D. *Chem Comm*. 2006, 2750–2752.
[126] Aleveque, O.; Frere, P.; Leriche, P.; Breton, T.; Cravino, A.; Roncali, J. *J Mater Chem*. 2009, 19, 3648–3651.
[127] Ostrick, J. R.; Dodabalapur, A.; Torsi, L.; Lovinger, A. J.; Kwock, E. W.; Miller, T. M.; Galvin, M.; Berggren, M.; Katz, H. E. *J Appl Phys*. 1997, 81, 6804–6808.
[128] Laquindanum, J. G. Katz, H. E.; Dodabalapur, A.; Lovinger, A. J. *J Am Chem Soc*. 1996, 118, 11331–11332.
[129] (a)Chesterfield, R. J.; Mckeen, J. C.; Newman, C. R.; Ewbank, P. C.; da Silva Filho, D. A.; Bredas, J.-L.; Miller, L. L.; Mann, K. R.; Frisbie, C. D. *J Phys Chem B* 2004, 108, 19281–19292. (b) Malenfant, P. R. L.; Dimitrakopoulos, C. D.; Gelorme, J. D.; Kosbar, L. L.; Graham, T. O.; Curioni, A.; Andreoni, W. *Appl Phys Lett*. 2002, 80, 2517–2519. (c) Horowitz, G.; Kouki, F.; Spearman, P.; Fichou, D.; Nogues, C.; Pan, X.; Garnier, F. *Adv Mater*. 1996, 8, 242–245.

[130] (a)Hiramoto, M.; Ihara, K.; Fukusumi, H.; Yokoyama, M. *J Appl Phys.* 1995, 78, 7153–7157. (b) Gundlach, D. J.; Pernstich, K. P.; Wilckens, G.; Gruter, M.; Haas, S.; Batlogg, B. *J Appl Phys*. 2005, 98, 064502/1–064502/8.

[131] Tatemichi, S.; Ichikawa, M.; Koyama, T.; Taniguchi, Y. *Appl Phys Lett.* 2006, 89, 112108/1–112108/3.

[132] Wen, Y.; Liu, Y.; Di, C.; Wang, Y.; Sun, X.; Guo, Y.; Zheng, J.; Wu, W.; Ye, S.; Yu, G. *Adv Mater*. 2009, 21, 1631–1635.

[133] Jones, B.; Ahrens, M. J.; Yoon, M.-H.; Facchetti, A.; Marks, T. J.; Wasielewski, M. R. *Angew Chem Int Ed.* 2004, 43, 6363–6366.

[134] Yoo, B.; Jung, T.; Basu, D.; Dodabalapur, A. *Appl Phys Lett.* 2006, 88, 082104/1–082104/3.

[135] Li, Y.; Tan, L.; Wang, Z.; Qian, H.; Shi, Y.; Hu, W. *Org Lett.* 2008, 10, 529–532.

[136] Ling, M.-M.; Erk, P.; Gomez, M.; Koenemann, M.; Locklin, J.; Bao, Z. *Adv Mater.* 2007, 19, 1123–1127. (b) Schmidt, R.; Oh, J. H.; Sun, Y.-S.; Deppisch, M.; Krause, A.-M.; Radacki, K.; Braunschweig, H.; Konemann, M.; Erk, P.; Bao, Z.; Wurthner, F. *J Am Chem Soc*. 2009, 131, 6215–6228.

[137] Schmidt, R.; Ling, M. M.; Oh, J. H.; Winkler, M.; Konemann, M.; Bao, Z.; Wurthner, F. *Adv Mater*. 2007, 19, 3692–3695.

[138] Katz, H. E.; Lovinger, A. J.; Johnson, J.; Kloc, C.; Siegrist, T.; Li, W.; Lin, Y.-Y. Dodabalapur, A. *Nature*, 2000, 404, 478–481.

[139] Shukla, D.; Nelson, S. F.; Freeman, D. C.; Rajeswaran, M.; Ahearn, W. G.; Meyer, D. M.; Carey, J. T. *Chem Mater.* 2008, 20, 7486–7491.

[140] Jones, B. A.; Facchetti, A.; Marks, T. J.; Wasielewski, M. R.; *Chem Mater*. 2007, 19, 2703–2705.

[141] Kruger, H.; Janietz, S.; Sainova, D.; Dobreva, D.; Koch, N.; Vollmer, A. *Adv Funct Mater.* 2007, 17, 3715–3723.

[142] Gao, X.; Qiu, W.; Yang, X.; Liu, Y.; Wang, Y.; Zhang, H.; Qi, T.; Liu, Y.; Lu, K.; Du, C.; Shuai, Z.; Yu, G.; Zhu, D. *Org Lett*. 2007, 9, 3917–3920. (b) Rger, C.; Wrthner, F. *J Org Chem.* 2007, 72, 8070–8075.

[143] Zhan, X.; Tan, Z.; Domercq, B.; An, Z.; Zhang, X.; Barlow, S.; Li, Y.; Zhu, D.; Kippelen, B.; Marder, S. R. *J Am Chem Soc*. 2007, 129, 7246–7247.

[144] (a)Chen, Z.; Zheng, Y.; Yan, H.; Facchetti, A. J Am Chem Soc. 2009, 131, 8–9. (b) Yan, H.; Chen, Z.; Zheng, Y.; Newman, C.; Quinn, J. R.; Dotz, F.; Kastler, M.; Facchetti, A. *Nature*, 2009, 457, 679–687. (c) Guo, X.; Watson, M. D. *Org Lett*. 2008, 10, 5333–5336.

[145] Ong, B. S.; Wu, Y.; Jiang, L.; Murti, K. Synth Met. 2004, 142, 49–52. (b) Ong, B. S.; Wu, Y.; Liu, P.; Gardner, S. *J Am Chem Soc*. 2004, 126, 3378–3379.

[146] (a)Roncali, J. Chem Rev. 1992, 92, 711–738. (b) Gurunathan, K.; Murugan, A. V.; Marimuthu, R.; Mulik, U. P.; Amalnerkar, D. P. *Mater Chem Phys*. 1999, 61, 173–191. (c) Yalcinkaya, S.; Tuken, T.; Yazici, B.; Erbil, M.; *Prog Org Coat*. 2008, 62, 236–244.

[147] (a)Yamamoto, T.; Morita, A.; Miyazaki, Y.; Maruyama, T.; Wakayama, H.; Zhou, Z.; Nakamura, Y.; Kanbara, T.; Sasaki, S.; Kubota, K. *Macromolecules* 1992, 25, 1214–1223. (b) Yamamoto, T. J *Organomet Chem.* 2002, 653, 195–199.

[148] (a)McCullough, R. D.; Lowe, R. D. *Chem. Comm.* 1992, 70–72. (b) Loewe, R. S.; Khersonsky, S. M.; McCullough, R. D. *Adv Mater.* 1999, 11, 250–253. (c) Bolognesi, A.; Porzio, W.; Bajo, G.; Zannoni, G.; Fannig, L. *Acta Polym.* 1999, 50, 151–155. (d) Chen, T.-A.; Rieke, R. D. *J Am Chem Soc.* 1992, 114, 10087–10088. (e) Chen, T.-A.; Wu, X.; Rieke, R. D. *J Am Chem Soc.* 1995, 117, 233–244. (f) Liversedge, I. A.; Higgins, S. J.; Giles, M.; Heeney, M.; McCulloch, I. *Tetra Lett.* 2006, 47, 5143–5146.
[149] Sirringhaus, H. Tessler, N.; Friend, R. H. *Science*, 1998, 280, 1741–1744.
[150] (a)Chang, J. F.; Sun, B.; Breiby, D. W.; Nielsen, M. M.; Solling, T. I.; Giles, M.; McCulloch, I.; Sirringhaus, H. *Chem Mater.* 2004, 16, 4772–4776. (b) Wang, G.; Swensen, J.; Moses, D.; Heeger, A. J. *J Appl Phys.* 2003, 93, 6137–6141. (c) Surin, M.; Leclere, P.; Lazzaroni, R.; Yuen, J. D.; Wang, G.; Moses, D.; Heeger, A. J.; Cho, S.; Lee, K. *J Appl Phys.* 2006, 100, 033712/1–033712/6.
[151] *Proceedings of the Society of Photo-optical Instrumentation Engineers.* 2007, 6658, 65810–65810.
[152] Huang, Y.; Wang, Y.; Sang, G.; Zhou, E.; Huo, L.; Liu, Y.; Li, Y. *J Phys Chem B* 2008, 112, 13476–13482.
[153] (a)Wang, Y.; Zhou, E.; Liu, Y.; Xi, H.; Ye, S.; Wu, W.; Guo, Y.; Di, C.; Sun, Y.; Yu, G.; Li, Y. *Chem Mater* 2007, 19, 3361–3363. (b) Lu, K.; Sun, X.; Liu, Y.; Di, C.; Xi, H.; Yu, G.; Gao, X.; Du, C. *J Poly Sci A Poly Chem.* 2009, 47, 1381–1392. (c) Hou, J.; Tan, Z.; Yan, Y.; He, Y.; Yang, C.; Li, Y. *J Am Chem Soc.* 2006, 128, 4911–4916. (d) Li, Y.; Zou, Y. *Adv Mater.* 2008, 20, 2952–2958.
[154] (a)McCulloch, I.; Bailey, C.; Giles, M.; Heeney, M.; Love, I.; Shkunov, M.; Sparrowe, D.; Tierney, S. *Chem Mater.* 2005, 17, 1381–1385. (b) Heeney, M.; Bailey, C.; Genevicius, K.; Shkunov, M.; Sparrowe, D.; Tierney, S.; McCulloch, I. *J Am Chem Soc.* 2005, 127, 1078–1079.
[155] McCulloch, I.; Heeney, M.; Bailey, C.; Genevicius, K.; Macdonald, I.; Shkunov, M.; Sparrowe, D.; Tierney, S.; Wagner, R.; Zhang, W.; Chabinyc, M. L.; Kline, R. J.; Mcgehee, M. D.; Toney, M. F. *Nature Mat.* 2006, 328–333.
[156] Hamadani, B. H.; Gundlach, D. J.; McCulloch, I.; Heeney, M. *Appl Phys Lett.* 2007, 91, 243512/1–243512/3.
[157] Zhu, Y.; Champion, R. D.; Jenekhe, S. A. *Macromolecules* 2006, 39, 8712–8719.
[158] Osaka, I.; Sauve, G.; Zhang, R.; Kowalewski, T.; McCullough, R. D. *Adv Mater.* 2007, 19, 4160–4165.
[159] Pan, H.; Li, Y.; Wu, Y.; Liu, P.; Ong, B. S.; Zhu, S.; Xu, G. *J Am Chem Soc.* 2007, 129, 4112–4113.
[160] Li, J.; Qin, F.; Li, C.; Bao, Q.; Chan-Park, M. B.; Zhang, W.; Qin, J.; Ong, B. S. *Chem Mater.* 2008, 20, 2057–2059.
[161] Fong, H. H.; Pozdin, V. A.; Amassian, A.; Malliaras, G. G.; Smilgies, D.-M.; He, M.; Gasper, S.; Zhang, F.; Sorensen, M. *J Am Chem Soc.* 2008, 130, 13202–13203.
[162] Coppo, P.; Turner, M. L. *J Mater Chem.* 2005, 15, 1123–1133.
[163] Zhang, M.; Tsao, H. N.; Pisula, W.; Yang, C.; Mishra, A. K.; Mullen, K. *J Am Chem Soc.* 2007, 129, 3472–3473.
[164] Letizia, J. A.; Salata, M. R.; Tribout, C. M.; Facchetti, A.; Ratner, M. A.; Marks, T. J. *J Am Chem Soc.* 2008, 130, 9679–9694.

[165] Guo, X.; Kim, F. S.; Jenekhe, S. A.; Watson, M. D. *J Am Chem Soc*. 2009, 131, 7206–7207.

[166] (a)Usta, H.; Facchetti, A.; Marks, T. J. *J Am Chem Soc*. 2008, 130, 8580–8581. (b) Uata, H.; Risko, C.; Wang, Z.; Huang, H.; Deliomeroglu, M. K.; Zhukhovitskiy, A.; Facchetti, A.; Marks, T. J. *J Am Chem Soc*. 2009, 131, 5586–5608.

[167] Chua, L.-L.; Zaumseil, J.; Chang, J.-F.; Ou, E. C.-W.; Ho, P. K.-H.; Sirringhaus, H.; Friend, R. H. *Nature*, 2005, 434, 194–199.

[168] (a)Babel, A.; Jenekhe, S. A. *J Am Chem Soc*. 2003, 125, 13656–13657. (b) Babel, A.; Jenekhe, S. A. *Adv Mater,* 2002, 14, 371–374.

[169] Shirota, Y. J Mater Chem. 2000, 10, 1–25. (b) Shirota, Y. *J Mater Chem*. 2005, 15, 75–93.

[170] (a)Veres, J.; Ogier, S. D.; Leeming, S. W.; Cupertino, D. C.; Khaffaf, S. M. *Adv Funct Mater*. 2003, 13, 199–204. (b) Veres, J.; Ogier, S.; Lloyd, G.; Leeuw, D. *Chem Mater*. 2004, 16, 4543–4555.

[171] Song, Y. B.; Di, C. A.; Yang, X. D.; Li, S. P.; Xu, W.; Liu, Y. Q.; Yang, L. M.; Shui, Z. G.; Zhang, D. Q.; Zhu, D. B. *J Am Chem Soc*. 2006, 128, 15940–15941.

[172] Bao, Z.; Lovinger, A. J.; Dodabalapur, A. *Appl Phys Lett*. 1996, 69, 3066–3068.

[173] de Boer, R. W. I.; Stassen, A. F.; Craciun, M. F.; Mulder, C. L. Molinari, A.; Rogge, S.; Morpurgo, A. F. *Appl Phys Lett*. 2005, 86, 262109/1–262109/3.

[174] Bao, Z. N.; Lovinger, A. J.; Brown, J. *J Am Chem Soc*. 1998, 120, 207–208.

[175] (a)Wang, H.; Song, D.; Yang, J.; Yu, B.; Geng, Y.; Yan, D. *Appl Phys Lett*. 2007, 90, 253510/1–253510/3. (b) Wang, H.; Zhu, F.; Yang, J.; Geng, Y.; Yan, D. *Adv Mater*. 2007, 19, 2168–2171. (c) Song, D.; Wang, H.; Zhu, F.; Yang, J.; Tian, H.; Geng, Y.; Yan, D. *Adv Mater*. 2008, 20, 2142–2144.

[176] Chen, Y. L.; Su, W.; Bai, M.; Jiang, J. Z.; Li, X. Y.; Liu, Y. Q.; Wang, L. X.; Wang, S. Q. *J Am Chem Soc*. 2005, 127, 15700–15701.

[177] (a)Park, Y. D.; Lim, J. A.; Lee, H. S.; Cho, K. *Materials today* 2007, 10, 46—54. (b) Yang, C.-Y.; Cheng, S.-S.; Ou, C.-W.; Chuang, Y.-C.; Wu, M.-C.; Dhananjay,; Chu, C.-W. *J Appl Phys*. 2008, 103, 094519/1–094519/6. (c) Ma, H.; Liu, M. S.; Jen, A. K-Y. *Polymer International*. 2009, 58, 594–619.

[178] Shtein, M.; Mapel, J.; Benziger, J. B.; Forrest, S. R. Appl Phys Lett. 2002, 81, 268–270. (b) Hayakawa, R.; Petit, M.; Chikyow, T.; Wakayama, Y. *Appl Phys Lett*. 2008, 93, 153301/1–153301/3.

[179] Choi, J.-M.; Jeong, S. H.; Hwang, D. K.; Im, S.; Lee, B. H.; Sung, M. M. *Org Elect*. 2009, 10, 199–204.

[180] Itaka, K.; Yamashiro, M.; Yamaguchi, J.; Haemori, M.; Yaginuma, S.; Matsumoto, Y.; Kondo, M.; Koinuma, H. *Adv Mater*. 2006, 18, 1713–1716.

[181] Yuan, G. C.; Xu, Z.; Gong, C.; Cai, Q. J.; Lu, Z. S.; Shi, J. S.; Zhang, F. J.; Zhao, S. L.; Xu, N.; Li, C. M. *Appl Phys Lett*. 2009. 94, 153308/1–153308/3.

[182] Cahyadi, T.; Kasim, J.; Tan, H. S.; Kulkarni, S. R.; Ong, B. S.; Wu, Y.; Chen, Z.-K.; Ng, C. M.; Shen, Z.-X.; Mhaisalkar, S. G. *Adv Funct Mater*. 2009, 19, 378–385.

[183] (a)Dimitrakopoulos, C. D.; Kymissis, I.; Purushothaman, S.; Neu-mayer, D. A.; Duncombe, P. R.; Laibowitz, R. B. *Adv Mater*. 1999, 11, 1372–1375. (b) Dimitrakopoulos, C. D.; Purushothaman, S.; Kymissis, J.; Callegari, A.; Shaw, J. M.

Science, 1999, 283, 822–824. (c) Yuan, J. F.; Zhang, J.; Wang, J.; Yan, X. J.; Yan, D. H.; Xu, W. *Appl Phys Lett*. 2003, 82, 3967–3969. (d) Fumagalli, L.; Natali, D.; Sampietro, M.; Peron, E.; Perissinotti, F.; Tallarida, G.; Ferrari, S. *Org Elect*. 2008, 9, 198–208.

[184] (a)Peng, X.; Horowitz, G.; Fichou, D.; Garnier, F. *Appl Phys Lett*. 1990, 57, 2013–2015. (b) Bao, Z.; Feng, Y.; Dodabalapur, A.; Raju, V. R.; Lovinger, A. *J. Chem Mater*. 1997, 9, 1299–1301. (c) Halik, M.; Klauk, H.; Zschieschang, U.; Schmid, G.; Dehm, C.; Schutz, M.; Maisch, S.; Effenberger, F.; Brunnbauer, M.; Stellacci, F. Nature, 2004, 431, 963–966. (d) Facchetti, A.; Yoon, M. H.; Marks, T. *J. Adv Mater*. 2005, 17, 1705–1725. (e) Jang, J.; Kim, J. W.; Park, N.; Kim, J.-*J. Org Elect*. 2008, 9, 481–486. (f) Roberts, M. E.; Queralto, N.; Mannsfeld, S. C. B.; Reinecke, B. N.; Knoll, W.; Bao, Z. *Chem Mater*. 2009, 21, 2292–2299.

[185] Yoon, M. H.; Kim, C.; Facchetti, A.; Marks, T. J. *J Am Chem Soc*. 2006, 128, 12851–12869.

[186] Fichou, D. *Handbook of Oligo- and Polythiophenes*. Wiley-VCH, New York, 1998.

[187] Burgi, L.; Richards, T. J.; Friend, R. H.; Sirringhaus, H. *J Appl Phys*. 2003, 94, 6129–6137.

[188] (a)Yun, D. J.; Lee, D. K.; Jeon, H. K.; Rhee, S. W.; *Org Elect*. 2007, 8, 690–694. (b) Ulman, A. *Chem Rev*. 1996, 96, 1533–1554. (c) Love, J. C.; Estroff, L. A.; Kriebel, J. K.; Nuzzo, R. G.; Whitesides, G. M. *Chem Rev*. 2005, 105, 1103–1169. (e) Di, C.; Yu, G.; Liu, Y.; Xu, X.; Wei, D.; Song, Y.; Sun, Y.; Wang, Y.; Zhu, D.; Liu, J.; Liu, X.; Wu, D. *J Am Chem Soc*. 2006, 128, 16418–16419.

[189] (a)Lee, J.; Hwang, D. K.; Choi, J. M.; Lee, K.; Kim, J. H.; Park, J. H.; Kim, E.; Im, S. *Appl Phys Lett*. 2005, 87, 023504/1–023504/3. (b) Hwang, D. K.; Park, J. H.; Lee, J.; Choi, J. Kim, J. H.; Kim, E.; Im, S. *Electrochem Solid-State Lett*. 2005, 8, G140–G142. (c) Choi, J. M.; Lee, J.; Hwang, D. K.; Kim, J. H.; Kim, E.; Im, S. *Appl Phys Lett*. 2006, 88, 043508/1–043508/3.

[190] (a)Kymissis, I.; Dimitrakopoulos, C. D.; Purushothaman, S. *IEEE Trans Electron Devices*. 2001, 48, 1060–1064. (b) Gundlach, D. J.; Jia, L. L.; Jackson, T. N. IEEE Electron Device Lett. 2001, 22, 571–573. (c) Rhee, S.-W.; Yun, D.-J. *J Mater Chem*. 2008, 18, 5437–5444.

INDEX

#

A

B

C

D

E

F

G

H

I

J

K

L

M

Q

R

S

T

U

V

W

X

Y

Z